McDonal

Happy Meal® Toys from the Eighties

Joyce & Terry Losonsky

4880 Lower Valley Road, Atglen, PA 19310 USA

Library of Congress Cataloging-in-Publication Data

Losonsky, Joyce.
McDonald's Happy Meal toys from the eighties/Joyce and Terry Losonsky.
p. cm.
ISBN 0-7643-0322-8 (softcover)
1. McDonald's Corporation--Collectibles--Catalogs.
2. Premiums (Retail trade)--Collectors and collecting--United states--Catalogs.
3. Lunchboxes--Collectors and collecting--United states--Catalogs.
I. Losonsky, Terry. II. Title.
NK6213.L6795 1998
688.7'2'0973075--dc21 98-21473
CIP

Designed by "Sue"
Type set in *Humanist521 BT*

ISBN: 0-7643-0322-8
Printed in China
1 2 3 4

Published by Schiffer Publishing Ltd.
4880 Lower Valley Road
Atglen, PA 19310
Phone: (610) 593-1777; Fax: (610) 593-2002
E-mail: Schifferbk@aol.com
Please write for a free catalog.
This book may be purchased from the publisher.
Please include $3.95 for shipping.

In Europe, Schiffer books are distributed by
Bushwood Books
6 Marksbury Avenue
Kew Gardens
Surrey TW9 4JF England
Phone: 44 (0)181 392-8585; Fax: 44 (0)181 392-9876
E-mail: Bushwd@aol.com

Please try your bookstore first.

We are interested in hearing from authors
with book ideas on related subjects.

Acknowledgements

The authors would like to take this opportunity to express our sincere appreciation and gratitude to all our collector friends who have contributed their time and knowledge to our research. In the event of oversight, we sincerely apology for any names and familiar faces left off our overgrowing list. It takes a collector group of friends to write a comprehensive price guide on McDonald's Happy Meal Toys From the Eighties. We would like to "Thank" the following:

ADVISORY BOARD

Ken Clee
Ron & Eileen Corbett
Jimmy & Pat Futch
John & Eleanor Larsen
McDonald's Corporation
Greg & Rhonda MacClaren
Bill & Pat Poe
E. J. Ritter
Rich & Laurel Seidelman
Bill Thomas
Meredith Williams

One reason for writing this book is to share our knowledge with other collectors who enjoy the thrill of seeking out McDonald's collectibles in the family toy boxes, thrift stores, and yard sales. The thrill of finding a "new" item is a sustaining factor among collectors. Since McDonald's is located in at least one hundred countries and distributes millions of toys every week, there are enough collectibles for all of us. Our desire to accurately record the history of McDonald's collectibles is the driving force behind our books, because understanding the history of the past provides insight into the future.

We would like to extend a special "hug" to our children and our families for their love and patient help over the last thirty years of collecting. To our children: Andrea, Natasha, Nicole, and Ryan and to children everywhere, we thank them for providing the incentive to visit McDonald's frequently. To our families: Stephen and Ann Zurko, Frank and Nancy Losonsky, Steve and Linda Zurko, Max Zurko, Phil and Alana Losonsky, Chris and Toni Losonsky and Aunt Ursula Shows, we sincerely thank them for all their encouragement and help over the years. We extend a heartfelt "Thanks" to the many McDonald's employees who over the years have fulfilled our request for items, especially for their cheerful, friendly, helpful, polite manners extended to collectors around the world. To Chris and Betsy, Kevin and Judy, the McDonald's owner/operators (O/O) and crew everywhere who say with a smile, "May I take your order, please?" we would like to express sincere appreciation. On behalf of all collectors, we expressly want to "Thank" McDonald's Corporation, Oak Brook, Illinois, for their assistance!

Lastly, a very special "Thank you and Hug" are extended to the many wonderful collector friends who have helped us in various ways over the last thirty years of collecting, from offering advice to sending material or photographs. These special friends have encouraged us to develop, update, and expand this book and our upcoming books. We wish to acknowledge assistance from the following individuals:

Helen Farrell - McDonald's Archives
Lois Dougherty - McDonald's Archives
Laura Kleiner - McDonald's Archives
Kathy List - Marketing/McDonald's
Wilma Weir - Marketing/McDonald's
Sam Apkarian
Dave Archer
Kathy Arne
Ron & Ethel Bacon
Linda Bailey
Richard & Crystal Banyon
Tom & Bonnie Becker
Terry Beedie
Becky Berger
Tom Borton
Bill & Marie Boyce
Harvey & Cleo Bradstreet
Bob & Mary Ann Brown
Sidney & Jeanne Bruce
Gerald & Helen Buchholz
Bert & Carolyn Buckler
Mark Carder
Carl & Rosemary Carlson
Maynard Carney
Karen Cavanaugh
Jim Challenger
Stephanie Chandler
Jim & Sally Christoffel
Ann Marie Clark
John & Brenda Clark
Kathy Clark
Judy Clark
Marilyn Clulow
Mark Coleman
David Cunningham
Clint Deale
Marvinette Dennis
Nate Downs
Darrell & Robyn Duncan
Gail Duzak
David Epstein
Gordon & Kath Fairgrieve
Leslie Fein
Fred Fiedler
Marjorie Fontana
Mike & Deanna Fountaine
Mike & Kathy Franze
Bonnie & Cheri Garnett
Jim & Linda Gegorski
Kay Geva
Brian Gildea
Mark & Carol Gillette
Bob & Gretchen Gipson
Lance Golba
Cindy Gore
Steve Gould
Pat & Martha Gragg
Shirley Graulich
Nick Graziano
Gary & Teena Greenberg
Chuck Gustafson
David Hale
Roberta Harris
Gary & Judy Heald
Gary & Shirley Henriques
Ed Hock
Roger Hordines
Sharon Iranpour
Steven & Ann Jackson
Dave Johnson
Brian Jones
Anne King
Joyce Klassen
Robert Lanier
Jerry Ledbetter
Pat Lonergan
Kent Longmire
Darrell Lulling
Lee Marsh
Thomas & Frankie Massey
Bob McClintock
Bill & Betty McCormick
Glen & Kathleen McElwee
Janet McGuire
Art McManis
Victor Medcalf
Don Metiva
George Miller
Howard Morris
Julius & Margaret Mortvedt
Stanley Mull
Pat Multz
Beulah Murphy
Rene & Anne Marie Naim
Steve & Margie Nation
Steven Jr. & Rebecca & Rachel Nation
Tom & Terry Nelson
Jim & Cooky Oberg
Harry Oberth
Roger Olshanski
Tom & Teresa Olszeski
Joe & Dolly Pascale
Mark Patterson
Garnett Pennington
Mark & Jane Petzel
Janet Phillips
Joe & Carol Pierce
Ray Podraza
Larry & Maynuella Poli
Edward & Jean Pomeroy
Jean Pomeroy
Mike Portzline
Charles & Connie Prater
Ron & Jane Prussiano
Fred Rauch
Jimmy Renella
Russell & Marie Rinehart
Alyce Roberts
Tom & Kathy Robusto
Natalie Royer
Ed Ruby
Chris & Julie Rucho
Emma Rush
Doug & Debbie Ryan
Barbara Saitta
Essie Saunders
Ed & Sharon Scarbrock
Pat Sentell
Bob Serighino
Jim Silva
Trudy Slaven
Scott Smiles
Dan Smith
Jerry & Lorraine Soltis
Rich & JoLyn Stack
Lorie Steele
Julie Stegeman
Peggy Stockard
David Stone
Richard & Marge Taibi
Debbie Taylor
Nigel Thomas
John & Virginia Thompson
Robert & Jackie Thompson
Ray & Dorothy Tognarelli
Frances Turey
Gary & Jill Turner
Lee Turpin
Dave Tuttle
Kees & Conny Versteeg
Taylor & Cindy Wagen
Lloyd & Nancy Washburn
Fred & Elaine Waterman
Ted Waters
Toni Welsh
Gary & Karen Wenzlaff
Robert Wilkey
Don Wilson
Jim & Rosalie Wolfe
Mike & Mary Ann Wooten
Ron & Eldra Word
Frank Work
Claire Zabo
Frank Zamarripa

Joyce and Terry Losonsky can be reached at:

7506 Summer Leave Lane
Columbia, Maryland 21046-2455
Tel: USA 1-410-381-3358
FAX: USA 1-410-381-1852
E-mail: JoyceUSA@aol.com

Introduction

Welcome to the world of McDonald's Happy Meal toy collecting from the beginning of the Happy Meal! The Happy Meal and Happy Meal toys, jingles, logos, slogans, and convention themes reflect an image of what America has experienced in the last forty-three years — McDonald's toy collecting is Americana at its best! The 1980s ushered in the era of McDonald's Happy Meal toys in grand style. From its meager beginning with the June 1979 "Circus Wagon Happy Meal" promotion to the Little Mermaid Happy Meal of Christmas 1989, this decade represents the epitome of fast food toy collecting. McDonald's has become an American institution producing the "gastronomical glue" which binds parents to children and grandparents to grandchildren in a common experience, all enjoying McDonald's food and collecting the Happy Meal toys.

Some collectors want to collect only Happy Meal toys from the eighties, others seek only Disney toys, while still others prefer only Garfield items or Batman items. Some collectors want only items with one specific slogan or mark, like the "Bisected Arches," era items, while other collectors use slogans or trademarks as their "cut off" point for collecting. That is, they collect nothing later than 1968 or they collect only "Archy" items or only mint on tree (MIT) items. The wishes and desires of the collectors are as infinite as the toys themselves.

McDonald's has tried and repeated many toy themes, slogans, logos, and trademarks throughout its forty-three year history and, as always, McDonald's collectibles provide many examples of each. If we asked a group of McDonald's customers what an advertising slogan is, they'd probably reply, "Just another way to get me to buy a product." While that is true for some, for McDonald's collectors, a slogan or trademark means much more. Identification of toys, slogans, logos, jingles, and convention themes is one of the best ways to date an advertising collectible! The age of an item, its condition, its rarity, and general distribution data on the item all tend to influence price. Everyone seems to want to know, "What is it worth?" They don't necessarily want to sell the item, they just want to have that warm, fuzzy feeling of knowing the item they possess has value. But in the end, price is determined between what amount a buyer is willing to pay and what amount the seller is willing to take. Prices of McDonald's items are constantly changing—up and down, regionally and globally—based on supply and demand.

"Where did it all start and where is it going?" is a common question among collectors. Before fully exploring the world of McDonald's Happy Meal Toys from the Eighties, whether they are the early paper items or the latest Disney venture toys, an understanding of the impressive history of the McDonald's success story is necessary. The following Losonsky Lists are provided to aid in the identification of items and to document the history of McDonald's toys, jingles, logos, slogans, and convention themes through the years. The lists are followed by helpful information for using this book and then a brief overview of McDonald's formative period during its first three decades (for a more detailed description of McDonald's early years, please see Joyce and Terry Losonsky's companion book, *McDonald's® Pre-Happy Meal® Toys in the Fifties, Sixties, and Seventies,* also available from Schiffer Publishing).

Losonsky's Identification Guides to McDonald's Collectibles

Losonsky List #1: Jingles and Slogans

1950s

1953 - "McDonald's HAMBURGERS...THERE'S ONE NEAR YOU!"

1955 - "Speedee" corporate logo
1955-62 - "Speedee, the Hamburger Man"
1957 - "COAST to COAST"

1959 - "HOME OF AMERICA'S 'GOODEST' HAMBURGER"
1959 - "ALL AMERICAN MEAL only 45 cents"

1960s

1960 - "Look for the Drive-in with the Arches"
1960 - Speedee says: "Look for me at McDonald's speedee drive -ins"
1960-65 - "Look for the Golden Arches" jingle
1960s - "McDonald's Speedee drive-ins - often imitated, never duplicated"
1960s - "Those who know—Go to McDonald's"

1961 - YOUR KIND OF PLACE
1961 - THE ALL AMERICAN MEAL - still only 45 cents

1962 - "Go for Goodness at McDonald's"
1962 - "...the Drive-in with the arches"
1962 - "...way of life coast to coast" (repeated)
1962 - "EVERYBODY'S FAVORITE Coast to Coast"
1962 - Home of America's Favorite Hamburger
1962 - Home of...America's Favorite Hamburgers... still only 15 cents

1963 - "Meet Ronald McDonald"
1963 - "EVERYBODY'S FAVORITE Coast to Coast" (repeated)

1964 - "Look for THE DRIVE-IN WITH THE (GOLDEN) ARCHES"
1964 - "Come as you are and eat in your car—it's always fun for the family to eat at McDonald's from coast-to-coast, McDonald's sells the most! " ("coast-to-coast" repeated)
1964 - "Come as you are and eat in your car"
1964 - ALL AMERICAN MEAL only 47 cents

1965 - "McDonald's—where quality starts fresh every day - look for the Golden Arches"
1965 - "The Sky's The Limit"

1966-68 - "Ronald McDonald and His Flying Hamburger"

1967 - "McDonald's is Our Kind of Place"
1967 - "Look For The Golden Arches - The Closest Thing To Home"
1967 - Welcome to McDonald's...The Closest Thing To Home

1968 - "McDonald's is Your Kind of Place" (modified & repeated)
1968 - "The Customer is #1...What have you done lately for the Customer?"

1969 - "You Deserve a Break Today - So Get Up and Get Away to McDonald's"
1969 - "The Challenge of Success"

1970s

1970 - "You Deserve A Break Today - So get Up And Get Away To McDonald's" (repeated)

1971 - "You Deserve A Break Today - So get Up And Get Away To McDonald's" (repeated)
1971 - "Success and Then Some"
1971 - "Grab a Bucket & Mop"

1972 - "Don't Forget to Feed the Wastebasket"
1972 - "Let's Face It"

1973 - "You Deserve a Break Today" (abbreviated ad)

1974 - "You Deserve A Break Today We're Close By...Right on Your Way"
1974 - "McFavorite Clown"

1975 - "Two All-Beef Patties Special Sauce"
1975-79 - "We Do It All For You"
1975 - "Twoallbeefpattiesspecialsaucelettuce cheesepicklesonionsonasesameseedbun"

1976 - "You, You're the One"
1976 - "The Challenging World of Number One"

1978 - "Keep Your Eyes on Your Fries"
1978 - "When We Work At It—It Works"
1978 - "Happy Meals Tickle My Tummy"

1979 - "Nobody Can Do It Like McDonald's Can"

1980s

1980 - "Our World...Today, Yesterday and Tomorrow"

1981 - "You Deserve a Break Today" (repeated)

1982 - "McDonald's and You" (Camp Nippersink)
1982 - "Delivering the Difference"

1983 - "McDonald's and You" (repeated)

1984 - "It's a Good Time for the Great Taste of McDonald's"
1984 - "When the USA Wins You Win"
1984 - "One of a Kind"

1985 - "Large Fries for Small Fries"
1985 - "The Hot Stays Hot and the Cool Stays Cool"

1986 - "Back To Our Future"

1987 - "McKids"

1988 - "Good Time, Great Taste of McDonald's"
1988 - "Good Time, Great Taste, That's Why This is My Place"
1988 - "Sharing the Dream"
1988 - "McKids" (repeated)

1990s

1990 - "Food, Folks, and Fun"
1990 - "Call to Action, Customer Satisfaction"

1991 - "Do You Believe in Magic?"

1992 - "What You Want Is What You Get [at McDonald's Today]"
1992 - "Together, we've got what it takes"

1993 - "McWorld"

1994 - "Great Expectations"

1995 - "Have You Had Your Break Today?"

1996 - "QSC and Me"

1997 - "My McDonald's"
1997 - "did somebody say McDonald's?"

1998 - "did somebody say McDonald's?" (repeated)
1998 - "Where the World's Best Come Together" (Olympics, 1998)
1998 - "Made for You"
1998 - "Made for You...At the Speed of McDonald's"
1998 - "the value of GOLD"

Losonsky List #2: Sign Identification

1948

1948 - Speedee is the McDonald brothers Company Symbol

1950s

1953 - Speedee sign with Single Arch and Speedee holding 15 cent sign

1955-62 - Speedee, the Hamburger Man

1957 - Double Arch replaces Single Arch with Speedee

1958 - ONE HUNDRED MILLION sold

1960s

1960 - 400 HUNDRED MILLION sold

1961 - 500 HUNDRED MILLION sold

1962 - 700 HUNDRED MILLION sold
1962 - Bisected arches M replaces "Speedee"
1962-63 - Bisected Arches with Half Arrow Head

1963 - ONE BILLIONTH Hamburger Served
1963 - Bisected Arches within a Ship's Wheel
1963 - Double Arches with "McDonald's Hamburgers" signs replace Single Arch signs with "McDonald's Hamburgers"

1964 - Over ONE BILLION SOLD

1966 - TWO BILLION SERVED

1967 - FOUR BILLION SERVED

1968 - Over 4 Billion Served
1968 - Billions served on signs
1962-68 - Bisected Arches replaces Speedee

1969 - "Billions Served" sign changed to "5 Billion Served"

1970s

1972 - 10th & 11th BILLION HAMBURGERS SOLD

1974 - 15th BILLIONTH SERVED

1976 - 20 BILLION SERVED

1978 - 25 BILLION SERVED

1979 - 30 BILLION SOLD

1980s

1983 - 45 BILLION SOLD

1984 - 50 BILLION SERVED

1985 - 55 BILLION SERVED

1986 - More Than 60 Billion Served

1987 - 65 BILLION SERVED

1988 - 70 BILLION SERVED

1989 - 75 BILLION SERVED

1990s

1990 - 80 BILLION SERVED

1992 - 90 BILLION SERVED

1992 - Signs changed to "McDonald's"

1995 - Signs on highways changed to Arches only (no McDonald's name)

1997 - 99 BILLION SERVED

Losonsky List #3: Character Introduction/Redesigns

1955

1955-62 Speedee character introduction

1960s

1963 - Ronald McDonald as Regional Spokesman - Bisected arches
1963-68 - Ronald McDonald with bisected arches costume

1964 - Archy McDonald (October - December)

1966 - Ronald McDonald as National Spokesman - bisected arches (October)

1969 - Ronald McDonald's costume changed to even pockets (no bisected arches)

1970s

1970 - Ronald McDonald introduced in McDonaldland (January)
1970 - Evil Grimace introduced on TV (Nov./Dec.)
1970 - Hamburglar introduced on TV (Nov./Dec.)
1970 - Mayor McCheese introduced on TV (Nov./Dec.)
1970 - Captain Crook introduced on TV (Nov./Dec.)
1970 - The Professor introduced on TV (Nov./Dec.)
1970 - Big Mac introduced on TV (Nov./Dec.)
1970 - Gobblins introduced on TV (Nov./Dec.)

1971 - Evil Grimace with four arms officially introduced
1971 - Hamburglar with unusual nose officially introduced
1971 - Mayor McCheese officially introduced
1971 - Captain Crook with skull and cross bones and green hair officially introduced
1971 - The Professor with long hair officially introduced
1971 - Big Mac officially introduced
1971 - Gobblins officially introduced

1973 - Evil Grimace redesigned to Grimace (two arms)
1973 - Hamburglar redesigned to long pointed nose
1973 - Mayor McCheese redesigned
1973 - Captain Crook redesigned
1973 - The Professor redesigned
1973 - Big Mac redesigned
1973 - Uncle O'Grimacey introduced

1973-81 - Ronald redesigned - Pockets have black lines

1974 - Grimace still called: The Grimace
1974 - Hamburglar still called: The Hamburglar

1975 - The Captain Crook's hair is no longer green

1979 - The Happy Meal Guys appear and disappear

1980s

1981 - Birdie, the Early Bird introduced

1982 - The Professor is redesigned
1982 - Gobblins were renamed the French Fry Guys
1982 - Design change of Captain Crook

1983 - McNugget Buddies introduced
1983 - (French) Fry Guys redesigned
1983 - Captain Crook redesigned to "The Captain"

1985 - Fry Girls redesigned

1987 - Mac Tonight introduced
1987 - Ronald redesigned (January)
1987 - Grimace redesigned again (January)
1987 - Birdie, the Early Bird redesigned (January)
1987 - Hamburglar redesigned to pudgy face, smiling character (January)

1988 - CosMc - The little space alien joins McDonaldland

1990s

1997 - Flubber (Disney/McDonald's) introduced

1998 - "Iam Hungry" computer animated character introduced

Helpful Information for Using this Book

Pricing

Note: Mint Price Range Only Listed for All Items

MINT - mint price value listed - Mint in the package range listed. $5.00-8.00 = Mint in the Package price range. The price range listed is for MINT IN THE PACKAGE.

LOOSE -Loose toys are 50% less than the low mint value listed. $2.50 = 50% off lowest price = Loose value. Damaged, chipped, or broken toys tend to have little value with a collector.

$—— - indicates no definitive price has been established for the item.

The real value of any collectible is what a buyer is willing to pay. This value may exceed the stated mint in the package (MIP) range. Likewise, since McDonald's makes millions of toys, value may be over inflated based on regional markets. Price ranges vary by regions, since some toys were distributed in specific regional markets.

Name and Numbering System

Premium Names - The premiums are listed by the names on the packaging whenever possible.

Box Names - The boxes were named by the authors with the accompanying identifying numbering system. Whenever possible, the names came from the front panel where the words "Happy Meal" are displayed.

Number System - The numbering system reflects the **country of origin/country of distribution,** followed by the first two letters of the Happy Meal or **first two letters of the Happy Meal** name or the generic representation of the Happy Meal items. The two letters are followed by the **year of distribution**, followed by a **numerical listing of the items**. The authors intention is to reflect a different alphabetical/numerical listing for each and every item distributed.

Example: USA Ci7909 = CIRCUS WAGON HAPPY MEAL, 1979

USA = Country of distribution/origin
Ci = [Ci] rcus Wagon Happy Meal
79 = Year of distribution
09 = Numerical listing of item

Example: USA Dt8865 = DUCK TALES I HAPPY MEAL, 1988

USA = Country of distribution/origin = USA
Dt = [D] uck [T] ales I Happy Meal, 1988
88 = Year of distribution
65 = Numerical listing of an item: Translite

Numerical Designator - Last two numbers of identification code.

BAG-Happy M	30-33
BANNER	27
BOX-Happy M	10-13
BUTTON	50
CEILING DANGLER	41
COUNTER CARD	42
COUNTER MAT	60
CREW Refer sheet	43
CREW POSTER	44
DISPLAY	26
HEADER CARD	62
LUG-ON	63
MESSAGE C INSERT	61
PIN	95
REGISTER TOPPER	45
TABLE TENT	56
TOYS	1-8
TRAYLINER	55
TRANSLITE/SM	64
TRANSLITE/LG	65

Whenever conflict in selecting the alphabetical/numerical designator arose, the first letter of the first two names of the Happy Meal was used and/or the generic alphabetic representation of the item was used. For example, Michael Jordan/Fitness Fun Happy Meal, 1992 and/or Fitness Fun/Michael Jordan Happy Meal is noted as Mj. Likewise, some Happy Meal promotions were repeated over the years. These were consistently assigned alphabetic listings: Attack Pack becomes Ap; Barbie becomes Ba; Batman becomes Bt; Cabbage Patch becomes Cp; Funny Fry Friends becomes Ff; Halloween becomes Ha; Hot Wheels becomes Hw; Tonka becomes Tk and so on. As time progresses, it is hoped these alpha/numeric listings will become standardized. The authors apologize for all past inconsistencies in developing a system which identifies each and every item with a separate alpha/numeric label. A Cross/Numbering listing can be found in the back of this text.

McDonald's Collecting Language

Hm = Happy Meal
MIP = Mint in Package
MOC = Mint on Card
MOT = Mint on Tree/plastic holder
Nd = No date on item
NP = Not packaged

Clean-up week - Open time period following a Happy Meal when no specific designed toy is distributed. The stock room backlog is given out.

Counter card - advertising or customer information card or board which sits on the counter.

Counter mat - advertising mat which sits on the counter; used in early years.

Display - advertising medium which holds/displays the toys being promoted and distributed during a specific time frame. These range from older bubble type to cardboard fold-up type. These are displayed in stand-up Ronald McDonald in the lobby.

Generic - item such as a box or a toy not specifically associated with a specific theme Happy Meal or promotion. The item(s) may be used in several different promotions over a period of time.

Header card - used in older Happy Meal promotions as advertising on top of the permanent display or ceiling dangler to display Happy Meal boxes or toys.

Insert card - advertising card within/along with the premium packaging.

Lug-on - sign added to the menu board.

McDonaldland - imaginary place where all the McDonald's cast of characters live and play; a playland area; a place in the hearts of children.

National - all stores in the USA distribute the same Happy Meal at the same time; supported with national advertising.

Register Topper - advertising item placed on the top of registers.

Regional - geographical distribution was limited to specific cities, states, stores, or marketing areas.

Self-liquidator - item intended to be sold over the counter which may or may not be included in the Happy Meal box.

Table tent - rectangle shaped advertising sign placed on the tables and counters in the lobby.

Translite - advertising transparent sign used on overhead or drive-through menu boards to illustrate the current promotion.

U-3 - under the age of 3 premiums; specifically designed for children under the age of 3. Packaging is typically in zebra stripes around the outside of the package. The colors of the stripes vary.

In the Beginning...

The Fifties

Ray Kroc, the founder of McDonald's, opened his first store in Des Plaines, Illinois on April 15, 1955, and the history of McDonald's began to unfold. Kroc was fifty-two years old at the time, a salesman for milk shake machines who saw a "golden" opportunity in the successful car-hop restaurants originally opened in California by Dick and Marice (Mac) McDonald in the late 1940s. Simplistically, Ray Kroc franchised the McDonald's Speedee Service System in 1955 and began constructing buildings using red and white candy striped tile walls and yellow neon arches going through the roof — thus the Golden Arches were born. It was Ray Kroc's vision of serving reasonably priced hamburgers, shakes, and french fries to the ever-growing population that propelled McDonald's into a legendary business of the twentieth century.

The first menu at the Des Plaines store was a check-off sheet with a limited selection of 15 cent hamburgers, 10 cent fries, and 20 cent milk shakes. It was the "All American Meal" for only 45 cents per person! The menu was promoted by the company symbol of "Speedee," the little hamburger man, and both speedy service and speedy growth soon followed. By the following year a dozen McDonald's restaurants were added in Illinois and California, and in 1959 the 100th McDonald's restaurant opened in Fond du Lac, Wisconsin.

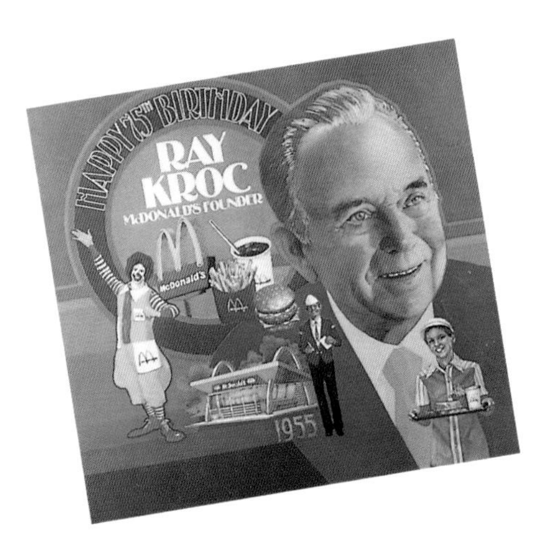

The Sixties

In 1960, "Look for the Golden Arches" jingle initiated McDonald's radio advertising, and the sounds of McDonald's calling were heard all around the country. Willard Scott became the symbol of Ronald McDonald, making his first public debut in Washington D.C. in October, 1963, and the Filet-O-Fish sandwich was officially added to the menu in 1964, priced at 24 cents. The Archy McDonald logo was used for a limited time in Ohio regional markets, along with the extremely successful Filet-O-Fish advertising. By 1966, McDonald's had become a national name and was voted "the growth company" of the year.

The Big Mac sandwich arrived in 1968, an item destined to be a dominant force in the fast food market for the next thirty years. The Big Mac Attack had begun! Along with the Big Mac sandwich, the Hot Apple Pie was added to the menu at the opening of the 1,000th McDonald's restaurant in Des Plaines, Illinois.

THE SIXTIES

The Seventies

"The best is yet to come" became McDonald's motto in the 1970s — certainly a prophetic statement! In 1970, McDonaldland was created, the home of Ronald McDonald and his colorful cast of characters: Big Mac, The Hamburglar, Grimace, Mayor McCheese, The Professor, The Gobblins, and Captain Crook. The characters, whose appearance would change little over the years, added another level of appeal to McDonald's concept and generated smiles wherever they appeared.

In 1972, the Quarter Pounder was added to the menu in conjunction with the opening of the 2,000th McDonald's restaurant in Des Plaines, Illinois. The breakfast menu was initially tested in California and instituted in all locations by 1976.

In the seventies, the Happy Meal concept was tested and promoted nationally; at the same time the Barrel of Fun "rolled" into the stores. McDonald's advertising focused on children and the "Collect All" theme began to take root. The sprout was green and rapidly growing with the introduction of the Circus Wagon Happy Meal in 1979. Ronald McDonald's theme ushered in the generation of fast food consumers and collectors alike:

Ronald McDonald's Theme

Nobody can do it like Ronald can.
Nobody else spreads laughter throughout the land.
He makes it fun, he's always got a joke in hand.
He's Ronald McDonald.
Nobody can do it like Ronald can.

Nobody else has a sunnier smile.
Nobody else has a funnier style.
Come on along, see what he has planned...

Happy Meal Officially Begins - June 1979

Circus Wagon Happy Meal, 1979

Boxes:

❑ ❑ USA Ci7930 **Hm Box - Capt Crook/Seal**... Why Isn't a Seal's Nose 12" Long, 1979. $25.00-40.00

❑ ❑ USA Ci7931 **Hm Box - Grimace/Elephant**... Why Do Elephants Have Such Wrinkled Knees, 1979. $25.00-40.00

Ci7930

Ci7931

Ci7932

❑ ❑ USA Ci7932 **Hm Box - Ronald/Hamb/Capt**...What Kind of Cat Find in Library, 1979. $25.00-40.00

❑ ❑ USA Ci7933 **Hm Box - Ronald/Goblins**...Half a Hamburger, 1979. $25.00-40.00

❑ ❑ USA Ci7934 **Hm Box - Ronald/Lion**...When Do Lions Have 8 Feet, 1979. $25.00-40.00

❑ ❑ USA Ci7935 **Hm Box - Ronald/Mayor**...What Do Zebras Have, 1979. $25.00-40.00

Premiums:

❑ ❑ USA Ci7905 **ID Bracelet - Big Mac,** 1979, Yel or Red with Alphabet Sticker Sheet/MIP in cellophane pkg.$10.00-15.00

❑ ❑ USA Ci7908 **ID Bracelet - Hamburglar,** 1979, Yel or Red with Alphabet Sticker Sheet/MIP in cellophane pkg. $10.00-15.00

❑ ❑ USA Ci7909 **ID Bracelet - Ronald,** 1979, Yel or Red with Alphabet Sticker Sheet/MIP in cellophane pkg.$10.00-15.00

Ci7933

Ci7935

Ci7934

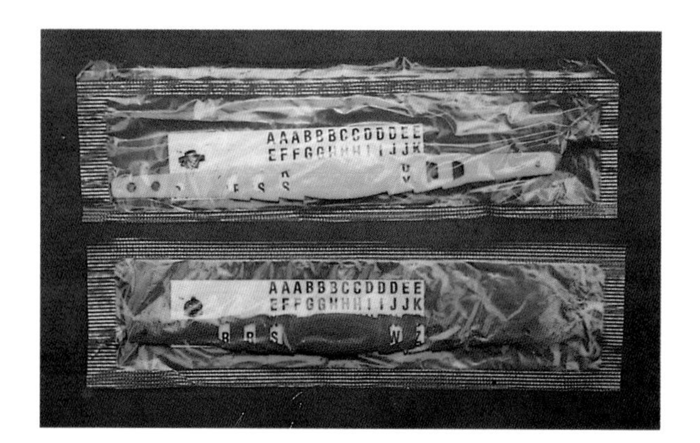

Top: Ci7908; Bottom: Ci7905

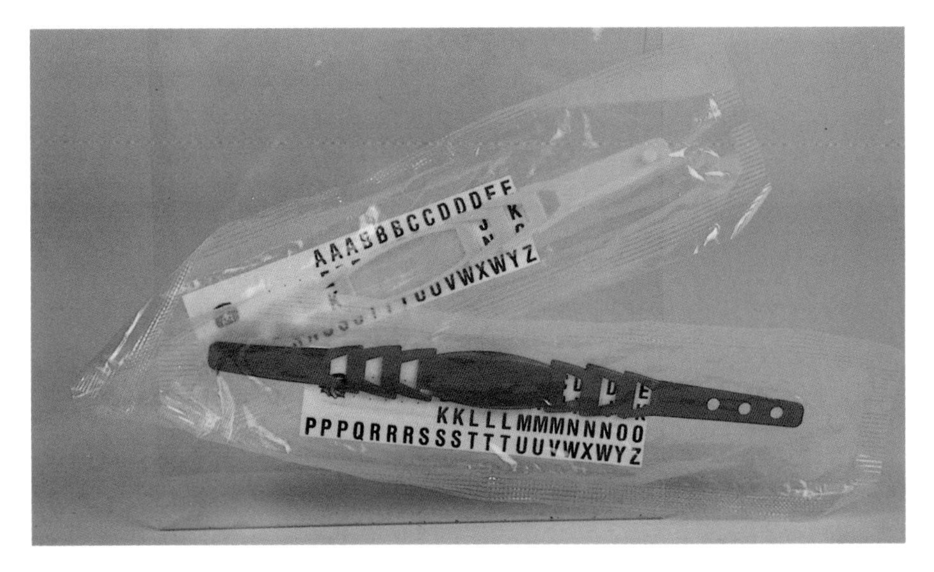

Top: Ci7908; Bottom: Ci7909

❑ ❑ USA Ci7906 **McDoodler Ruler - Ronald**, 1979, Red or Blu or Yel/2p/MIP in cellophane pkg. $20.00-25.00

❑ ❑ USA Ci7907 **McPuzzle Lock - Ronald,** 1979, Red or Blu or Yel/5p MOT on Plastic Tree. $8.00-10.00

❑ ❑ USA Ci7926 **Floor Display**, 1979. $100.00-150.00

❑ ❑ USA Ci7941 **Ceiling Dangler,** 1979. $50.00-75.00

Ci7907

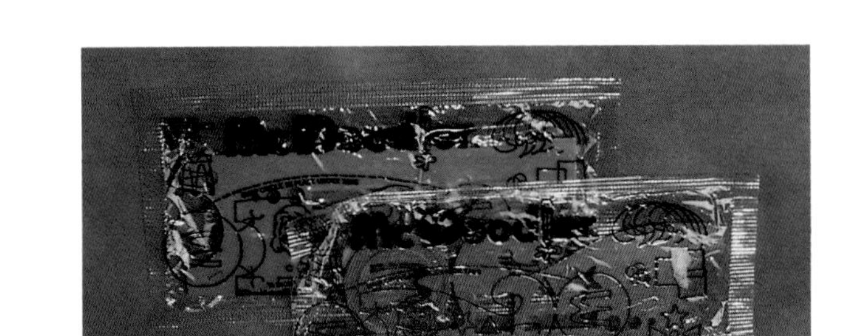

Ci7906

Ci7906

Ci7941

Ci7954

❑ ❑ USA Ci7954 **Crew Badge,** 1979, Paper Blu/Turq Hm Box with Pinback. $15.00-20.00

❑ ❑ USA Ci7965 **Translite/6 Boxes/Lg,** 1979. $50.00-75.00
❑ ❑ USA Ci7966 **Translite/1 Box/Lg,** 1979. $45.00-50.00

Comments: National Distribution: USA - June 11 - July 31, 1979. **Circus Wagon Happy Meal is considered to be the First National Happy Meal Promotion in the USA, June 11 - July 31, 1979. This promotion is considered to be the first of hundreds of Happy Meals to come. A mere fifteen years later, "Ronald McDonald Celebrates Happy Birthday Happy Meal, 1994" is celebrated as the commemoration of the Fifteenth Anniversary of the Happy Meal in the USA, October 28 - December 1, 1994.**

Minor note: The Losonsky authors have NO EVIDENCE (at this time) indicating the Diener figurines — clown, poodle, horse, ape, or others — were distributed with this Happy Meal. On a regional basis, the Diener figurines could have been distributed when demand exceeded supply. This could be an example of a "send me anything you have" option, not a concerted effort to distribute pre-planned premiums.

Ci7965

Ci7966

Space Theme Meal Happy Meal, 1979

Boxes:
❑ ❑ USA Sp7900 **Hm Box - Big Mac/Six Martians/Flight Log/ Space Quiz,** 1979. $35.00-50.00
❑ ❑ USA Sp7901 **Hm Box - Grimace/Thick Shake,** 1979. $35.00-50.00
❑ ❑ USA Sp7902 **Hm Box - Space Creatures/Ron Feeding Creatures,** 1979. $35.00-50.00
❑ ❑ USA Sp7903 **Hm Box - McDonaldland Friends in Plant-A-Tarium,** 1979. $35.00-50.00
❑ ❑ USA Sp7904 **Hm Box - Ron/Stowaway Spacemen "Six Spacemen ..",** 1979. $35.00-50.00
❑ ❑ USA Sp7905 **Hm Box - Space Zoo/Ronald and Space Zoo Animals,** 1979. $35.00-50.00

Sp7900

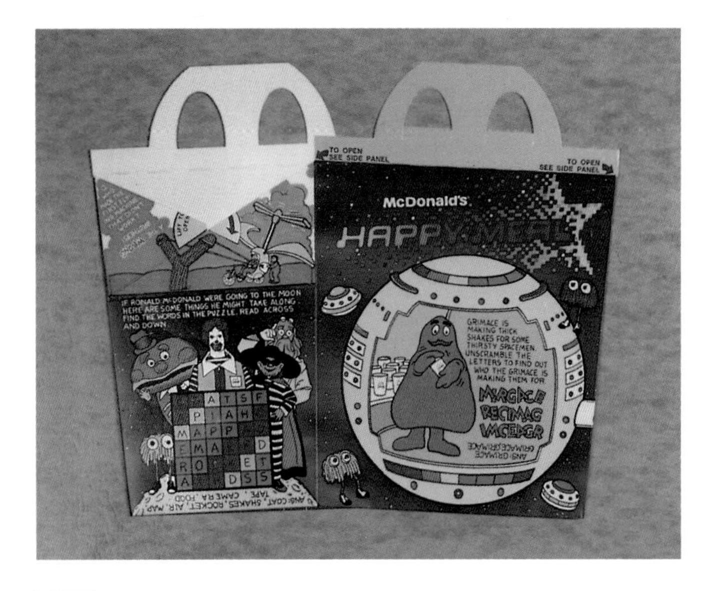

Sp7901

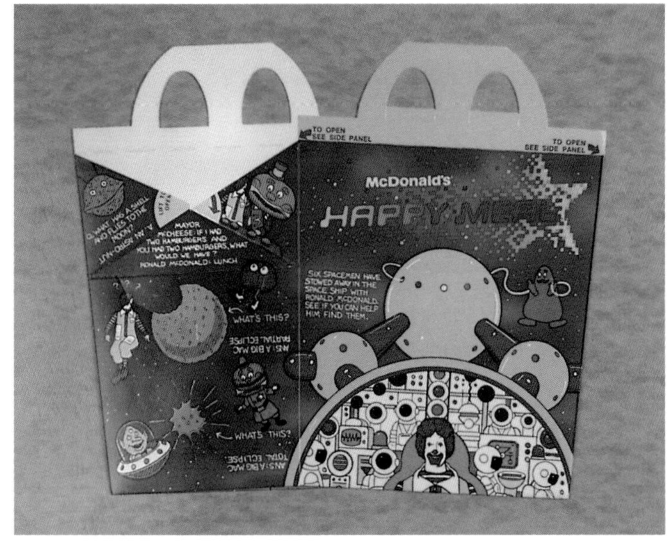

Sp7904

Sp7902

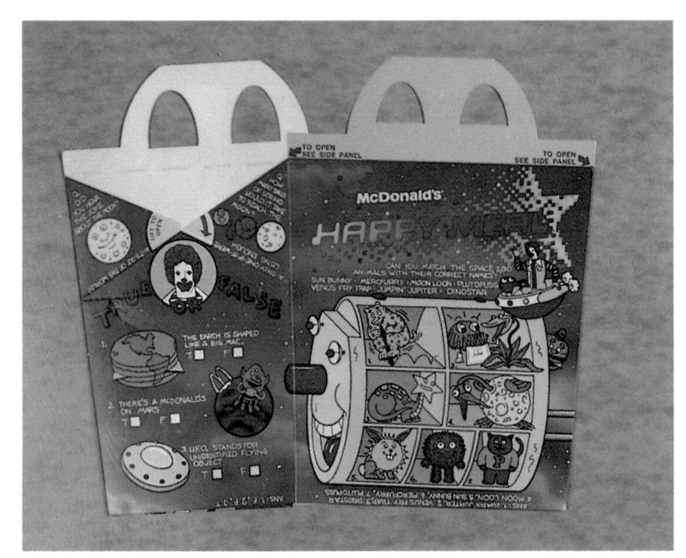

Sp7905

Sp7903

1979 Space Meal.

From left: Sp7990, Sp7991, Sp7992, Sp7993, Sp7994, Sp7995, Sp7996, Sp7997.

Sp7941

Premiums:

❑ ❑ USA Sp7990 **Gill Face Creature,** Nd, Org/Pur/Yel/Brn/Grn/Brn. $1.00-2.00

❑ ❑ USA Sp7991 **Horned Cyclops,** Nd, Org/Pur/Yel/Brn/Grn/Blu. $1.00-2.00

❑ ❑ USA Sp7992 **Insectman,** Nd, Org/Pur/Yel/Brn/Grn/Blu. $1.00-2.00

❑ ❑ USA Sp7993 **Lizard Man,** Nd, Org/Pur/Yel/Brn/Grn/Blu. $1.00-2.00

❑ ❑ USA Sp7994 **Tree Trunk Monster,** Nd, Org/Pur/Yel/Brn/Grn/Blu. $1.00-2.00

❑ ❑ USA Sp7995 **Vampire Bat Creature**, Nd, Org/Pur/Yel/Brn/Grn/Blu. $1.00-2.00

❑ ❑ USA Sp7996 **Veined Cranium,** Nd, Org/Pur/Yel/Brn/Grn/Blu. $1.00-2.00

❑ ❑ USA Sp7997 **Winged Amphibian,** Nd, Org/Pur/Yel/Brn/Grn/Blu. $1.00-2.00

❑ ❑ USA Sp7965 **Translite/Space Aliens,** 1979. $50.00-75.00

❑ ❑ USA Sp7966 **Translite/Alien Creatures,** 1979. $50.00-75.00

❑ ❑ USA Sp7967 **Translite/Happy Meal,** 1979. $50.00-75.00

❑ ❑ USA Sp7942 **Counter Card.** $40.00-50.00

❑ ❑ USA Sp7941 **Ceiling Dangler.** $45.00-75.00

Comments: Regional Distribution: USA - July - December 1979. Premium Markings - "Diener Ind." Space Raiders were soft rubber figurines, Space Aliens were hard rubber figurines in 1979. In later years, Diener made both soft and hard rubber figurines. These Space Aliens/Raiders Premiums could have been given out in earlier happy meal offerings. For example, in 1978 during HM Test in Kansas City, some stores had Diener figurines. Again, it is an example of a "send me anything you have" request from the stores to the distribution centers, which create this unusual distribution of toys/premiums. Note: USA Sp7990-97 Diener Soft Rubber Figure Versions are still available at retail stores in the 1990s. No McDonald's markings on loose premiums.

Minor Note: "Space Theme Meal" name is a revised name for the Happy Meal, based on new research.

Star Trek Meal Happy Meal, 1980/1979

Boxes:

❑ ❑ USA St7920 **Hm Box - Bridge - Draw the Alien,** 1979. $7.00-9.00

❑ ❑ USA St7921 **Hm Box - Bridge - Planet Faces,** 1979. $7.00-9.00

St7920

St7921

❑ ❑ USA St7922 **Hm Box - Federation,** 1979. $7.00-9.00
❑ ❑ USA St7923 **Hm Box - Klingons,** 1979. $7.00-9.00
❑ ❑ USA St7924 **Hm Box - Spacesuit - Spock's Code,** 1979. $7.00-9.00
❑ ❑ USA St7925 **Hm Box - Transporter Room - Klingon Match,** 1979. $7.00-9.00

St7924

St7922

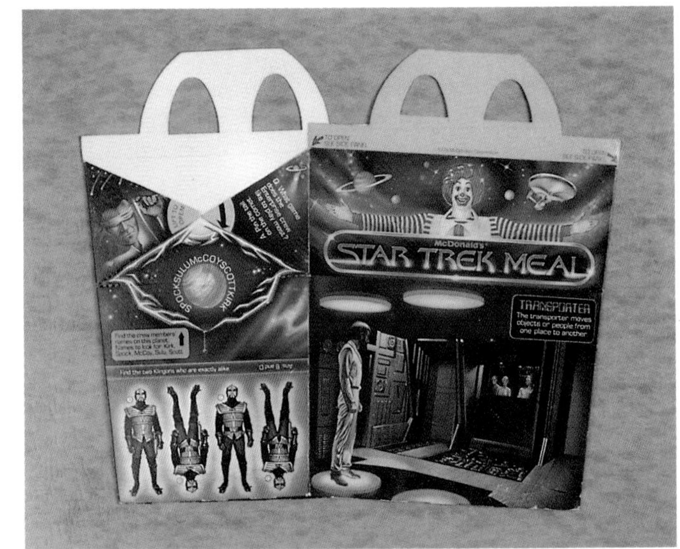

St7925

St7923

Star Trek
Blue Book

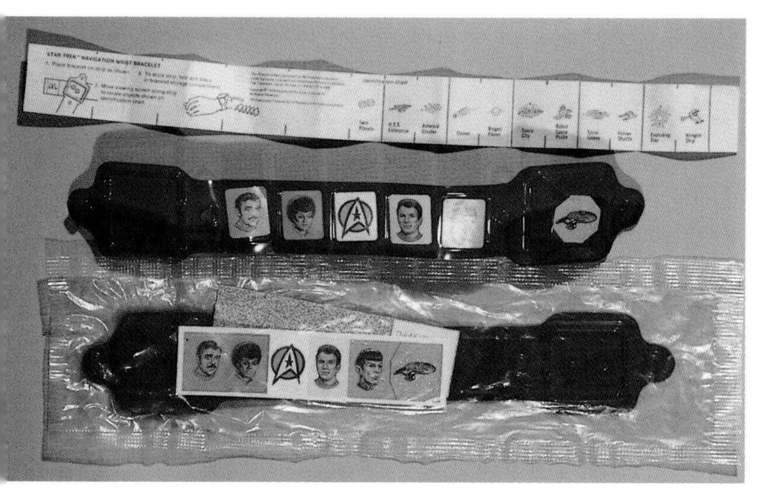

St7915

Premiums:

❑ ❑ USA St7915 **Navigation Wrist Bracelet,** 1979, Navigators Chart/Galaxy 6 Decals/8 3/4"/Blu. $25.00-35.00

❑ ❑ USA St7916 **Game,** 1979, Paper Board with Removable Pieces/5" x 10 1/4". $15.00-20.00

❑ ❑ USA St7901 **Video Communicator,** 1979, Gry or Blk/ Comic #1 - Star Trek Stars. $25.00-40.00

❑ ❑ USA St7902 **Video Communicator,** 1979, Gry or Blk/ Comic #2 - a Pill Swallows the Enterprise. $25.00-40.00

❑ ❑ USA St7903 **Video Communicator,** 1979, Gry or Blk/ Comic #3 - Time & Time & Time Again. $25.00-40.00

❑ ❑ USA St7904 **Video Communicator,** 1979, Gry or Blk/ Comic #4 - Votec's Freedom. $25.00-40.00

❑ ❑ USA St7905 **Video Communicator,** 1979, Gry or Blk/ Comic #5 - Starlight Starfight. $25.00-40.00

❑ ❑ USA St7906 **Ring - Capt Kirk,** 1979, **Yellow.** $30.00-40.00

❑ ❑ USA St7906 **Ring - Capt Kirk,** 1979, **Blue.** $25.00-35.00

❑ ❑ USA St7906 **Ring - Capt Kirk,** 1979, **Red.** $50.00-75.00

❑ ❑ USA St7907 **Ring - U.S.S. Enterprise,** 1979, **Yellow.** $30.00-40.00

❑ ❑ USA St7907 **Ring - U.S.S. Enterprise,** 1979, **Blue.** $25.00-35.00

❑ ❑ USA St7907 **Ring - U.S.S. Enterprise,** 1979, **Red.** $50.00-75.00

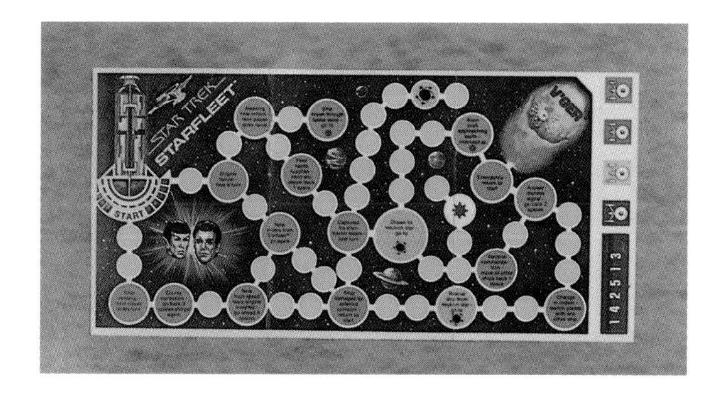

St7916

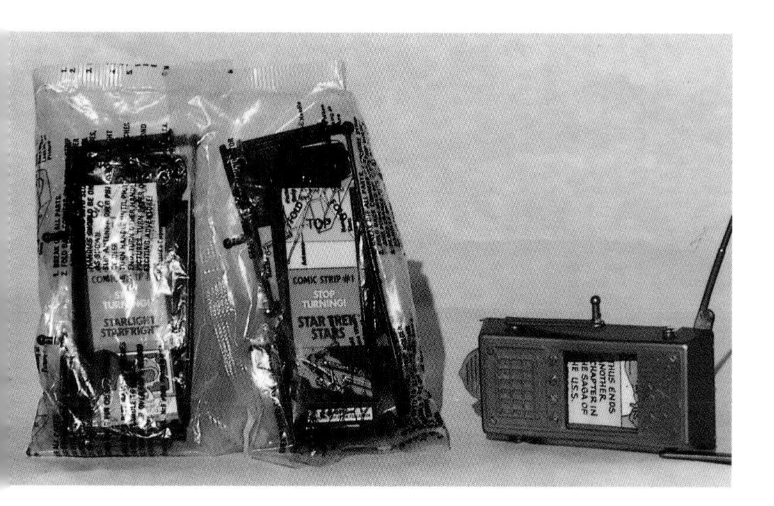

St7905 St7901 St7902

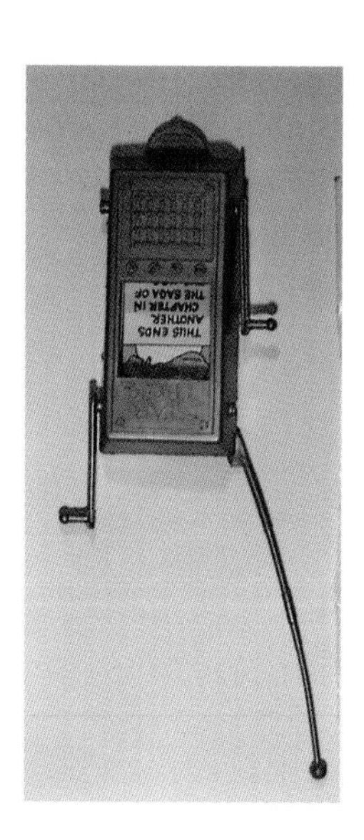

St7904

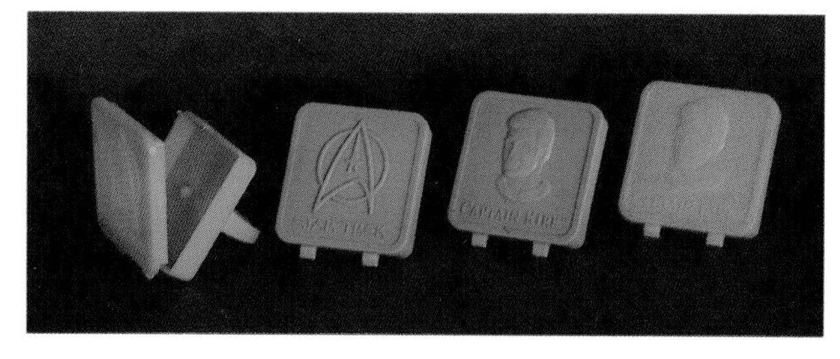

St7909 St7906

❑ ❑ USA St7908 **Ring - Spock,** 1979, **Yellow.** $30.00-40.00
❑ ❑ USA St7908 **Ring - Spock,** 1979, **Blue.** $25.00-35.00
❑ ❑ USA St7908 **Ring - Spock,** 1979, **Red.** $50.00-75.00

❑ ❑ USA St7909 **Ring - Star Trek Logo,** 1979, **Yellow.** $30.00-40.00
❑ ❑ USA St7909 **Ring - Star Trek Logo,** 1979, **Blue.** $25.00-35.00
❑ ❑ USA St7909 **Ring - Star Trek Logo,** 1979, **Red.** $50.00-75.00

❑ ❑ USA St7910 **Iron-On Transfer - Capt Kirk,** 1979. $20.00-25.00
❑ ❑ USA St7911 **Iron-On Transfer - Dr. McCoy,** 1979. $20.00-25.00
❑ ❑ USA St7912 **Iron-On Transfer - Lt. Ilia,** 1979. $20.00-25.00
❑ ❑ USA St7913 **Iron-On Transfer - Mr. Spock,** 1979. $20.00-25.00

❑ ❑ USA St7934 **Crew Badge/Paper Stick-On.** $25.00-40.00
❑ ❑ USA St7935 **Entry Form/Star Trek the Motion Picture,** 1980, Paper. $10.00-15.00

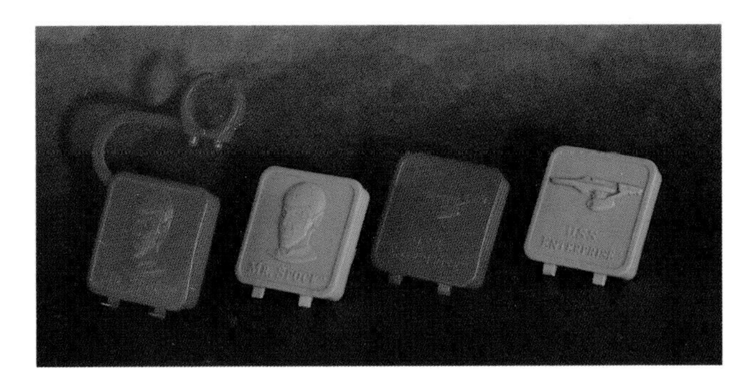

St7908 St7907

St7910

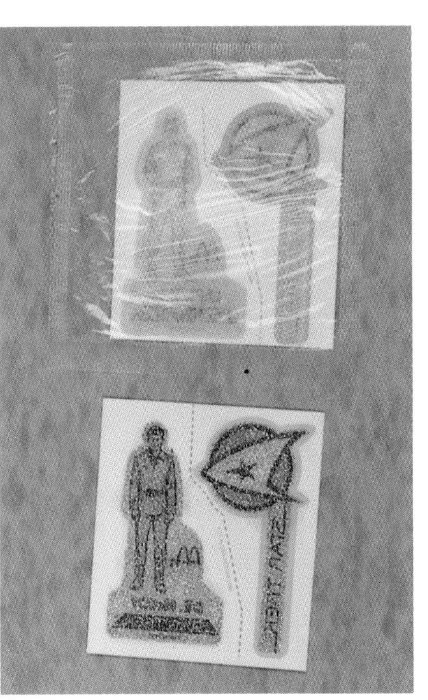
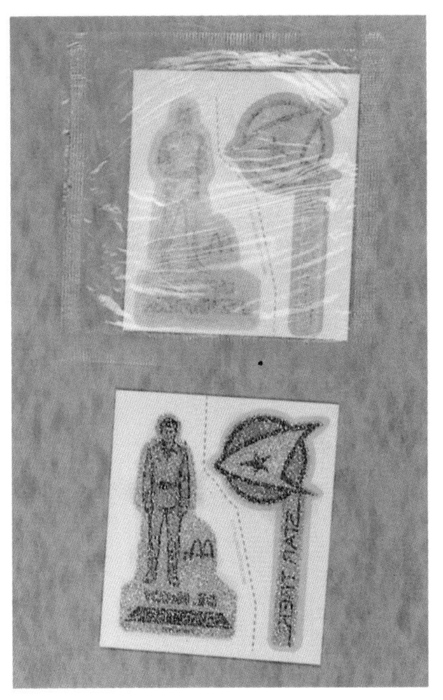

St7911

Top: St7913; Bottom: St7912

St7934

St7935

St7936

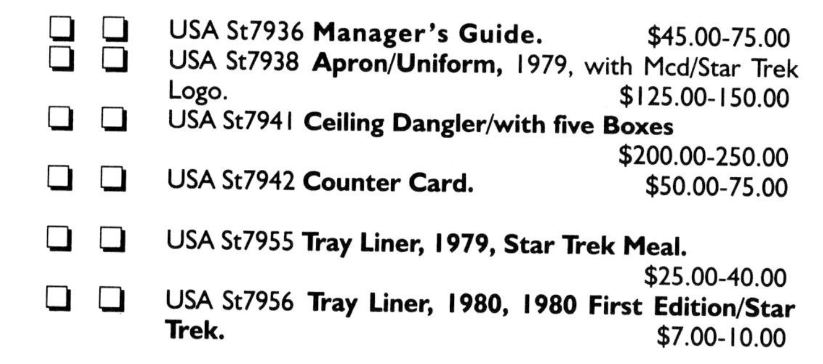

❑ ❑ USA St7936 **Manager's Guide.** $45.00-75.00

❑ ❑ USA St7938 **Apron/Uniform,** 1979, with Mcd/Star Trek Logo. $125.00-150.00

❑ ❑ USA St7941 **Ceiling Dangler/with five Boxes** $200.00-250.00

❑ ❑ USA St7942 **Counter Card.** $50.00-75.00

❑ ❑ USA St7955 **Tray Liner, 1979, Star Trek Meal.** $25.00-40.00

❑ ❑ USA St7956 **Tray Liner, 1980, 1980 First Edition/Star Trek.** $7.00-10.00

St7938

St7942

St7955

St7941

St7956

❑ ❑ USA St7965 **Translite/1 Large Star Trek Hm Box.** $175.00-200.00

❑ ❑ USA St7966 **Translite/Hm Box/Ham/Fries/Shake/Pin Holes.** $175.00-200.00

Comments: National Distribution: USA - December 26, 1979 - February 3, 1980. This was USA Happy Meal, where premiums were designed for a national motion picture release, *Star Trek: The Motion Picture*. Communicators came with one comic strip having two back-to-back different titles/listed. Some regions developed their own promotions/premiums. Generic premiums could have been given with the Happy Meal. No under 3 (U-3) premiums were furnished. Some stores offered entry forms "Win Tickets for Two" to the "Star Trek: the Motion Picture" (1980).

The name, "Star Trek MEAL Happy Meal" followed "Space MEAL Happy Meal." The word "Meal" was dropped in 1980, with the following promotions emphasizing, "HAPPY MEAL." The Star Trek Meal included one of six boxes, one of five prizes, and a sampler packet of McDonaldland Cookies. Five basic box designs were used, with the sixth box only slightly different due to cost and efficiency constraints in the printing process (information from the Manager's Guide).

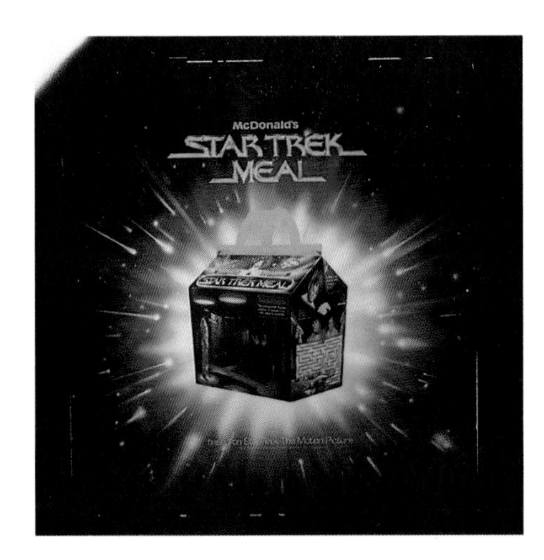

St7965

St7966

The Eighties

McDonald's "famous menu" was promoted in the 1980s by the national advertising slogan, "Nobody can do it like McDonald's can." During the 1980s, nobody could.

The full introduction of the breakfast menu in the eighties meant McDonald's was becoming more things to more people. Slowly, McDonald's went from drive-in's to inside seating to inside seating plus drive-thru's and playland parks. The impact of the playland parks outside and inside McDonald's would be felt for decades to come.

The concept of the Happy Meal for children was being refined during the eighties. In testing stages, the Happy Meal was tested with different toys around the country. In 1982, the focus shifted to collect a "Prize" in every box, not just "Collect the box."

By 1984, in spite of the death of Ray Kroc, the celebration candle did not die out. Instead, the candle burned brighter with the 25th Anniversary and the 30th Anniversary celebrations continuing to refocus McDonald's towards communities. Community involvement through the Ronald McDonald Houses and sports activities involved the surrounding communities and resulted in increased sales in the local stores, the golden objective. Through market testing regional toys, the advertising and marketing departments decided on a national as opposed to a regional focus. National advertising tied to movie promotions brought the most successful ventures in Happy Meal promotions. Disney-McDonald's joint promotions include Bambi, Duck Tales, and Oliver & Company Happy Meal promotions. Joint marketing efforts were exclusive outside the USA during this period, with McDonald's teaming with Disney. Within the USA, the team approach was still selective and cautious. That is, McDonald's rode a steady course until the competition began to appear in the burger business. In the late 1980s Burger King, Wendy's, KFC, Hardee, Sonic, White Castle, Taco Bell, and others began to make their presence known and McDonald's took immediate note by picking up the advertising pace.

The restaurant of the eighties was expanded to include drive-thru service and playgrounds. The impact of both concepts moved customers quickly towards McDonald's. That magic moment, when the decision on where to eat was being made, was based on McDonald's consistent reputation for Quality, Service, Value, and Cleanliness, as well as entertainment for the children.

Hamburglar Statue

Ronald McDonald Statue

Captain Crook Statue

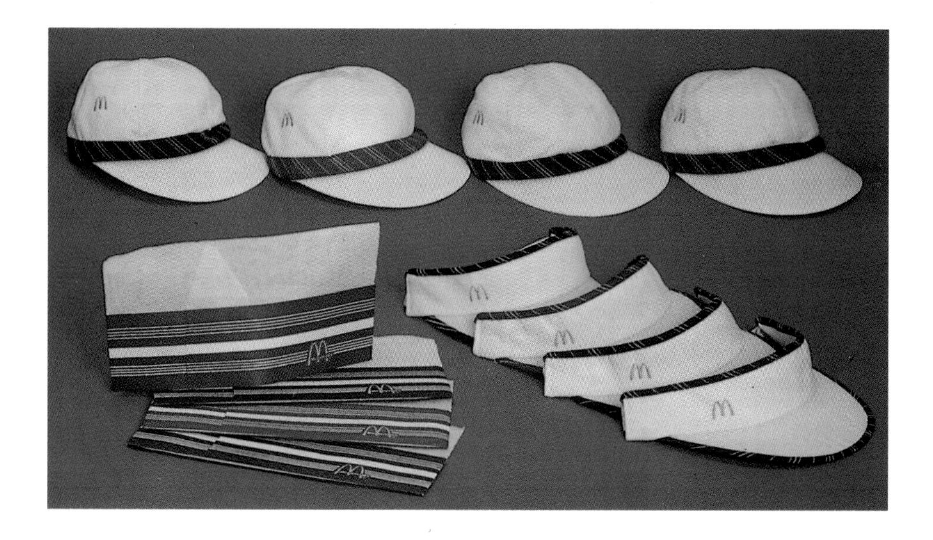

Apple Pie Tree

PLAYLAND RULES AND REGULATIONS SIGN

The Playland Rules and Regulations Sign is recommended for all Playlands by McDonald's Corporation. This sign has a solid Aluminum frame for attaching to building or poles. The sign face is protected by a clear Lexan sheet and comes complete with a set of numbers for changeable Playland hours. Size 36" x 48".

Playland Sign

THE APPLE PIE TREE

The Apple Pie Tree is a friendly attraction for children, whether inside or outside. When combined with children's

Big Mac Statue

Mayor McCheese Statue

Grimace Statue

Da8036

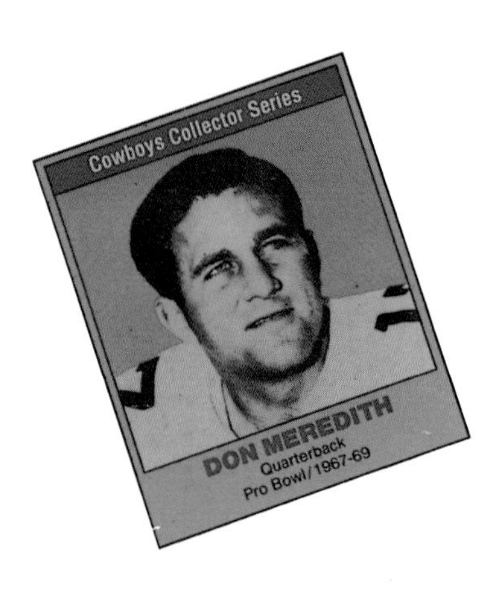

1980

Dallas Cowboys Super Box Happy Meal, 1980
Happy Hotcakes Promotion, 1980
Look Look Books Happy Meal, 1980
Olympic [Medal] Happy Meal (Canceled), 1980
Safari Adventure Meal Happy Meal, 1980
Undersea Happy Meal, 1980
USA Generic Promotions, 1980

- **McHappy Day promotion**
- **25th Silver Anniversary of McDonald's**
- **6th National O/O Convention held**
- **"Our World...Today, Yesterday and Tomorrow" O/O theme**
- **Birdie, The Early Bird introduced unofficially**

Dallas Cowboys Super Box Happy Meal, 1980

❑ ❑ USA Da8035 **Hm Super Box I Chuck Howley/Don Perkins**, 1980. $200.00-250.00
❑ ❑ USA Da8036 **Hm Super Box II Don Meredith/Bob Lilly**, 1980. $200.00-250.00
❑ ❑ USA Da8037 **Hm Super Box III Roger Staubach/W. Garrison**, 1980. $200.00-250.00

Comments: Regional Distribution: USA - 1980 in Dallas, Texas. Two cut-out cowboys collector series football cards on each box.

Happy Hotcakes Promotion, 1980

❑ ❑ USA Ho8010 **Pancakes Wrapper - Batter up for Breakfast**, 1980. $15.00-20.00
❑ ❑ USA Ho8011 **Pancakes Wrapper - Flyin' High**, 1980. $15.00-20.00
❑ ❑ USA Ho8012 **Pancakes Wrapper - Too Pooped to Pedal**, 1980. $15.00-20.00

Ho8010

Top: Ho8012; Bottom: Ho8011

Premiums:
- ❑ ❑ USA Ho8001 **Cup: Happy Milk - Ron with Net/Red Band**, 1980. $3.00-5.00
- ❑ ❑ USA Ho8002 **Cup: Happy Milk - Ron with Binoculars/ Blue Band**, 1980. $3.00-5.00
- ❑ ❑ USA Ho8003 **Cup: Happy Milk - Ron with Pillow/Brn Band**, 1980. $3.00-5.00
- ❑ ❑ USA Ho8004 **Cup: Happy Milk - Ron with Birdie Flying/ Grn Band**, 1980. $3.00-5.00

Comments: Regional Distribution: USA - 1980 in South Carolina and Ohio.

Ho8001 Ho8002 Ho8003 Ho8004

Look Look Books Happy Meal, 1981/1980

Premiums: St. Louis, Missouri Area Books (1980):
- ❑ ❑ USA Lo8000 **Book: Animals of the Sea,** 1980, Logo on back/Paper Booklet. $40.00-50.00
- ❑ ❑ USA Lo8001 **Book: Animals That Fly,** 1980, Logo on back/ Paper Booklet. $40.00-50.00
- ❑ ❑ USA Lo8002 **Book: Cats in the Wild,** 1980, Logo on back/ Paper Booklet. $40.00-50.00
- ❑ ❑ USA Lo8003 **Book: The Biggest Animals,** 1980, Logo on back/Paper Booklet. $40.00-50.00

Premiums: Marine World/California Area Books (1981):
- ❑ ❑ USA Lo8005 **Book: Animals of the Sea,** 1980, Logo on front/Paper Booklet. $40.00-50.00
- ❑ ❑ USA Lo8006 **Book: Animals That Fly,** 1980, Logo on front/ Paper Booklet. $40.00-50.00
- ❑ ❑ USA Lo8007 **Book: Cats in the Wild,** 1980, Logo on front/ Paper Booklet. $40.00-50.00
- ❑ ❑ USA Lo8008 **Book: The Biggest Animals,** 1980, Logo on front/Paper Booklet. $40.00-50.00

- ❑ ❑ USA Lo8042 **Counter Card**/Dated April 20, 1980. $50.00-75.00

Comments: Regional Distribution: USA - St. Louis, Missouri in 1980 and Southern California by Marine World in 1981. Offer good until April 20, 1980 in St. Louis set, July 5, 1981 in California set. Both sets of books are exactly the same size, 5" x 4" with 12 pages. MIP St. Louis books came packaged in cellophane. MIP California books came loose, not packaged in cellophane. Books were produced by Western Publishing with the McDonald's logo on the back of the book on St. Louis set and on the front of the book on California set. Historically, this promotion is, "McDonald's Happy Meal" giving away Look Look Books.

Lo8005-08

Lo8002

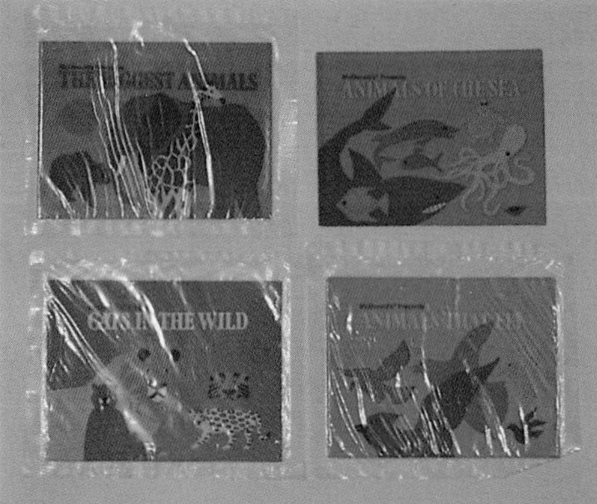

Lo8000-03

Lo8042

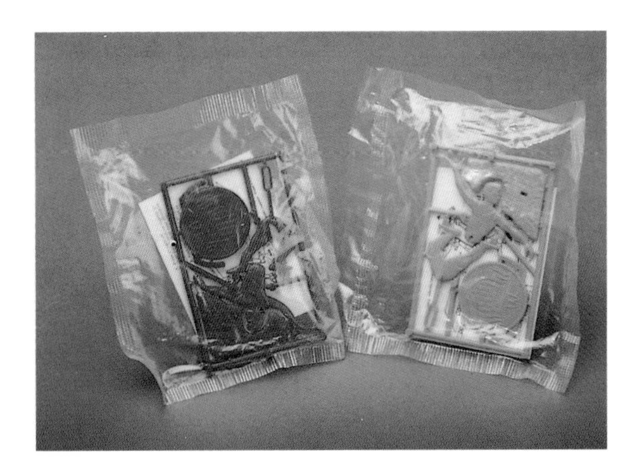

O18005 O18003

Olympic [Medal] Happy Meal (Canceled), 1980

Premiums:

❑ ❑ USA O18001 **Male Basketball Player,** 1980, red or yellow or blue/3p. $20.00 -35.00

❑ ❑ USA O18002 **Male Swimmer/Diver,** 1980, red or yellow or blue. $20.00 -35.00

❑ ❑ USA O18003 **Female Swimmer/Diver,** 1980, with paper diving medal in red or yellow or blue. $20.00 -35.00

❑ ❑ USA O18004 **Female Gymnast,** 1980, red or yellow or blue. $20.00-35.00

❑ ❑ USA O18005 **Male Javelin Thrower,** 1980, red or yellow or blue. $20.00-35.00

❑ ❑ USA O18006 **Male Pole Vaulter,** 1980, red or yellow or blue. $20.00-35.00

❑ ❑ USA O18007 **Male Soccer Player,** 1980, with paper soccer goal in red or yellow or blue. $20.00-35.00

Comments: National Distribution planned: USA June 2 - July 6, 1980. This Happy Meal was planned in conjunction with the Summer Olympics to be held in Moscow, Russia. President Jimmy Carter protested the Soviet's invasion of Afghanistan in December 1979 and subsequently canceled the USA's participation in the games. McDonald's canceled the Happy Meal and replaced the time slot with Safari Adventure Happy Meal. Each premium came with an insert card which either included instructions or paper accessories.

Sa8015

Safari Adventure Meal Happy Meal, 1980

Boxes:

❑ ❑ USA Sa8015 **Hm Box - Ronald and Grimace**, 1980. $25.00-40.00

❑ ❑ USA Sa8016 **Hm Box - Ronald and Hyena,** 1980. $25.00-40.00

❑ ❑ USA Sa8017 **Hm Box - Ronald/Monkey/Fry Guy**, 1980. $25.00-40.00

❑ ❑ USA Sa8018 **Hm Box - Ronald/Vines/Carrot/Hippo**, 1980. $25.00-40.00

Sa8017

Sa8016

Premiums:

- ❑ ❑ USA Sa8001 **Alligator,** 1980, Blu or Brn or Gry or Org or Pnk or Pur or Yel or Grn. $1.00-2.00
- ❑ ❑ USA Sa8002 **Ape,** 1980, Blu or Brn or Gry or Org or Pnk or Pur or Yel or Grn. $1.00-2.00
- ❑ ❑ USA Sa8003 **Elephant,** 1980, Blu or Brn or Gry or Org or Pnk or Pur or Yel or Grn. $1.00-2.00
- ❑ ❑ USA Sa8004 **Hippo,** 1980, Blu or Brn or Gry or Org or Pnk or Pur or Yel or Grn. $1.00-2.00
- ❑ ❑ USA Sa8005 **Lion,** 1980, Blu or Brn or Gry or Org or Pnk or Pur or Yel or Grn. $1.00-2.00
- ❑ ❑ USA Sa8006 **Monkey,** 1980, Blu or Brn or Gry or Org or Pnk or Pur or Yel or Grn. $1.00-2.00
- ❑ ❑ USA Sa8007 **Rhino,** 1980, Blu or Brn or Gry or Org or Pnk or Pur or Yel or Grn. $1.00-2.00
- ❑ ❑ USA Sa8008 **Tiger,** 1980, Blu or Brn or Gry or Org or Pnk or Pur or Yel or Grn. $1.00-2.00

- ❑ ❑ USA Sa8019 **Comb - Ronald,** 1980, Red or Yel or Blu. $1.00-2.00
- ❑ ❑ USA Sa8020 **Comb - Grimace,** 1980, Red or Yel or Blu. $1.00-2.00

- ❑ ❑ USA Sa8021 **Sponge - Grimace,** Nd, Grimace Standing/Purple. $5.00-8.00
- ❑ ❑ USA Sa8022 **Sponge - Ronald,** Nd, Ronald Sitting with Arms/Legs Crossed. $3.00-5.00
- ❑ ❑ USA Sa8023 **Sponge - Hamburglar Floater,** Nd, Org 5 1/8" Circle Disc. $15.00-20.00

- ❑ ❑ USA Sa8024 **Cookie Mold - Ronald,** 1980, Red or Yel. $1.00-1.50
- ❑ ❑ USA Sa8025 **Cookie Mold - Grimace,** 1980, Red or Yel. $1.00-1.50

Sa8001 Sa8002   Sa8003 Sa8004   Sa8005 Sa8006   Sa8007 Sa8008

Sa8019

Sa8021 Sa8023 Sa8022

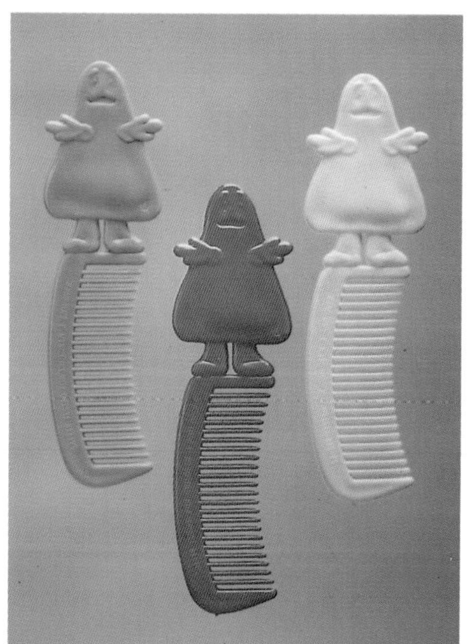

Sa8020

Left: Sa8024
Right: Sa8025

Clockwise from left: Sa8027, Sa8030, Sa8029, Sa8026, Sa8028.

❑ ❑ USA Sa8026 **Dome Game - Big Mac Double Balls in Rings,** 1979, Yel Base. $10.00-12.00

❑ ❑ USA Sa8027 **Dome Game - Captain 3 Balls in Slot,** 1979, Yel Base. $10.00-12.00

❑ ❑ USA Sa8028 **Dome Game - Hamburglar Single 2 Rings Toss,** 1979, Yel Base. $10.00-12.00

❑ ❑ USA Sa8029 **Dome Game - Mayor 2 Balls in Ring Toss,** 1979, Yel Base. $10.00-12.00

❑ ❑ USA Sa8030 **Dome Game - Ronald Double Ring 2 Rings Toss,** 1979, Yel Base. $10.00-12.00

❑ ❑ USA Sa8031 **Light Switch Cover,** 1980, Ron/Stick-On/ 4" x 2 1/2"/Glow-In-Dark. $3.00-5.00

❑ ❑ USA Sa8032 **Ring - Ronald Whistle,** 1980, Red. $8.00-10.00

❑ ❑ USA Sa8033 **Pennant - Ronald,** 1980. $15.00-20.00

❑ ❑ USA Sa8034 **Pennant - Grimace,** 1980. $15.00-20.00

❑ ❑ USA Sa8035 **Bead Game - Tip 'N' Tilt,** 1980, Triangle Bead Game. $50.00-75.00

❑ ❑ USA Sa8036 **Hamburglar Hockey,** 1980, plastic game with sticker sheet/org. $4.00-6.00

Sa8031

Sa8032

Sa8033
Sa8034

Sa8035

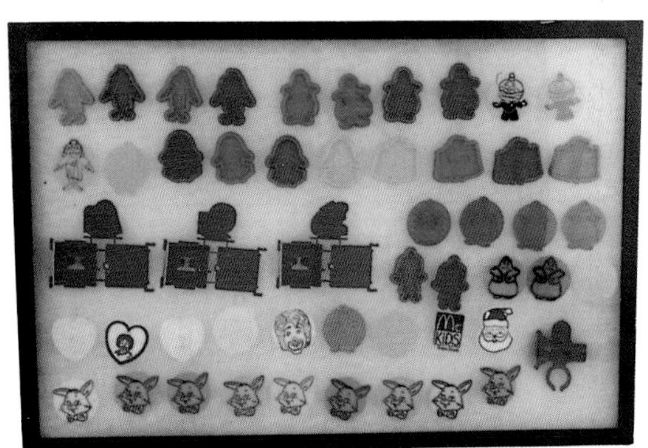

Sa8036

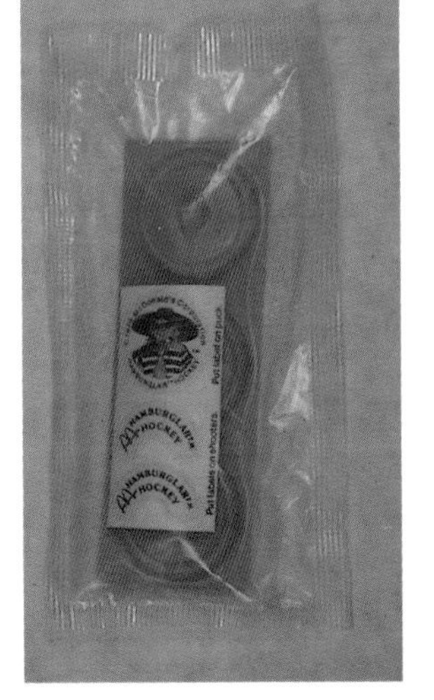

❑ ❑ USA Sa8037 **Letterland Stationery - Ronald/Professor,** 1980. $3.00-4.00

❑ ❑ USA Sa8038 **Letterland Stationery - Capt/Big Mac/Grimace/Fry Guy,** 1980. $3.00-4.00

❑ ❑ USA Sa8039 **Game - Fishin' Fun,** 1980, Paper. $8.00-10.00

❑ ❑ USA Sa8040 **Game - Who's in the Zoo?** 1980, Paper. $8.00-10.00

❑ ❑ USA Sa8043 **McDonaldland Cookie Sampler,** 1979, Knock, knock. Who's there? $3.00-4.00

❑ ❑ USA Sa8044 **McDonaldland Cookie Sampler,** 1979, It's hands are always on it's face. What is it? $3.00-4.00

❑ ❑ USA Sa8045 **McDonaldland Cookie Sampler,** 1979, What animal keeps the time? $3.00-4.00

❑ ❑ USA Sa8046 **McDonaldland Cookie Sampler,** 1979, What runs without feet? $3.00-4.00

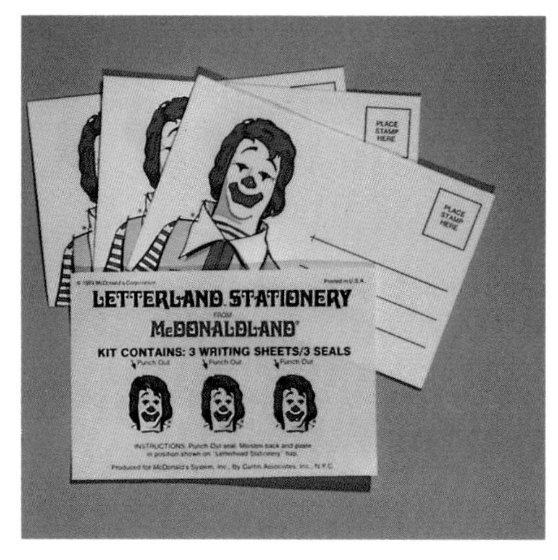

Sa8037

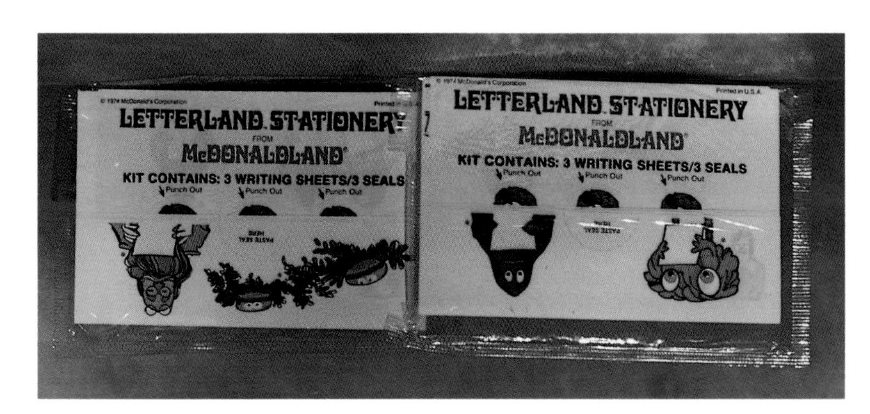

Sa8038

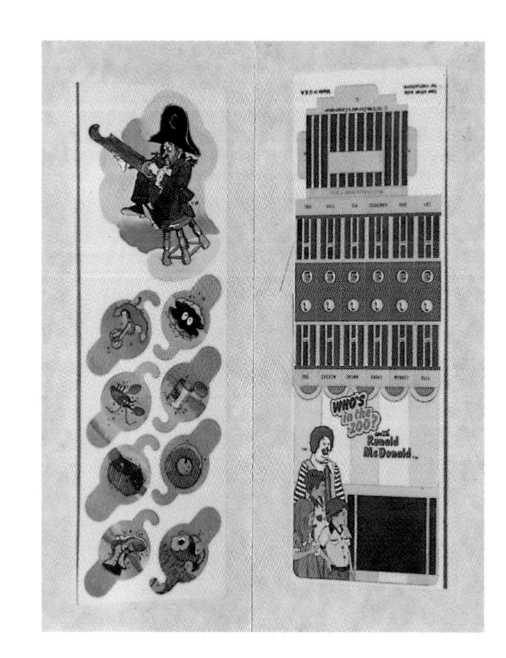

Sa8039 Sa8040

Sa8045

Sa8043

Sa8044

Sa8065

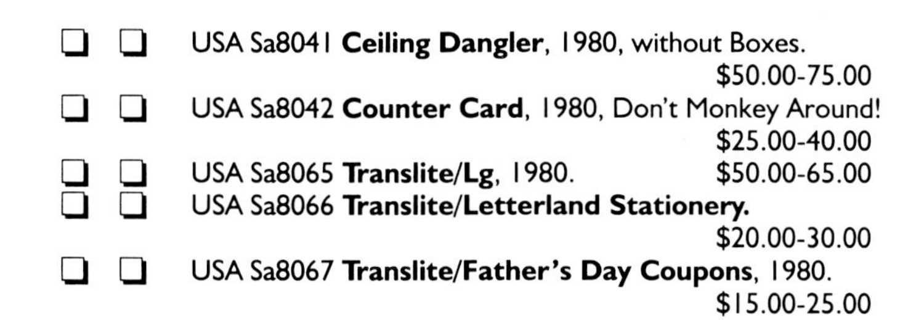

- ❑ ❑ USA Sa8041 **Ceiling Dangler**, 1980, without Boxes. $50.00-75.00
- ❑ ❑ USA Sa8042 **Counter Card**, 1980, Don't Monkey Around! $25.00-40.00
- ❑ ❑ USA Sa8065 **Translite/Lg**, 1980. $50.00-65.00
- ❑ ❑ USA Sa8066 **Translite/Letterland Stationery.** $20.00-30.00
- ❑ ❑ USA Sa8067 **Translite/Father's Day Coupons**, 1980. $15.00-25.00

Comments: National Promotion: USA June 2-29, 1980. Figurine premium markings - "Diener Ind." Listed figurines were rubber and loose. Regions developed their own promotions/premiums during 1980. Some stores gave out hard rubber safari figurines later (post 1982) wrapped in plastic with the warning, "Safety Tested for Children 3 and over." These figurines with the safety test markings have an estimated value of $40.00 each, MIP. They were distributed later and not with the Safari Happy Meal in 1980. Safety test information on the packaging came about after the Playmobile Happy Meal recall in 1982. McDonaldland Cookie Samplers were given with each Happy Meal.

Sa8066

Sa8067

Undersea Happy Meal, 1980

Boxes:

- ❑ ❑ USA Un8035 **Hm Box - Captain Crook & Ronald in Fish Mouth**, 1980. $35.00-50.00
- ❑ ❑ USA Un8036 **Hm Box - Grimace Squid & Scuba Diver**, 1980. $35.00-50.00

Un8035

Un8036

☐ ☐ USA Un8037 **Hm Box - Grimace on Porpoise Squid,** 1980. $35.00-50.00

☐ ☐ USA Un8038 **Hm Box - Sea Shell Secret Message**, 1980. $35.00-50.00

☐ ☐ USA Un8039 **Hm Box - Ronald in Submarine,** 1980. $35.00-50.00

☐ ☐ USA Un8040 **Hm Box - Whale of a Tale/Eels/Dog/ Chicken**, 1980. $35.00-50.00

☐ ☐ USA Un8001 **Alligator,** Nd, "Diener" Molded into Figurine. $1.00-2.00

☐ ☐ USA Un8002 **Dolphin/Porpoise,** Nd, "Diener" Molded into Figurine. $1.00-2.00

☐ ☐ USA Un8003 **Seal,** Nd, No Company Name. $1.00-2.00

☐ ☐ USA Un8004 **Shark/Great White,** Nd, No Company Name. $1.00-2.00

☐ ☐ USA Un8005 **Shark/Hammerhead,** Nd, "Hammerhead" Molded into Figurine. $1.00-2.00

☐ ☐ USA Un8006 **Shark/Tiger,** Nd, "Tiger Shark" Molded into Figurine. $1.00-2.00

☐ ☐ USA Un8007 **Shark/Whale,** Nd, "Whaleshark" Molded into Figurine. $1.00-2.00

☐ ☐ USA Un8008 **Turtle/Sea,** Nd, "Diener" Molded into Figurine. $1.00-2.00

☐ ☐ USA Un8009 **Walrus,** Nd, "Diener" Molded into Figurine. $1.00-2.00

☐ ☐ USA Un8010 **Whale,** Nd, "Diener" Molded into Figurine. $1.00-2.00

Un8037

Un8038

Un8039

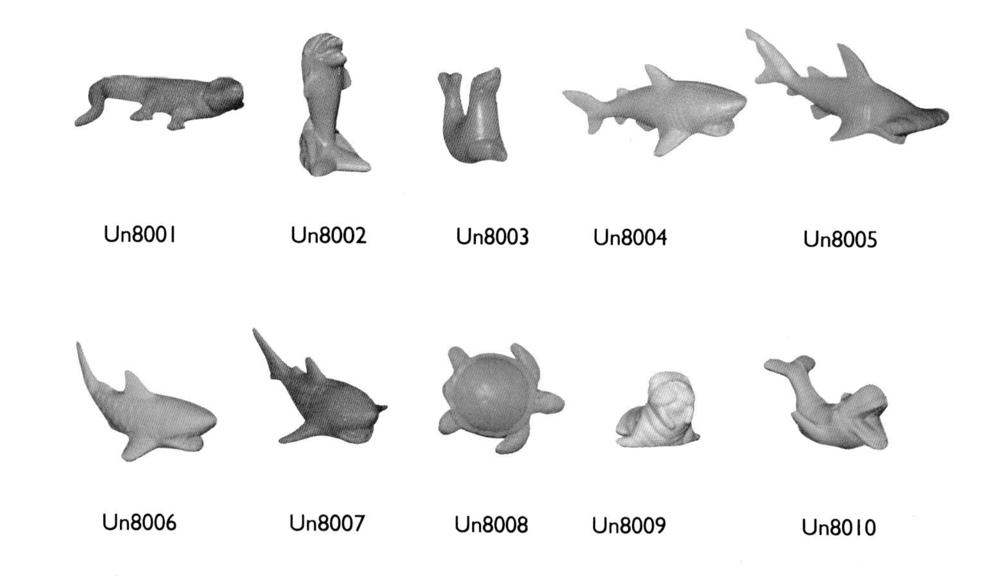

Un8001 Un8002 Un8003 Un8004 Un8005

Un8006 Un8007 Un8008 Un8009 Un8010

Un8011 Un8012

- ❑ ❑ USA Un8011 **Eggart Whapper - Grimace,** Nd. $10.00-15.00
- ❑ ❑ USA Un8012 **Eggart Whapper - Ronald,** Nd. $10.00-15.00

- ❑ ❑ USA Un8041 **Ceiling Dangler** without Boxes, 1980. $65.00-90.00
- ❑ ❑ USA Un8042 **Counter Card**, 1980. $25.00-35.00
- ❑ ❑ USA Un8065 **Translite/Lg**, 1980. $50.00-65.00

Comments: National Optional Distribution: USA - March 24-April 6, 1980 (in the Spring). All figurines were soft rubber and came in assorted colors. Regions could accept the national promotion or develop their own promotions. Generic premiums could have been given with this Happy Meal. The Happy Meal name "undersea" is not on the front of the boxes, but is on the Counter Card and Translite. "Undersea" is really the location of a "Happy Meal" which is surfacing. Grammatically, "undersea" is spelled with lower case letters and not capitalized, in reference to McDonald's selected name.

In the Boston regional area, Eggart Whapper with Ronald and Grimace were given with this promotion. These Eggart Whappers were covers for hard boiled eggs, given out during the Easter season. McDonaldland Cookie Sampler packages, like in the Safari Adventure Meal Happy Meal, were also given with each Happy Meal.

Un8042

USA Generic Promotions, 1980

- ❑ ❑ USA Ge8001 **Kite: McDonald's Fry'n the Sky.** $75.00-125.00
- ❑ ❑ USA Ge8003 **Jalopy Car - Birdie,** 1980, Org or Red or Yel/5p. $15.00-20.00
- ❑ ❑ USA Ge8006 **Jalopy Car - Hamburglar,** 1980, Org or Red or Yel/5p. $15.00-20.00
- ❑ ❑ USA Ge8009 **Jalopy Car - Ronald,** 1980, Org or Red or Yel/5p. $15.00-20.00

- ❑ ❑ USA Ge8012 **Icicle Stick Mold,** 1980, Yel/4 1/4"/2p. $1.00-2.00
- ❑ ❑ USA Ge8013 **Pen - Grimace,** 1980, Purp with Corded Thread. $5.00-7.00
- ❑ ❑ USA Ge8014 **Pen - Big Mac,** 1980, Brn with Corded Thread. $10.00-12.00
- ❑ ❑ USA Ge8015 **Pen - Hamburglar,** 1980, Blk with Corded Thread. $5.00-7.00
- ❑ ❑ USA Ge8016 **Pen - Ronald**, 1980, Yel/Red/Wht with Corded Thread. $5.00-7.00

Ge8001

Ge8016 Ge8014 Ge8015 Ge8013

Ge8012 Ge8009 Ge8003 Ge8006

❑ ❑ USA Ge8017 **Pencils/ Collector Series: Ronald McDonald (Bubble Print), Mayor McCheese, Big Mac, Grimace, Hamburglar, Captain Crook.** Brown band around eraser. In early 1980, McDonald's was still trying for character identification with its premiums. Happy Meals were six months into the program. September brought collector series of pencils illustrating the characters. Each pencil has a brown band around the gold cap holding the eraser. Names are in capital letters. Birdie, The Early Bird was not an official McDonaldland character, at this time. Set of 6 $18.00-24.00

❑ ❑ USA Ge8018 **Ronald Styro-glider, red/white.** Styro-glider airplanes appeared again in 1980. These airplanes were easily flown and crashed. Dated 1980. $7.00-10.00

❑ ❑ USA Ge8019 **Space Cruiser, USA 648-50, red/white,** no date. $7.00-10.00

❑ ❑ USA Ge8020 **Ronald McDonald Airlines, yellow/red,** no date. $7.00-10.00

❑ ❑ USA Ge8021 **Ronald McDonald airplane, florescent red/ yellow, yellow nose weight,** no date. $7.00-10.00

❑ ❑ USA Ge8022 **Sunglasses: Ronald, Grimace, Hamburglar or BIRDIE**. A pair of 1980 sunglasses illustrating Birdie, The Early Bird, arriving ahead of schedule was given out regionally. $4.00-5.00

❑ ❑ USA Ge8023 **Valentine Cookie Sleeves:** Four 5" x 5" Cookie Box Sleeves, Chicago regional: a. Valentine, I hoped from the START (maze); b. PSSST! Here's a great play; c. Color me any color but blue; d. I'd have to search the whole world through. Each $5.00-8.00

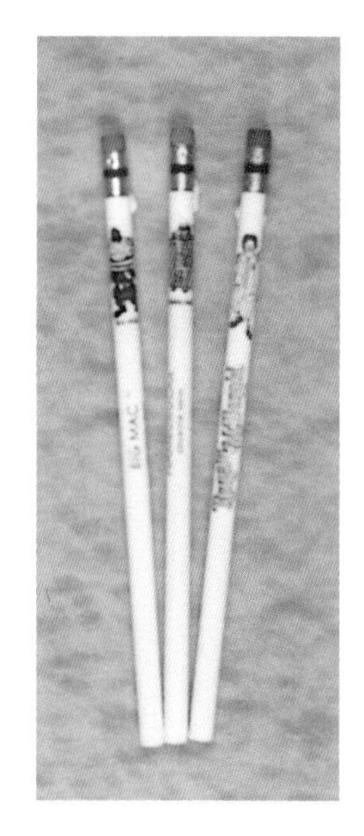

Ge8017

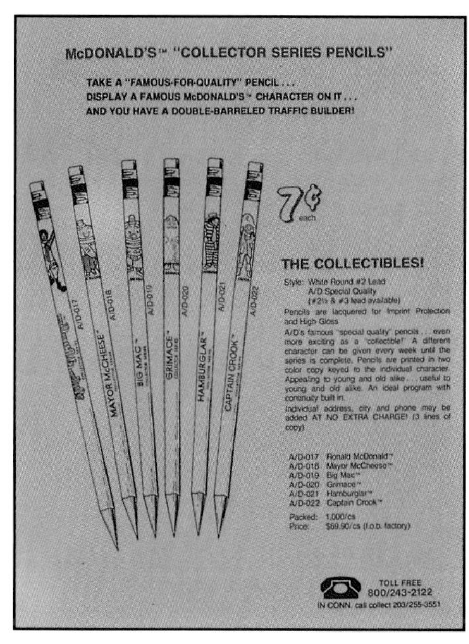

Ge8017

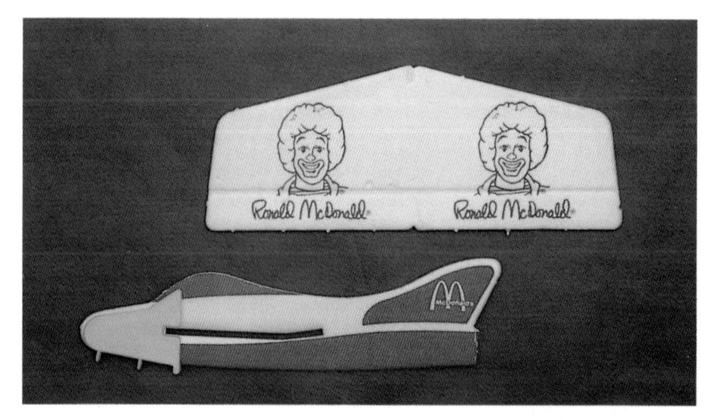

Ge8018

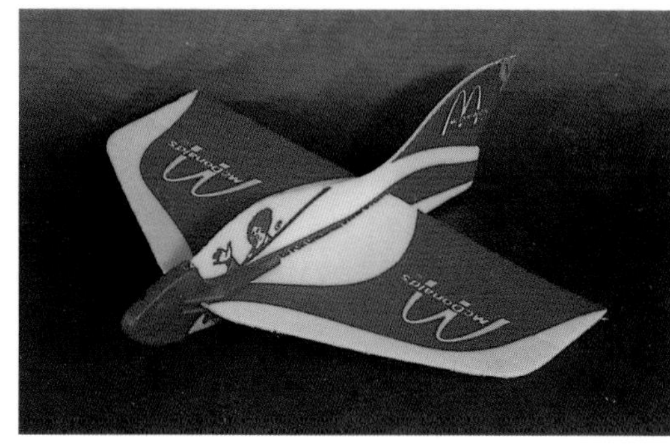

Ge8023

Ge8023

Ge8026 Ge8024 Ge8027 Ge8025

❑ ❑ USA Ge8024 **White Plastic Cup: Grimace playing Tug of War with pink elephant, 5". 1980 McDonaldland ZOO Happy Cups - 5" WHITE plastic cup.** Action graphics with characters at the Zoo. Copyright McDonald's System, Inc. 1978 on the base of cup. Same as yellow 5" cups. **White.** $3.00-4.00

❑ ❑ USA Ge8025 **White Plastic Cup: Ronald jumping rope held by two apes,** 5" white. $3.00-4.00

❑ ❑ USA Ge8026 **White Plastic Cup: Captain Crook sitting in boat near white seal,** 5" white. $3.00-4.00

❑ ❑ USA Ge8027 **White Plastic Cup: Hamburglar with Zebra,** 5" white. $3.00-4.00

❑ ❑ USA Ge8028 **Yellow Plastic Cup: Ronald McDonald Saves the Falling Star, 4 3/4". 1980 McDonaldland Adventure Series, 4 3/4" yellow plastic cup.** Same graphics as the GLASS McDonaldland Adventure Series. Copyright 1980 McDonald's Corporation at base. Six cups in the series. $3.00-4.00

❑ ❑ USA Ge8029 **Yellow Plastic Cup: Grimace Climbs a Mountain,** 4 3/4". $3.00-4.00

❑ ❑ USA Ge8030 **Yellow Plastic Cup: Hamburglar Hooks the Hamburgers,** 4 3/4". $3.00-4.00

❑ ❑ USA Ge8031 **Yellow Plastic Cup: Captain Crook Sails the Bounding Main,** 4 3/4". $3.00-4.00

❑ ❑ USA Ge8032 **Yellow Plastic Cup: Big Mac Nets the Hamburglar,** 4 3/4". $3.00-4.00

❑ ❑ USA Ge8033 **Yellow Plastic Cup: Mayor McCheese Rides a Runaway Train,** 4 3/4". $3.00-4.00

Ge8028 Ge8029 Ge8030

Ge8031

❑ ❑ USA Ge8034 **Yellow Plastic Cup: Ronald McDonald and Grimace in front of restaurant, "Happy Cup" under logo,** No date. Logo on back. 5". $4.00-5.00

❑ ❑ USA Ge8035 **Ronald McDonald Sugar Cookie,** large size. $15.00-25.00

❑ ❑ USA Ge8036 **Record: Ronald McDonald visits America with story book, 78 RPM.** A child's tour of the 50 states with Ronald-includes book & record. $10.00-15.00

❑ ❑ USA Ge8037 **Record: Ronald McDonald & Friends, 45 RPM.** Share a Song From Your Heart and "F-R-I-E-N-D-S". Casablanca Record and FilmWorks, Inc. $5.00-8.00

Ge8035

Ge8034

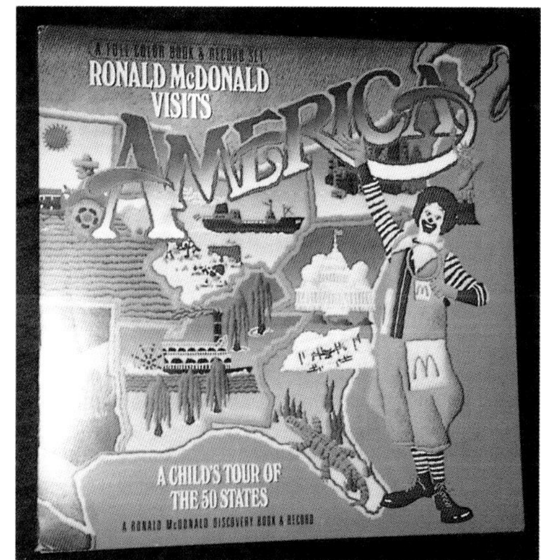

Ge8036

Ge8035

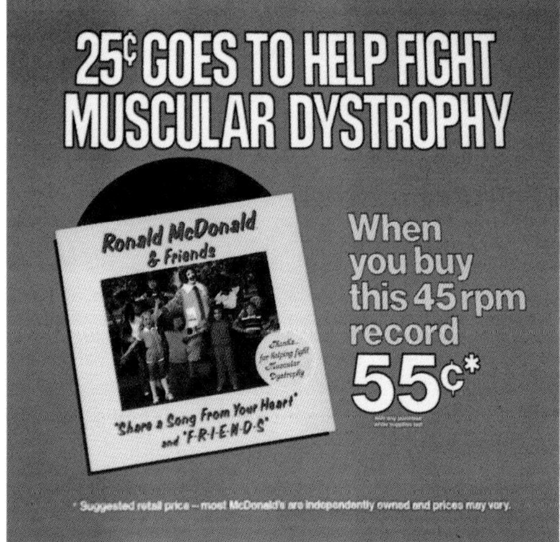

Ge8037

Ge8037

Ge8038 Ge8039

- ❑ ❑ USA Ge8038 **Record: Kids Radio Birthday Party Starring Ronald McDonald, Dee-Jay - yellow jacket, 78 RPM,** 1 of a 2 record set. $10.00-15.00
- ❑ ❑ USA Ge8039 **Record: Kids Radio Rainy Day Fun Starring Ronald McDonald, Dee-Jay - blue jacket, 78 RPM,** 1 of a 2 record set. $10.00-15.00
- ❑ ❑ USA Ge8040 **Super Door Alarm,** red plastic 2 piece with balloon. One stretches the balloon over the two pieces snapped together, blows up the balloon, twist ties it to keep the air from escaping, and waits for the "SOUNDS" of laughter! This premium was not widely distributed due to its design. $15.00-20.00
- ❑ ❑ USA Ge8041 **1980 Gift Wrap with Ribbons/ Seals/Tags/French Fry Coupon: SEASON'S GREETINGS.** $7.00-10.00
- ❑ ❑ USA Ge8042 **1980 Calendar: Ronald McDonald Coloring Calendar Book of Fun Facts...** $5.00-8.00
- ❑ ❑ USA Ge8043 **Stop Watch.** Yellow with paper decal. $20.00-25.00

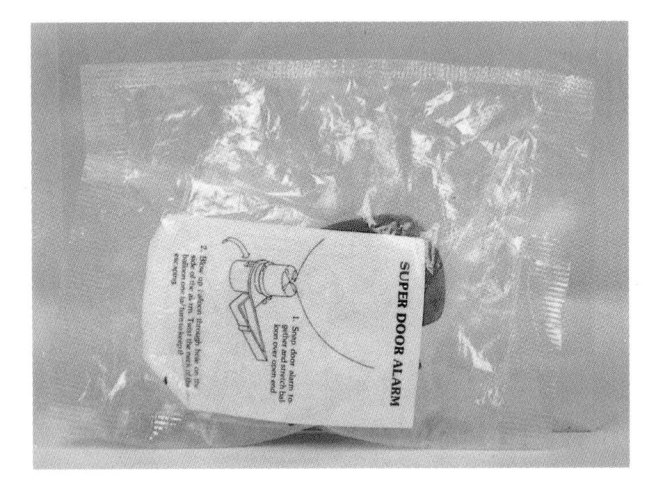

Ge8040

Ge8040

Ge8041

Ge8042

Ge8042

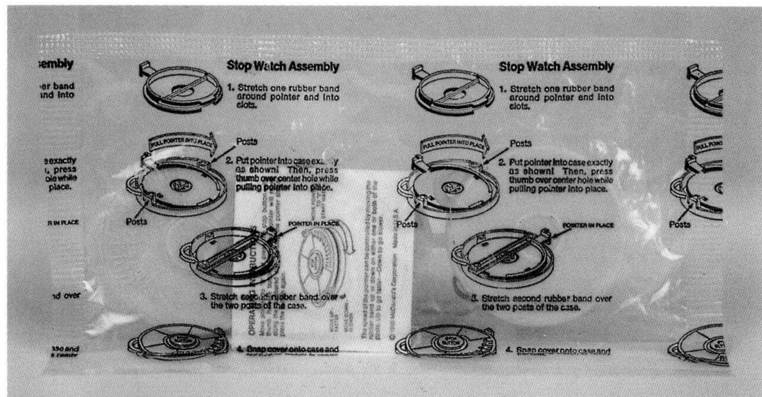

Ge8043

Comments: The 6th National O/O Convention was held in Toronto, Canada (Ontario). The occasion was also the celebration of the 25th Anniversary of the founding of the McDonald's operations. "Our World...Today, Yesterday and Tomorrow" was the advertising theme. Gift wrap was sold in the stores during the holiday period.

Ge8043

McDonaldland Fun Times Magazines, 1980

McDonaldland Fun Times magazine was initially a quarterly magazine designed especially for children. Each sixteen page issue features educational articles, games, word puzzles, and riddles for the children to enjoy. Over the years, the focus age of children changed from 5-7 years of age to pre-teens 9-13.

Ronald McDonald and the McDonaldland characters were an integral part of the magazine. It was designed to be the magazine premium that kids take home, play with, and read with their parents. It was initially designed to be without product advertising, without reference to specific promotions or regional orientation. This lofty approach has changed drastically over the years. Foremost, the magazine is a quality premium for a store's marketing plan aimed at children.

Initially, *McDonaldland Fun Times* was published each spring, summer, fall, and winter, beginning spring of 1979 in Canada. The magazine was first distributed in the USA in 1980. The Special Issue Summer 1980 Premier issue was the same as Vol. 1 No. 4 Spring 1980 from Canada, except that the Maple Leaf logo was deleted from the last page. In 1981, approximately 7,000,000 copies were circulated in Canada, the United States (including Hawaii and the Virgin Islands), and Singapore. Ireland received its first shipment in the fall of 1981.

CAN Ft7901

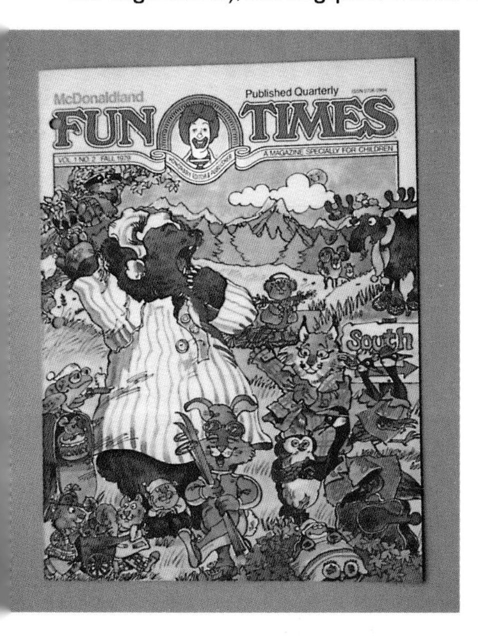

CAN Ft7902

CAN Ft7903

CAN Ft7904

❑ ❑ USA Ge8044 **Fun Times Magazine: Vol. 1 No. 4 Spring, 1980,** NO Maple leaf under McDonald's logo. $75.00-100.00

Vol. 2 No. 1 Fall 1980 is called the first nationally distributed issue of *Fun Times* in the USA. Its Fall cover illustrates Birdie, The Early Bird arriving ahead of schedule, with no specific mention of the character's name. If one is collecting only Birdie, The Early Bird items, then the collection would start in late 1980, even though McDonald's did not officially introduce the character until 1981.

❑ ❑ USA Ge8045 **Fun Times Magazine: Vol. 2 No. 1 Fall 1980 Fun Times.** $4.00-8.00

❑ ❑ USA Ge8046 **Fun Times Magazine: Vol. 2 No. 2 Winter 1980 Fun Times.** $4.00-8.00

Comments: McDonaldland Fun Times Magazines in the USA: Made in the USA: 1980-1998+.

Ge8045

Ge8046

1981

- **"You Deserve a Break Today" (jingle repeated)**
- **"Birdie, The Early Bird" officially introduced**
- **Cookie Boxes stress character identification**

Birdie, The Early Bird

Birdie, The Early Bird officially joins Ronald McDonald's cast of characters in the Spring of 1981 on the Breakfast Menu advertising, four years after the complete breakfast line was added to the national menu. This new character added to Ronald's cast of characters is intended to draw the attention of the children. The "You Deserve a Break Today" advertising campaign reaches the hearts and minds of millions. Unofficially, Birdie, The Early Bird, had already joined the McDonald's characters. But, in 1981, Birdie, The Early Bird made plans to become an OFFICIAL McDonaldland resident!

From March 3 to June 2, 1981, McDonald's introduced Birdie, The Early Bird. She appeared in one introductory commercial entitled "Early Bird" and two sustaining commercials - one entitled "On Target for Breakfast" and one entitled "Get Out of Bed, Sleepyhead."

"Early Bird" (:60/:30) picks up where Ronald and Grimace are bird watching and spot "Birdie, The Early Bird." After a misguided landing and introduction, Birdie and her new friends (Ronald and Grimace) head for McDonald's and "get off the ground" with a good breakfast.

In **"On Target for Breakfast"** (:60/:30), all of McDonaldland watches as Birdie tries to land for breakfast. Only with Ronald's help does she finally succeed and they all go off for a nutritious breakfast at McDonald's.

"Get Out of Bed, Sleepyhead" (:30) illustrates how Birdie, after several attempts, finally succeeds in waking Ronald for a breakfast at McDonald's.

Birdie, The Early Bird, 1990s.

In all these commercials Birdie, The Early Bird interacts with Ronald McDonald and the characters and displays her flying abilities and landing mishaps. The Rise'n Shiner, Birdie expects the best from everyone -- including herself.

The 1981 Cookie Box commemorating the introduction of Birdie, The Early Bird states: "Birdie (pictured) the Early Bird sure makes you giggle. Her landings are a smile (pictured) and she flies (pictured) with a wiggle. So peppy and bright she makes morning (pictured) time fun. If you want a friend for breakfast (pictured) our Birdie's the 1 (pictured)."

In 1986, Birdie, The Early Bird was redesigned to look younger. By the 1990s, Birdie, The Early Bird was a bright, cheerful, enthusiastic bird. Being a bird, she loves to fly. Unfortunately, she hasn't mastered the art of landing yet. Her attempts are delightfully comical, even after "training for almost 10 years [other characters were introduced in 1971]." Birdie, The Early Bird still puts a lot of energy into everything she does -- sports, dancing, exercise, and especially flying. She is always smiling, happy, bright-eyed and optimistic. And, as her name indicates, her favorite time of the day is morning. Birdie, The Early Bird character ad-

vertising was a tremendous success for McDonald's and ushered in the Happy Meal concept to a safe landing! In 1981, the Happy Meal emphasis was still on "Collect all 6 boxes," not the premiums, but this did not last for long (see Dinosaur Days Happy Meal, 1982/1981).

The Cookie Box description of Ronald McDonald in 1981 reflects: It's #1 clown's style, to make you smile (pictured). He'll dance and juggle (pictured) and spin, just to make you grin. He's everybody's friend, BIG or small. And he's known as Ronald McDonald (pictured) to all!

Birdie, The Early Bird was illustrated on the Happy Times wrapping paper. Her appearance was in a pink flight suit and white scarf with goggles on her head. Her flight suit carries a McDonald's "M" with the sunshine behind the arches -- The Breakfast Sunshine.

McDonald's animation cells.

3-D Happy Meal, 1981

Boxes:

❑ ❑ USA Th8155 **Hm Box - Bugsville/Hungry Funnies**, 1981. $25.00-40.00

Above and right: Th8155

❑ ❑ USA Th8156 Hm Box - **High Jinx/Clownish Capers**, 1981. $25.00-40.00

❑ ❑ USA Th8157 Hm Box - **Locomotion/Laughing Stock**, 1981. $25.00-40.00

❑ ❑ USA Th8158 Hm Box - **Space Follies/Gurgle Gags**, 1981. $25.00-40.00

Premium:

❑ ❑ USA Th8150 **3-D Paper Eye Glasses,** 1981, Cardboard with Blu/Red Cellophane with Arches. $15.00-20.00

❑ ❑ USA Th8165 **Translite/Lg**, 1981. $75.00-100.00

Comments: Regional Distribution: USA - May/June/Nov 1981. Test Market Regional Promotion.

Th8156

Th8157

Th8158

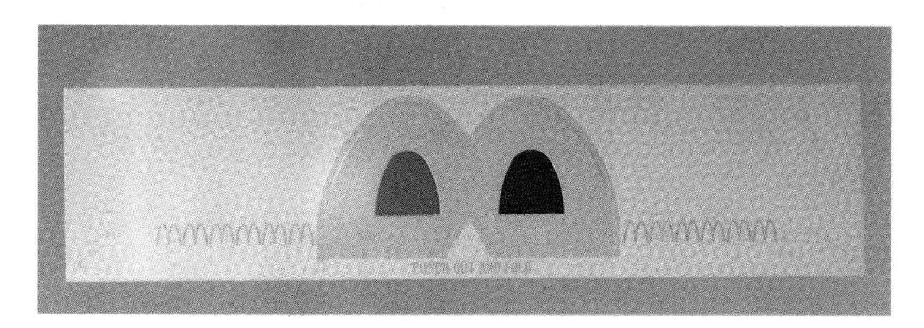

Th8150

Adventures of Ronald McDonald Happy Meal, 1981

Boxes:

- ❑ ❑ USA Ad8114 **Hm Box - Find way to end of the Rainbow/ Ronald/Grimace/Hamburglar**, 1981. $25.00-30.00
- ❑ ❑ USA Ad8115 **Hm Box - Rodeo/Ronald Bull Riding**, 1981. $25.00-30.00
- ❑ ❑ USA Ad8116 **Hm Box - Speech/Mayor Can't Read Speech**, 1981. $25.00-30.00
- ❑ ❑ USA Ad8117 **Hm Box - Express/Hamburglar Stealing Cheeseburger**, 1981. $25.00-30.00
- ❑ ❑ USA Ad8118 **Hm Box - Hide N' Seek/Ronald and Friends**, 1981. $25.00-30.00
- ❑ ❑ USA Ad8119 **Hm Box - Kaboom/Mayor Finding Light**, 1981. $25.00-30.00

Ad8114

Ad8115

Ad8116

Ad8117

Ad8118

Ad8119

Ad8101-06

Premiums:

- ❑ ❑ USA Ad8101 **Figurine - Big Mac,** 1981, 2" Soft Rubber. $10.00-15.00
- ❑ ❑ USA Ad8102 **Figurine - Captain,** 1981, 2" Soft Rubber. $10.00-15.00
- ❑ ❑ USA Ad8103 **Figurine - Grimace,** 1981, 2" Soft Rubber. $10.00-15.00
- ❑ ❑ USA Ad8104 **Figurine - Hamburglar,** 1981, 2" Soft Rubber. $10.00-15.00
- ❑ ❑ USA Ad8105 **Figurine - Mayor,** 1981, 2" Soft Rubber. $10.00-15.00
- ❑ ❑ USA Ad8106 **Figurine - Ronald,** 1981, 2" Soft Rubber. $10.00-15.00

Ad8103 Ad8106 Ad8103 Ad8102 Ad8104 Ad8113 Ad8107

- ❑ ❑ USA Ad8107 **Figurine - Big Mac,** 1981, 2" Hard Rubber/Yel/Grn/Org. $10.00-15.00
- ❑ ❑ USA Ad8107 **Figurine - Big Mac,** 1981, 2" Hard Rubber Blue. $15.00-20.00

- ❑ ❑ USA Ad8108 **Figurine - Captain,** 1981, 2" Hard Rubber/Yel/Grn/Org. $10.00-15.00
- ❑ ❑ USA Ad8108 **Figurine - Captain,** 1981, 2" Hard Rubber Blue. $15.00-20.00

- ❑ ❑ USA Ad8109 **Figurine - Grimace,** 1981, 2" Hard Rubber/Yel/Grn/Org. $10.00-15.00
- ❑ ❑ USA Ad8109 **Figurine - Grimace,** 1981, 2" Hard Rubber Blue. $15.00-20.00

- ❑ ❑ USA Ad8110 **Figurine - Hamburglar,** 1981, 2" Hard Rubber/Yel/Grn/Org. $10.00-15.00
- ❑ ❑ USA Ad8110 **Figurine - Hamburglar,** 1981, 2" Hard Rubber Blue. $15.00-20.00

- ❑ ❑ USA Ad8111 **Figurine - Mayor,** 1981, 2" Hard Rubber/Yel/Grn/Org. $10.00-15.00
- ❑ ❑ USA Ad8111 **Figurine - Mayor,** 1981, 2" Hard Rubber Blue. $15.00-20.00

- ❑ ❑ USA Ad8112 **Figurine - Ronald,** 1981, 2" Hard Rubber/Yel/Grn/Org. $10.00-15.00
- ❑ ❑ USA Ad8112 **Figurine - Ronald,** 1981, 2" Hard Rubber Blue. $15.00-20.00

- ❑ ❑ USA Ad8113 **Figurine - Birdie,** 1981, 2" Hard Rubber/Yel/Grn/Org. $10.00-15.00
- ❑ ❑ USA Ad8113 **Figurine - Birdie,** 1981, 2" Hard Rubber Blue. $15.00-20.00

- ❑ ❑ USA Ad8126 **Display/Ronald Standing**, 1981. $350.00-500.00
- ❑ ❑ USA Ad8165 **Translite/Lg**, 1981. $50.00-75.00

Ad8165

❑ ❑ USA Du8226 **Floor Display/Boss Hogg**, 1982. $——

❑ ❑ USA Du8241 **Ceiling Dangler/Cars**, 1982. $——

❑ ❑ USA Du8265 **Translite/Cars/Lg**, 1982. $50.00-75.00

❑ ❑ USA Du8266 **Translite/Cups/Lg**, 1982. $20.00-35.00

Comments: Limited Regional Distribution: USA May 21 - July 4, 1982 in St. Louis, Missouri. The 16 oz. Plastic Cups were used as an additional premium or free with the purchase of a large drink. Premiums were the Vacuum Formed Happy Meal Containers with sticker sheet. Auction prices for the Vacuum Form Containers have exceeded book value.

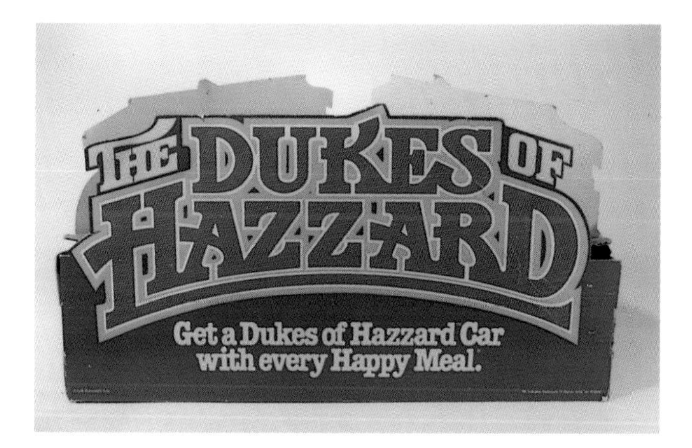

Du8241

Du8226

Giggles and Games Happy Meal, 1982

Boxes:

❑ ❑ USA Gi8200 **Hm Box - Bumper Car Tag**, 1982. $25.00-40.00

❑ ❑ USA Gi8201 **Hm Box - Make a Face Chase**, 1982. $25.00-40.00

Du8265

Gi8200

Gi8201

Gi8202

❏ ❏ USA Gi8202 **Hm Box - Monster Marathon**, 1982. $25.00-40.00

❏ ❏ USA Gi8203 **Hm Box - Outer Space Battle**, 1982. $25.00-40.00

❏ ❏ USA Gi8204 **Hm Box - Road Rally**, 1982. $25.00-40.00

❏ ❏ USA Gi8205 **Hm Box - Sunken Treasure**, 1982. $25.00-40.00

Premiums:

❏ ❏ USA Gi8220 **Sponge - Tickle Feather**, 1982, orange sponge. $25.00-40.00

❏ ❏ USA Gi8221 **Slate - Tic-Tac-Teeth**, 1982. $2.00-3.00

Gi8205

Gi8203

Gi8220

Gi8204

Gi8221

Gi8222

Gi8222

❑ ❑ USA Gi8222 **Whistle - Gobblin Caller,** 1982, Blue or Red or Green. $5.00-8.00

❑ ❑ USA Gi8223 **Comb - Gobblin Groomer,** 1982, Org or Grn. $2.00-3.00

❑ ❑ USA Gi8224 **Frisbee Flying Wheel - Flinger,** 1981, Frisbee/Ronald/Yel with Red. $8.00-10.00

❑ ❑ USA Gi8261 **MC Insert/Cardboard,** 1982. $45.00-60.00
❑ ❑ USA Gi8262 **Header Card,** 1982. $25.00-40.00
❑ ❑ USA Gi8265 **Translite/Lg**, 1982. $50.00-65.00

Comments: National Optional Happy Meal: USA - June 28 - August 29, 1982. Generic premiums could have been given with the Happy Meal. No under 3 (U-3) premiums were furnished. The green colored Gi8222 Gobblin Caller, dated 1982 without the name "Gobblin Caller" on the premium is an example of a giveaway during this period. It could have been distributed as a Happy Meal premium in some locations. Small booklets, "Open for giggles and screams" could have been regionally distributed by Circus World in Florida, for the Giggles and Games Happy Meal.

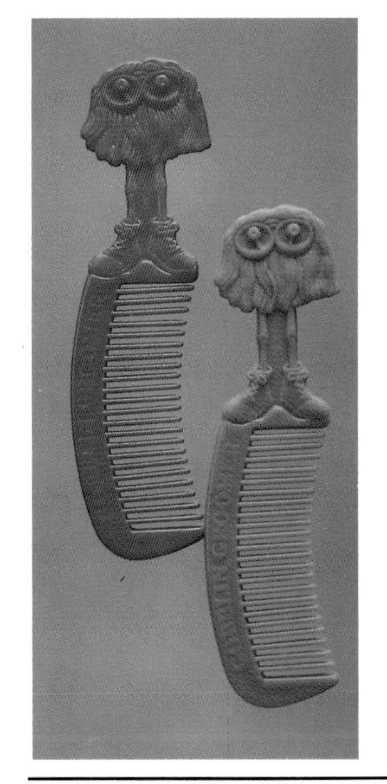

Gi8223

Gi8224

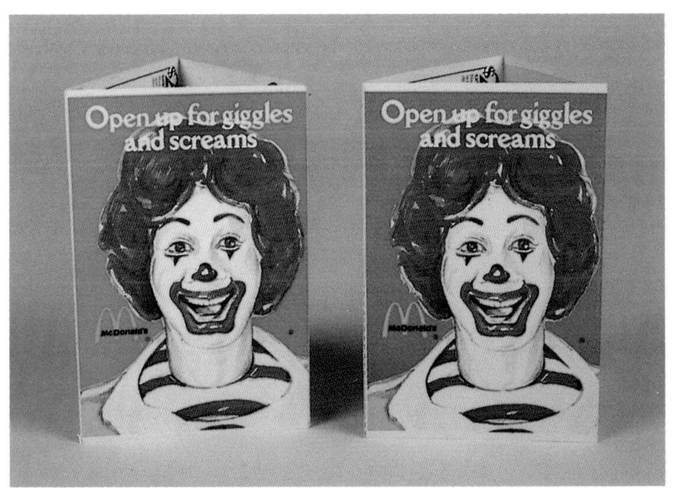

Go8201

Going Places Happy Meal, 1982

Boxes:
❑ ❑ USA Go8201 **Hm Box - Biplane/Ronald**, 1981. $15.00-20.00
❑ ❑ USA Go8202 **Hm Box - Dune Buggy/Ronald,** 1981. $15.00-20.00
❑ ❑ USA Go8203 **Hm Box - Elephant,** 1981. $15.00-20.00

Go8203

Go8202

❑ ❑ USA Go8204 **Hm Box - Fire Engine**, 1981. $15.00-20.00

❑ ❑ USA Go8205 **Hm Box - Paddle Wheeler**, 1981. $15.00-20.00

❑ ❑ USA Go8206 **Hm Box - Steam Engine**, 1981. $15.00-20.00

❑ ❑ USA Go8220 **Sponge,** 1981, Jolly Jet. $15.00-20.00

❑ ❑ USA Go8222 **Game - McDonaldland Hockey,** 1982, Red or Blu or Yel. $5.00-8.00

❑ ❑ USA Go8223 **Game - Gobblins Horseshoes,** 1982, Org or Lt Blue or Grn/2 Goblins with Stars/2 Gob with Diamonds/ 6p. $5.00-8.00

Go8204

Go8220

Go8205

Go8222

Go8206

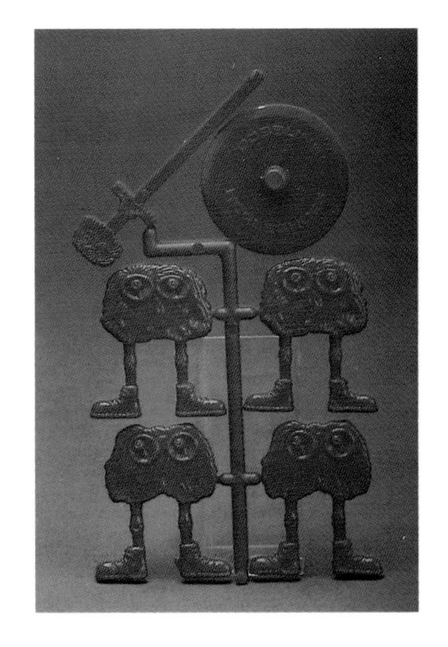

Go8223

❑ ❑ USA Go8224 **Game - Gobblins Bowling,** 1981, Grn or Yel or Purp Ball with 6 Gobblins Pins/7p. $4.00-5.00

❑ ❑ USA Go8225 **Scissors - Ronald, 1981 date,** Yellow. $4.00-5.00

❑ ❑ USA Go8261 **MC Insert Cardboard**, 1981.$25.00-40.00
❑ ❑ USA Go8265 **Translite/Lg**, 1981. $20.00-25.00

Comments: Regional Distribution: USA - February 21-April 28, 1982. Regions developed their own promotions/premiums. Generic and/or regional premiums could have been given with this Happy Meal. These Hm Boxes were regionally distributed again with the 1983 Going Places/Hot Wheels promotion, along with Hot Wheels Cars. A Going Places Plastic Cup was distributed in 1983. Red/Blue Scissors were prototype and not nationally/regionally distributed. Two other pairs of Yellow Scissors exist. Yellow Go8225 scissors dated 1982 were giveaways during 1982. The two different pairs are both dated 1982; one with Ronald McDonald's name and the other with "Safety Test" information. Only the Yellow Scissors dated 1981 were used with this Happy Meal.

Going Places Happy Meal in 1982 introduced a change in advertising focus, "A Prize in every box. Get all six." The promotion emphasized a prize and a box along with a food purchase. The premiums prizes soon became the focus of children's attention on every visit.

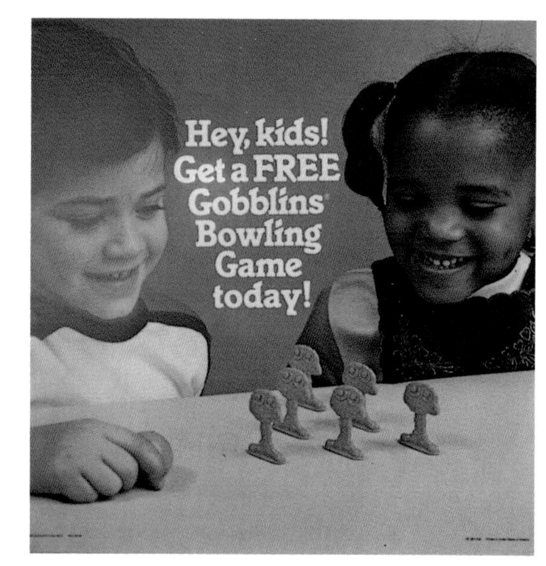

Go8224

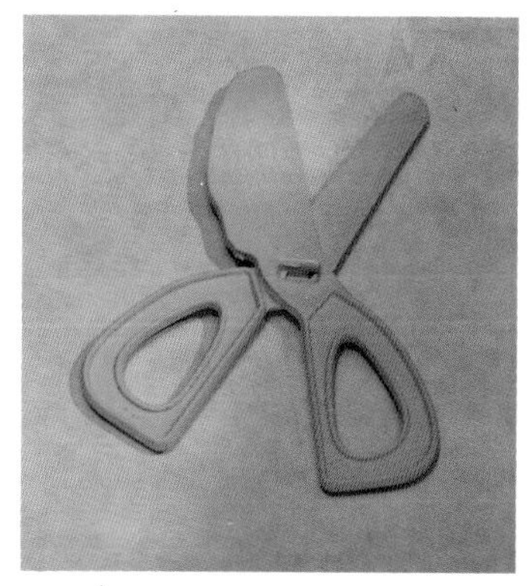
Go8225

USA Go8265

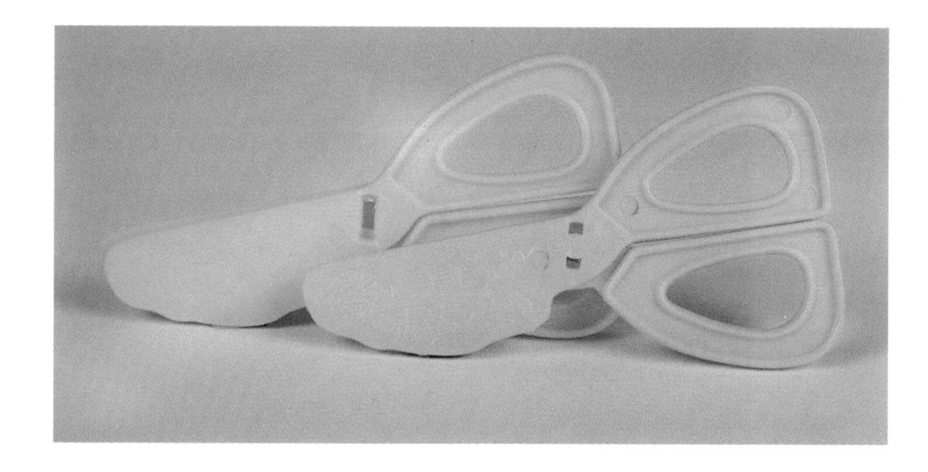

Li8258

Little Golden Book Happy Meal, 1982

Box:

❑ ❑ USA Li8258 **Hm Box - Ronald and Friends**, 1982. $10.00-15.00

Premiums: Books:

❑ ❑ USA Li8250 **Book - Benji Fastest Dog in the West**, 1982.. $2.00-3.00

❑ ❑ USA Li8251 **Book - The Country Mouse and the City Mouse**, 1982. $2.00-3.00

❑ ❑ USA Li8252 **Book - The Monster at the End of this Book**, 1982. $2.00-3.00

❑ ❑ USA Li8253 **Book - The Poky Little Puppy**, 1982. $2.00-3.00

❑ ❑ USA Li8254 **Book - Tom and Jerry's Party**, 1982. $2.00-3.00

Li8252

Li8251

Li8250

Li8253

Li8254

Li8252

Li8251

❑ ❑ USA Li8273 **Crew Badge - Benji,** 1982, Paper/Stick on. $7.00-10.00

❑ ❑ USA Li8274 **Crew Badge - Country Mouse**, 1982, Paper/Stick on. $7.00-10.00

❑ ❑ USA Li8275 **Crew Badge - Monster at the End**, 1982, Paper/Stick on. $7.00-10.00

❑ ❑ USA Li8276 **Crew Badge - Poky Little Puppy**, 1982, Paper/Stick on. $7.00-10.00

❑ ❑ USA Li8277 **Crew Badge - Tom and Jerry**, 1982, Paper/Stick on. $7.00-10.00

❑ ❑ USA Li8265 **Translite/Father with 2 Girls**, 1982. $15.00-25.00

❑ ❑ USA Li8271 **Message Center/Cardboard**, 1982. $25.00-40.00

❑ ❑ USA Li8272 **Counter Card/Girl with 5 Books**, 1982. $25.00-40.00

Comments: National Distribution: USA - July 16- August 23, 1982.

Li8273 Li8274

Li8265

Li8275 Li8277

Li8272

McDonaldland Express and Little Golden Books blue book.

USA Ex8209 USA Ex8208 USA Ex8207 USA Ex8206
USA Ex8203 USA Ex8202 USA Ex8201 USA Ex8200

McDonaldland Express Happy Meal, 1982

Premiums: Vacuum Formed Containers:

❑ ❑ USA Ex8200 **Train Engine,** 1982, Vacuum Formed/Red or Blue. $20.00-25.00
❑ ❑ USA Ex8201 **Coach Car,** 1982, Vacuum Formed/Blue or Orange. $20.00-25.00
❑ ❑ USA Ex8202 **Freight Car,** 1982, Vacuum Formed/Green or Orange. $20.00-25.00
❑ ❑ USA Ex8203 **Caboose,** 1982, Vacuum Formed/Red or Green. $25.00-35.00

❑ ❑ USA Ex8206 **Sticker Sheet - Engine,** 1982. $5.00-8.00
❑ ❑ USA Ex8207 **Sticker Sheet - Coach,** 1982. $5.00-8.00
❑ ❑ USA Ex8208 **Sticker Sheet - Freight,** 1982. $5.00-8.00
❑ ❑ USA Ex8209 **Sticker Sheet - Caboose,** 1982. $5.00-8.00

❑ ❑ USA Ex8225 **Display with premiums going through the mountain/w Premiums**, 1982. $800.00+
❑ ❑ USA Ex8226 **Display with premiums going over the mountain/Premiums**, 1982. $200.00-250.00+
❑ ❑ USA Ex8241 **Ceiling Dangler**, 1982. $75.00-100.00
❑ ❑ USA Ex8250 **Button**, 1982, Engine. $15.00-20.00
❑ ❑ USA Ex8261 **MC Insert/Cardboard**, 1982. $25.00-40.00
❑ ❑ USA Ex8262 **Header Card,** 1982. 40.00-50.00
❑ ❑ USA Ex8265 **Translite/Lg**, 1982. $25.00-40.00

Comments: National Distribution: USA - June 11-July 15, 1982. Vacuum formed train cars served as the food container and premium.

Ex8206

Ex8225

Ex8226

Ex8265

Ex8250

Playmobil II Happy Meal, 1982

Boxes:

❑ ❑ USA PI8260 **Hm Box - Barn**, 1982. $20.00-25.00
❑ ❑ USA PI8261 **Hm Box - Log Cabin**, 1982. $20.00-25.00
❑ ❑ USA PI8262 **Hm Box - School House**, 1982. $20.00-25.00
❑ ❑ USA PI8263 **Hm Box - Trading Post**, 1982. $20.00-25.00

Premiums:

❑ ❑ USA PI8250 **Set 1 Sheriff,** 1982, with Chair/Rifle/Hat/Cape. $8.00-10.00
❑ ❑ USA PI8251 **Set 2 Indian,** 1982, with Shield/Sticker/Gun/Spear/Peace Pipe/Headdress. $8.00-10.00
❑ ❑ USA PI8252 **Set 3 Horse,** 1982, Brn with Attached Saddle with Rect Trough. $10.00-15.00
❑ ❑ USA PI8253 **Set 4 Umbrella Girl,** 1982, with 2p Umbrella/Yel Suitcase/Red Dress/Blk Hair. $8.00-10.00
❑ ❑ USA PI8254 **Set 5 Farmer,** 1982, with Grn Shirt/Pants/Grey Scyle/Rake/Yel Hat/Dog. $15.00-20.00

See more items from this set on next two pages.

PI8260

PI8262

PI8261

PI8263

PI8250 PI8251 PI8252 PI8253 PI8254

playmobil
INDIAN

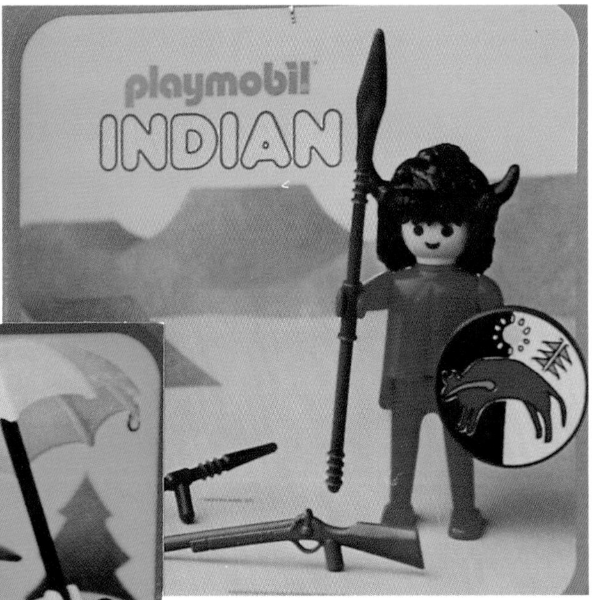

❏ ❏ USA PI8226 **Display/Premiums**, 1982 $150.00-200.00
❏ ❏ USA PI8227 **Playmobile Accessory Set,** 1982. $100.00-125.00
❏ ❏ USA PI8255 **Trayliner.** $3.00-5.00
❏ ❏ USA PI8265 **Translite/Lg**, 1982. $25.00-40.00
❏ ❏ USA PI8266 **Translite/Cardboard/Lg**, 1982. $25.00-40.00
❏ ❏ USA PI8267 **Blue Book: Playmobil II.** $40.00-50.00

PI8227

PI8226

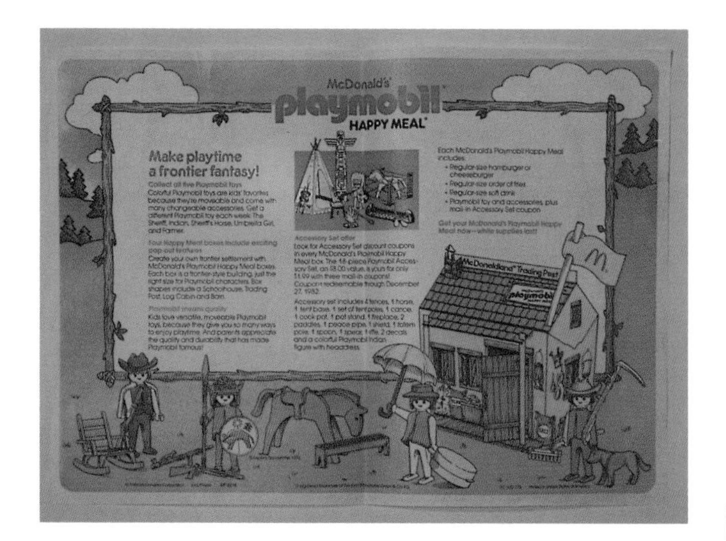

PI8255

PI8265

PI8267

❑ ❑USA Pl8271 **Counter Card/Recalled Notice**, 1982. $20.00-25.00

❑ ❑ USA Pl8272 **Button,** 1982, Star with Playmobil Meal Deal. $15.00-25.00

Pl8272

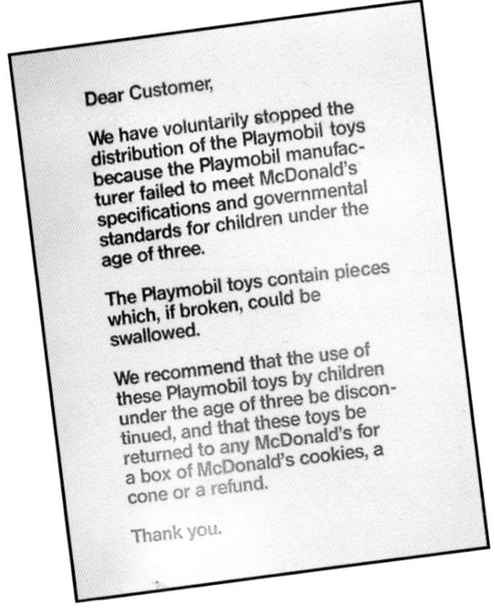

Dear Customer,

We have voluntarily stopped the distribution of the Playmobil toys because the Playmobil manufacturer failed to meet McDonald's specifications and governmental standards for children under the age of three.

The Playmobil toys contain pieces which, if broken, could be swallowed.

We recommend that the use of these Playmobil toys by children under the age of three be discontinued, and that these toys be returned to any McDonald's for a box of McDonald's cookies, a cone or a refund.

Thank you.

Pl8271

Comments: National Distribution: USA - October 2-November 28, 1982. Sets 1 and 2 were basically the only nationally distributed premiums. Because the Playmobil Happy Meal premiums contained small pieces and were potentially dangerous for small children, **McDonald's voluntarily recalled and stopped distribution of the Playmobil toys; this was McDonald's first national recall. They shifted focus from the box to the toy, but stumbled in the process. Successfully rebounding, McDonald's began the Under 3 (U-3) program, toys specifically designed for children under the age of 3.** A $1.98 accessory kit was offered by a coupon in the distributed packages. The accessory kit was not widely recognized, due to limited distribution of the coupon. The Playmobil People were marked "1974 Geobra." The dog in USA Pl8254 - Set 5 came in two versions - with and without mouth open. The Indian Shield was distributed in two versions; one being in the original Happy Meal and the second being in the discount stores liquidation packages which resold the toys. The liquidation packages were boxed. The Indian Shield pictured is the one in the October/November Happy Meal Promotion. The accessory kit was available for $1.98 when accompanied by additional coupons available in the Playmobile Happy Meal. Note that the shield in the accessory kit is the same as in the Happy Meal promotion.

Sky-Busters Happy Meal, 1982

Premiums: Rubber Airplanes:

❑ ❑ USA Sk8201 **Mig-21,** Yel or Blu or Brn or Pnk or Grn or Org Matchbox (Lesney). $3.00-4.00

❑ ❑ USA Sk8202 **Mirage F1,** Yel or Blu or Brn or Pnk or Grn or Org Matchbox (Lesney). $3.00-4.00

❑ ❑ USA Sk8203 **Phantom F4E,** Yel or Blu or Brn or Pnk or Grn or Org Matchbox (Lesney). $3.00-4.00

❑ ❑ USA Sk8204 **Sky Hawk A4F,** Yel or Blu or Brn or Pnk or Grn or Org Matchbox (Lesney). $3.00-4.00

❑ ❑ USA Sk8205 **Tornado,** Yel or Blu or Brn or Pnk or Grn or Org Matchbox (Lesney). $3.00-4.00

❑ ❑ USA Sk8206 **United DC-10,** Yel or Blu or Brn or Pnk or Grn or Org Matchbox (Lesney). $3.00-4.00

❑ ❑ USA Sk8264 **Translite/Sm,** 1982. $45.00-60.00

Sk8206

Sk8205

Sk8204

Sk8203

Sk8202

Sk8201

Comments: Regional Distribution: USA - 1982. Initially, hard rubber premiums were distributed. Later, some regions may have distributed soft rubber airplanes. Airplanes are marked, "Lesney."

Spaceship Happy Meal, 1982

Premiums: Vacuum Formed Containers:

❑ ❑ USA Sp8230 **Spaceship #1- 8 Windows,** 1981, Blu or Grn or Red or Yel with Circle M with Decals. $10.00-15.00

❑ ❑ USA Sp8231 **Spaceship #2 - Round/Circular with Rear Engine,** 1981, Blu or Grn or Red or Yel with Decals. $10.00-15.00

Sp8230

Sp8231

❑ ❑ USA Sp8232 **Spaceship #3 - Pointed Front,** 1981, Blu or Grn or Red or Yel with Decals. $10.00-15.00

❑ ❑ USA Sp8233 **Spaceship #4 - 4 Knobs,** 1981, Blu or Grn or Red or Yel with Decals. $10.00-15.00

❑ ❑ USA Sp8204 **Sticker Sheet #1 - 8 Windows.** $4.00-7.00

❑ ❑ USA Sp8205 **Sticker Sheet #2 - Round Front.** $4.00-7.00

❑ ❑ USA Sp8206 **Sticker Sheet #3 - Pointed Front.** $4.00-7.00

❑ ❑ USA Sp8207 **Sticker Sheet #4 - Four Knobs.** $4.00-7.00

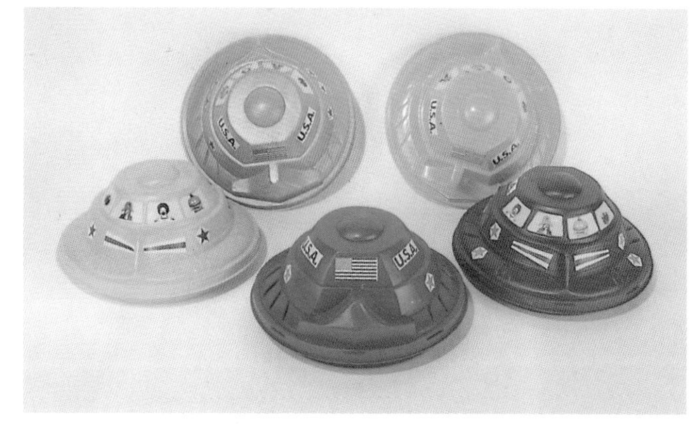

Sp8232

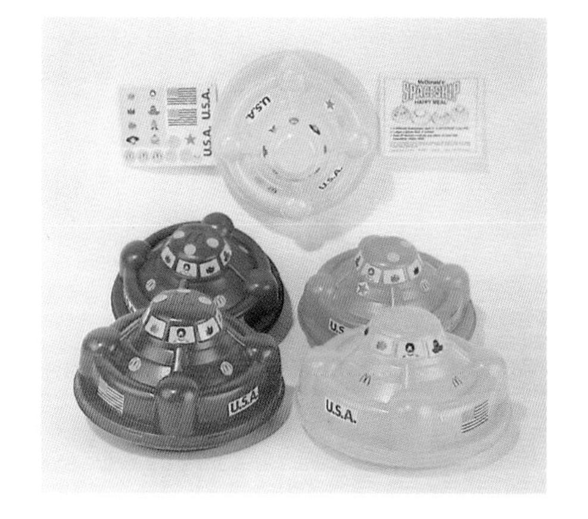

Sp8233

Sp8205

Sp8204

Sp8205

Sp8206

Sp8207

Sp8206

Sp8207

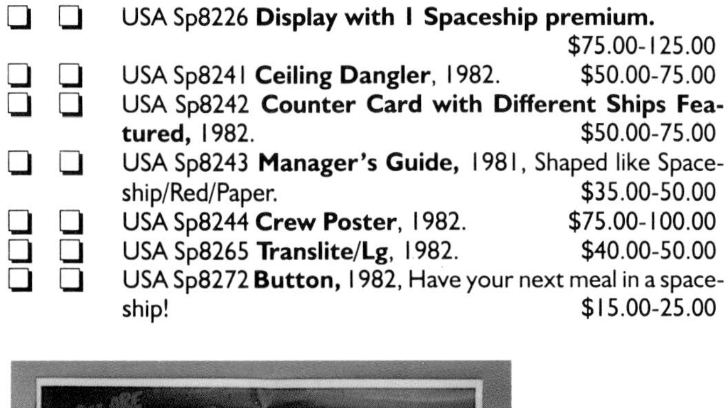

- ❑ ❑ USA Sp8226 **Display with 1 Spaceship premium.** $75.00-125.00
- ❑ ❑ USA Sp8241 **Ceiling Dangler**, 1982. $50.00-75.00
- ❑ ❑ USA Sp8242 **Counter Card with Different Ships Featured,** 1982. $50.00-75.00
- ❑ ❑ USA Sp8243 **Manager's Guide,** 1981, Shaped like Spaceship/Red/Paper. $35.00-50.00
- ❑ ❑ USA Sp8244 **Crew Poster**, 1982. $75.00-100.00
- ❑ ❑ USA Sp8265 **Translite/Lg**, 1982. $40.00-50.00
- ❑ ❑ USA Sp8272 **Button,** 1982, Have your next meal in a spaceship! $15.00-25.00

Sp8226

Sp8244

Sp8243

Sp8272

Sp8265

Comments: National Distribution: USA - January 22-February 28, 1982. Spaceship Happy Meal is sometimes mistakenly referred to as the "Unidentified" Happy Meal test marketed in the Kansas City, Missouri and St. Louis areas in 1981. Spaceship Happy Meal 1982 set of containers had a shiny finish; with a molded "M" on top or a molded "M" within a molded circle. The top and bottom were held together by eight lugs. The 1981 Unidentified set had only four lugs. The two piece Spaceship container held food and served as the premium. Color variations of the Spaceships exist. These same Spaceships were distributed in Canada and internationally. This worldwide distribution could account for the color variations.

As the years passed, other space-oriented Happy Meal promotions were presented and not utilized in the USA. One interesting promotion is the McDonaldland Space Patrol Happy Meal Promotion.

Spaceships and Playmobil toys, packaged in retail sales blue boxes, were used at McDonald's convention gatherings. Playmobil figures were recalled and later sold in retail stores for the over-3 market. Spaceships were designed for all ages.

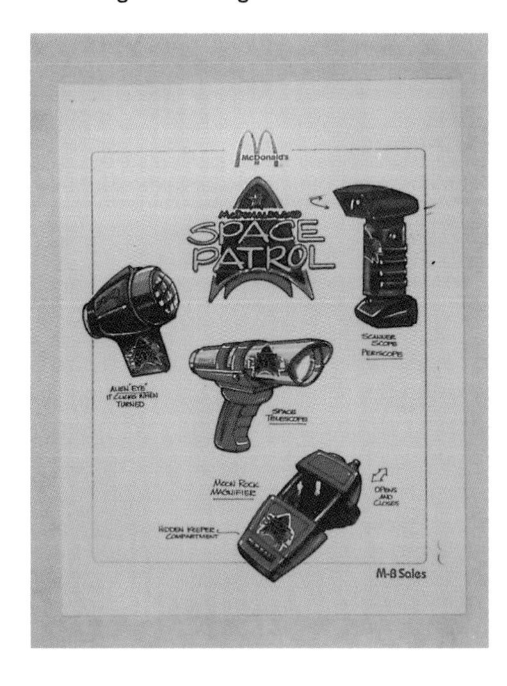

Wacky Happy Meal, 1982

Boxes:

❑ ❑ USA Wa8201 **Hm Box - Wacky Farm**, 1981. $40.00-50.00

❑ ❑ USA Wa8202 **Hm Box - Wacky Forest,** 1981. $40.00-50.00

❑ ❑ USA Wa8203 **Hm Box - Wacky Party**, 1981. $40.00-50.00

❑ ❑ USA Wa8204 **Hm Box - Wacky Parade,** 1981. $40.00-50.00

Wa8204

Wa8202

❑ ❑ USA Wa8205 **Hm Box - Wacky Picnic,** 1981. $40.00-50.00

❑ ❑ USA Wa8206 **Hm Box - Wacky Zoo**, 1981. $40.00-50.00

Premiums: 6 Drink Cups:

❑ ❑ USA Wa8210 **Cup - Airplane Hanger,** Nd, 8 oz. Plastic Cup. $35.00-50.00

❑ ❑ USA Wa8211 **Cup - Country Club,** Nd, 8 oz. Plastic Cup. $35.00-50.00

❑ ❑ USA Wa8212 **Cup - Figure Skating,** Nd, 8 oz. Plastic Cup. $35.00-50.00

❑ ❑ USA Wa8213 **Cup - Jungle Gym,** Nd, 8 oz. Plastic Cup. $35.00-50.00

❑ ❑ USA Wa8214 **Cup - Monkey Business,** Nd, 8 oz. Plastic Cup. $35.00-50.00

❑ ❑ USA Wa8215 **Cup - Traffic Jam,** Nd, 8 oz. Plastic Cup. $35.00-50.00

❑ ❑ USA Wa8262 **Header Card,** 1982. $40.00-50.00

❑ ❑ USA Wa8265 **Translite/Lg,** 1982. $50.00-65.00

Comments: Limited Regional Distribution: USA March 1-May 23, 1982. Boxes with cut-outs were the premiums. The cups may have been given as an additional premium in St. Louis, Missouri.

Wa8205

Wa8206

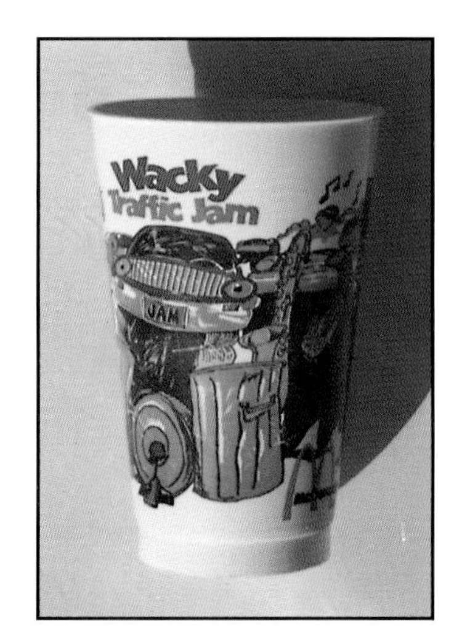

Wa8215

Wa8210 Wa8212

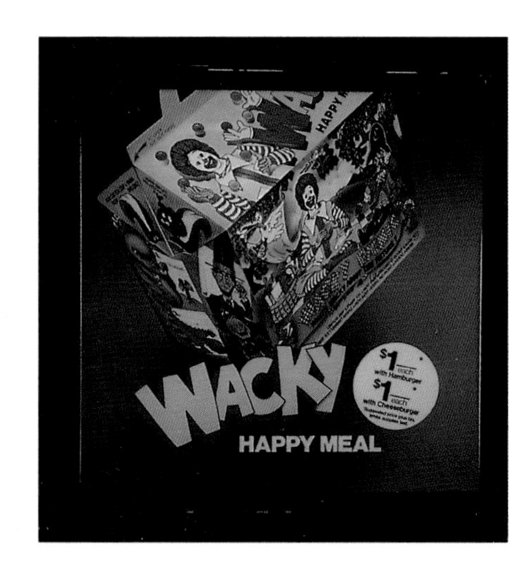

Wa8265

USA Generic Promotions, 1982

❑ ❑ USA Ge8201 **Basketball Hoop Shooter - Grimace or Ronald,** 1982, White or bright green Hoop/Shooter/Ball/6p. $10.00-12.00

❑ ❑ USA Ge8202 **Faller Zig Zag - Birdie,** 1982, Purp/3p. $10.00-12.00

❑ ❑ USA Ge8226 **Faller Zig Zag - Grimace,** 1982, Purp/3p. $10.00-12.00

❑ ❑ USA Ge8203 **French Fry Grabber - Captain,** 1982, Blu or Org or Pnk or Red/9p. $12.00-15.00

❑ ❑ USA Ge8204 **French Fry Grabber - Fry Kid,** 1982, Blu or Org or Pnk or Red/9p. $12.00-15.00

❑ ❑ USA Ge8205 **French Fry Grabber- Hamburglar,** 1982, Blu or Org or Pnk or Red/9p. $12.00-15.00

❑ ❑ USA Ge8206 **French Fry Grabber - Ronald,** 1982, Blu or Org or Pnk or Red/9p. $12.00-15.00

❑ ❑ USA Ge8207 **Gymnastic - Ronald,** 1982, Grn or Org/Parallel Bars/5p. $25.00-35.00

❑ ❑ USA Ge8208 **Gyro Top - Birdie,** 1982, Red or Gold or Turq Gyro with Zip Strip/4p. $12.00-15.00

❑ ❑ USA Ge8225 **Gyro Top - Grimace,** 1982, Red or Gold or Turq Gyro with Zip Strip/4p. $12.00-15.00

❑ ❑ USA Ge8209 **Gyro Top - Ronald,** 1982, Red or Gold or Turq Gyro with Zip Strip/4p. $12.00-15.00

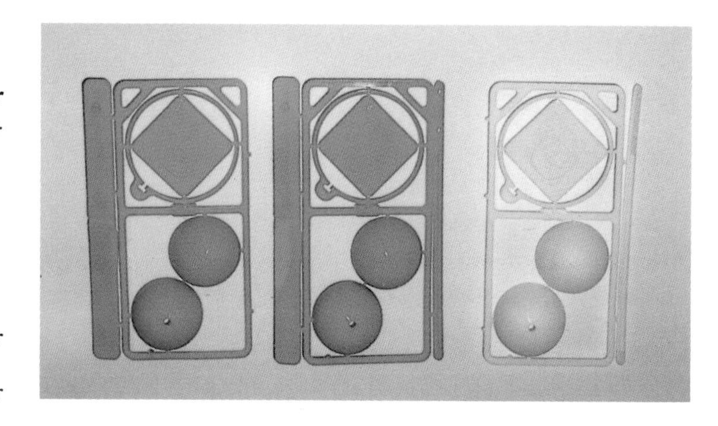

Ge8201

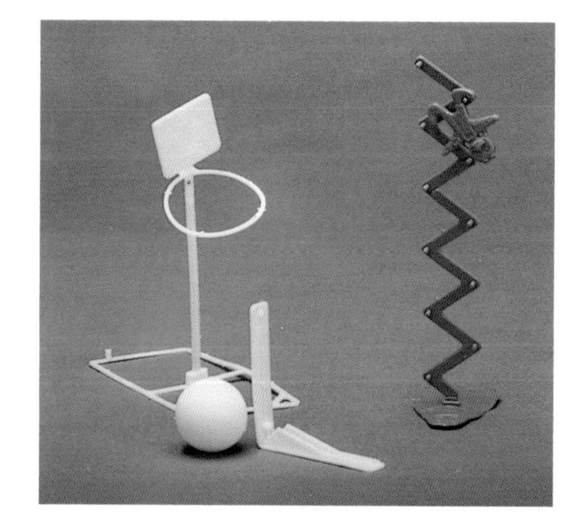

Ge8201 Ge8202

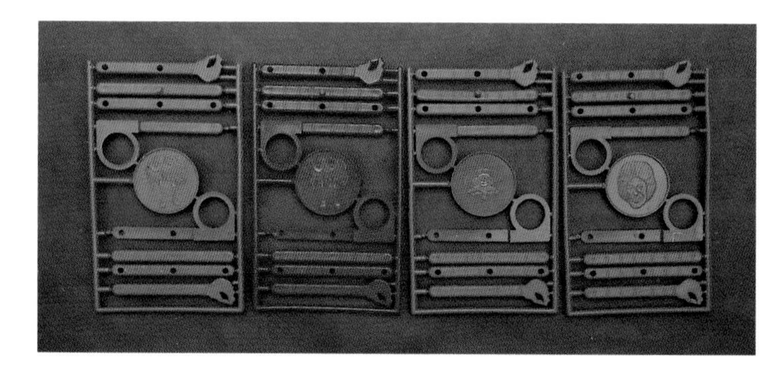

Ge8204 Ge8206

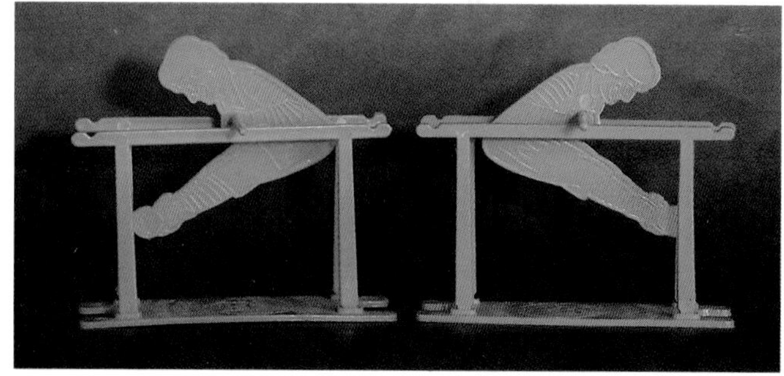

Ge8207

Ge8205

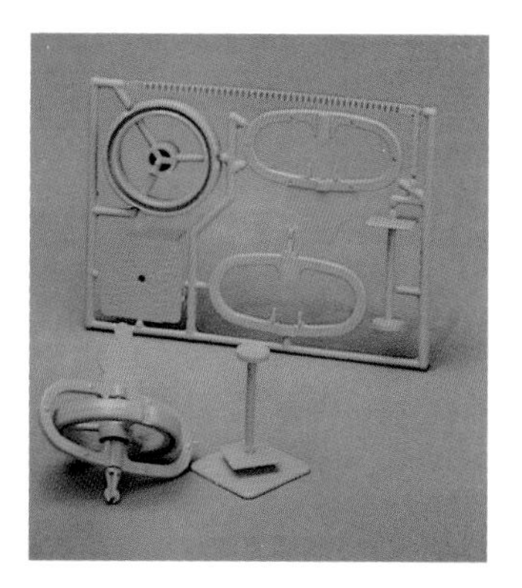

Ge8208

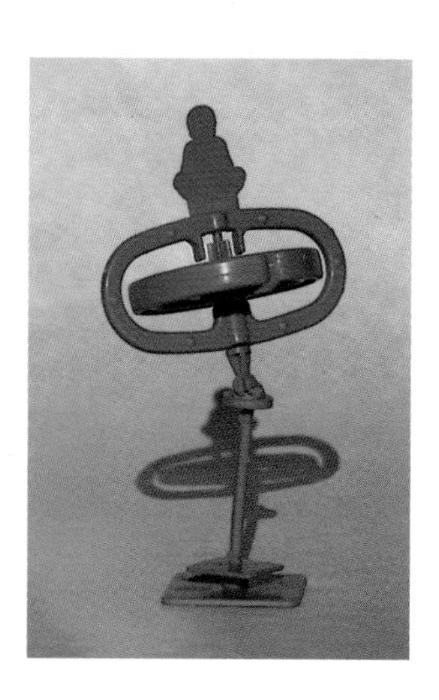

Ge8209

☐ ☐ USA Ge8210 **Hang Glider - Capt Crook,** 1982, Yel or Red or Blu Char with Pnk/Wht Styro Wings. $12.00-15.00

☐ ☐ USA Ge8211 **Hang Glider - Professor,** 1982, Yel or Red or Blu Char with Pnk/Wht Styro Wings. $12.00-15.00

☐ ☐ USA Ge8212 **Hang Glider - Ronald,** 1982, Yel or Red or Blu Char with Pnk/Wht Styro Wings. $12.00-15.00

☐ ☐ USA Ge8213 **Hardworking Burger Bulldozer,** 1982, Grn or Org. $10.00-15.00

☐ ☐ USA Ge8214 **Hardworking Burger Dump Truck,** 1982, Grn or Org. $10.00-15.00

☐ ☐ USA Ge8215 **Fly, Fly, Birdie Launcher - Birdie,** 1982, Pink or Wht or Yel Launcher/3p. $10.00-15.00

☐ ☐ USA Ge8216 **Pen - Mayor,** 1982, Pink Pen/Purp Cord. $8.00-12.00

☐ ☐ USA Ge8224 **Pop Car - Birdie,** 1982, Blu or Grn or Org/air compression launcher/6p. $12.00-15.00

☐ ☐ USA Ge8217 **Pop Car - Hamburglar,** 1982, Blu or Grn or Org/air compression launcher/6p. $12.00-15.00

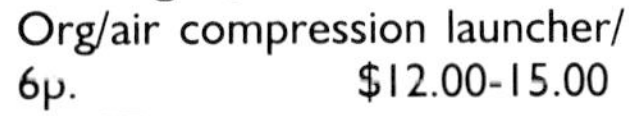

☐ ☐ USA Ge8218 **Pop Car - Ronald,** 1982, Blu or Grn or Org/air compression launcher/6p. $12.00-15.00

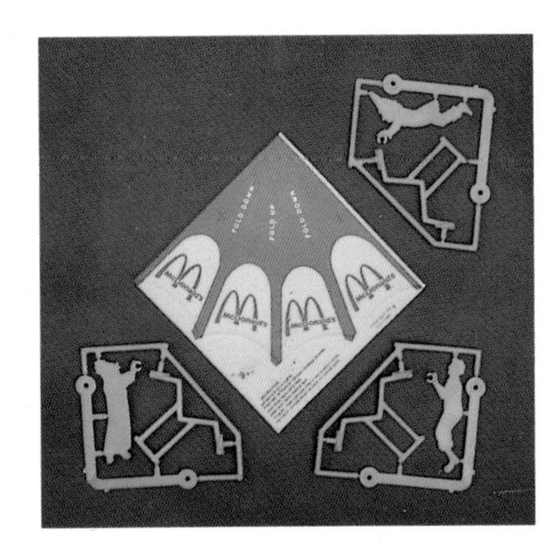

Ge8210-12

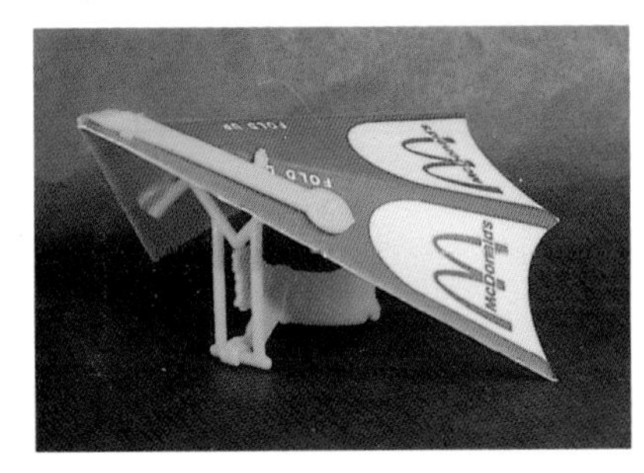

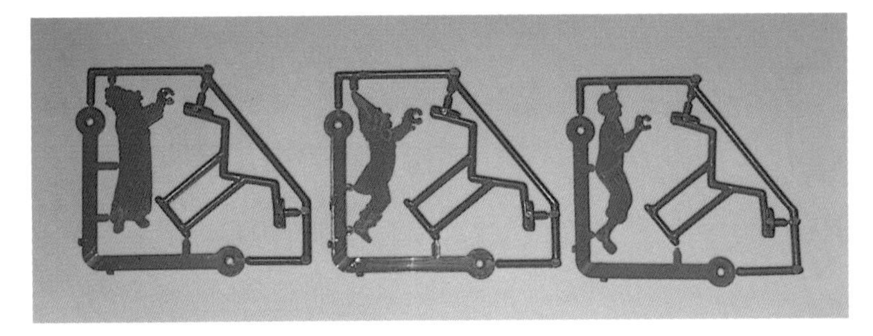

Ge8211 Ge8210 Ge8212

Ge8214 Ge8213

Ge8215

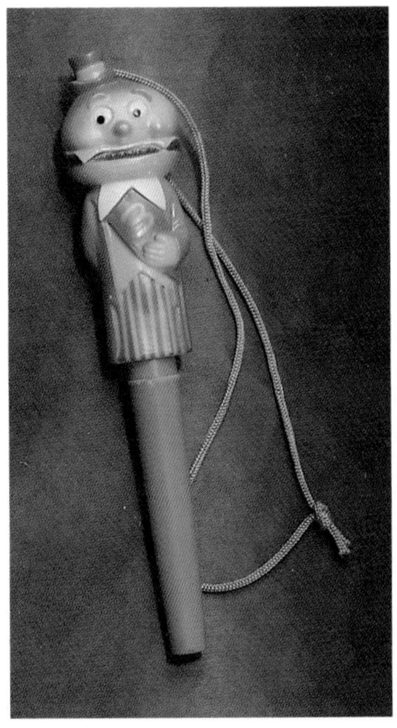

Ge8216

Ge8217

Ge8218

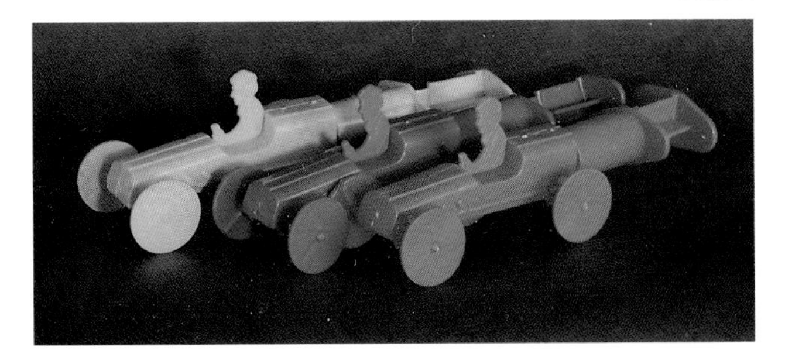

- ❑ ❑ USA Ge8219 **Puzzle Guess N' Glow,** 1982, Ronald with Ghost. $4.00-6.00

- ❑ ❑ USA Ge8220 **Sardine Can - Birdie,** 1982, Blu or Yel or Pur/3p. $12.00-15.00
- ❑ ❑ USA Ge8221 **Sardine Can - Grimace,** 1982, Blu or Yel or Pur/3p. $12.00-15.00
- ❑ ❑ USA Ge8222 **Sardine Can - Ronald,** 1982, Blu or Yel or Pur/3p. $12.00-15.00

- ❑ ❑ USA Ge8227 **Mystical Scrambler Kaleidoscope - Grimace,** 1982, Dk Blu or Dk Grn or Purp/3p. $12.00-15.00
- ❑ ❑ USA Ge8228 **Mystical Scrambler Kaleidoscope - Hamburglar,** 1982, Dk Blu or Dk Grn or Purp/3p. $12.00-15.00
- ❑ ❑ USA Ge8229 **Mystical Scrambler Kaleidoscope - Ronald,** 1982, Dk Blu or Dk Grn or Purp/3p. $12.00-15.00

- ❑ ❑ USA Ge8230 **Spin Pipe - Grimace,** 1982, Blu or Bright Green or Org/3p. $5.00-6.00
- ❑ ❑ USA Ge8231 **Spin Pipe - Hamburglar,** 1982, Blu or Bright Green or Org/3p. $5.00-6.00
- ❑ ❑ USA Ge8232 **Spin Pipe - Ronald,** 1982, Blu or Bright Green or Org/3p. $5.00-6.00

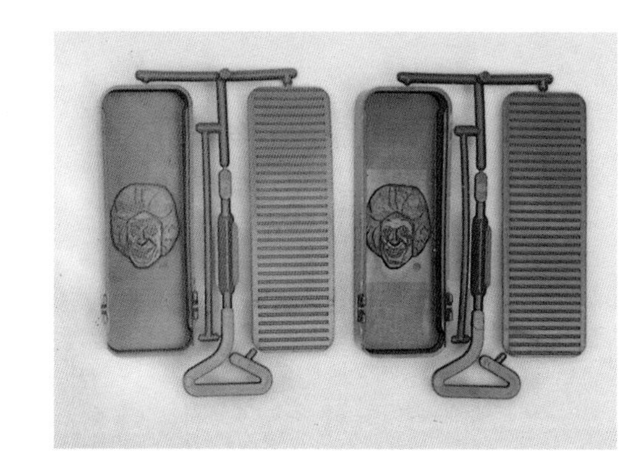

Ge8222

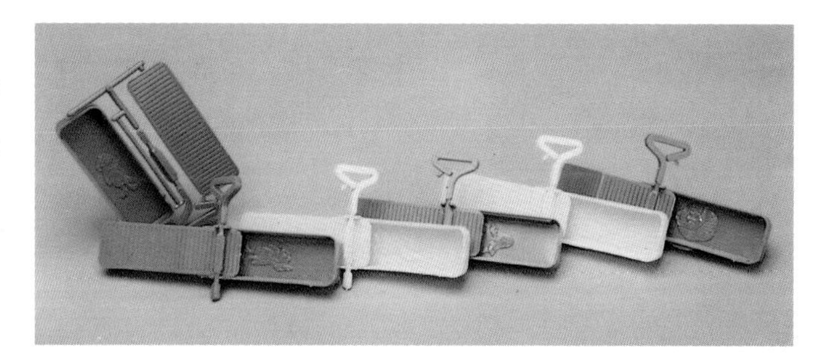

Ge8220, Ge8221, and Ge8222

Ge8227 Ge8228

Ge8219

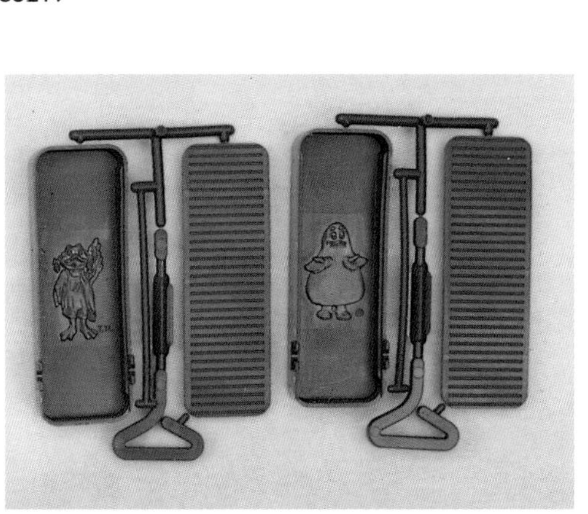

Ge8220 Ge8221

Ge8236

❑ ❑ USA Ge8236 **Photo Card: Ronald in Hot Air Balloon,** 1982. $15.00-20.00

❑ ❑ USA Ge8237 **White Plastic Cup: Ronald holding birthday cake with Grimace,** Hamburglar & Fry Guys with balloons, 1982 Ronald McDonald Happy Birthday Cup, 5" WHITE. $3.00-4.00

❑ ❑ USA Ge8246 **McDonaldland Cookies/Small packets:** Hamburglar. $1.00-1.50

❑ ❑ USA Ge8247 **Record: The Ronald McDonald All Star Party, 78 RPM,** 18 fun songs, 6 new games. $7.00-10.00

❑ ❑ USA Ge8238 **Record: (The) Magic Record #1, 45 RPM** with blue jacket. Song or story appear and disappear while the record is playing. $65.00-75.00

❑ ❑ USA Ge8239 **Record: (The) Magic Record #2, 45 RPM** with blue jacket. Song or story appear and disappear while the record is playing. $40.00-50.00

❑ ❑ USA Ge8240 **Fun Times Magazine: Vol. 3 No. 3 Spring, 1982**. In Vol. 4 No. 2 Winter 1982 the Gobblins were renamed the French Fry Guys in the Fun Times magazine. Vol. 4 No. 3 Mystery Issue dropped seasonal issues and focused on each Happy Meal promotion. Vol. 4 No. 4 Transportation issue showed the first design change of Captain Crook on the masterhead. $4.00-5.00

❑ ❑ USA Ge8241 **Fun Times Magazine: Vol. 3 No. 4 Summer, 1982.** $4.00-5.00

Ge8247

Ge8238 Ge8239

Ge8238 Ge8239

❑ ❑ USA Ge8242 **Fun Times Magazine: Vol. 4 No. 1 Fall, 1982.** $4.00-5.00
❑ ❑ USA Ge8243 **Fun Times Magazine: Vol. 4 No. 2 Winter, 1982.** $4.00-5.00

❑ ❑ USA Ge8244 **1982 Calendar: RONALD McDONALD STICKER FUN COLORING CALENDAR.** $5.00-8.00
❑ ❑ USA Ge8245 **1982 Ornament, brass/Rockwell/two children with tree.** $4.00-7.00

Comments: Regional Distribution: USA - 1982 during Clean-Up weeks. Some of these generic premiums were sold in retail stores. Many of these premiums were distributed during the Second and Third Quarter of 1982 from the Premium Distribution Program. This is a sampling of generic premiums given away during 1982. The distribution dates of bead games Ge8233-35 could vary between 1978 - 1984. These bead games were regionally distributed in parts of Kentucky.

The Professor's costume was redesigned in 1982.

The 7th National O/O Convention was held in California at the Jack Murphy Stadium in San Diego. The convention was held in conjunction with Ray Kroc's eightieth birthday party. The theme of the party and the convention was: "Delivering the Difference."

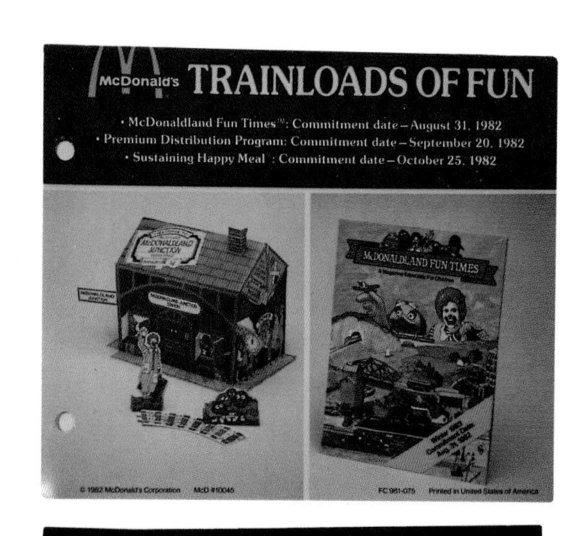

Ge8245

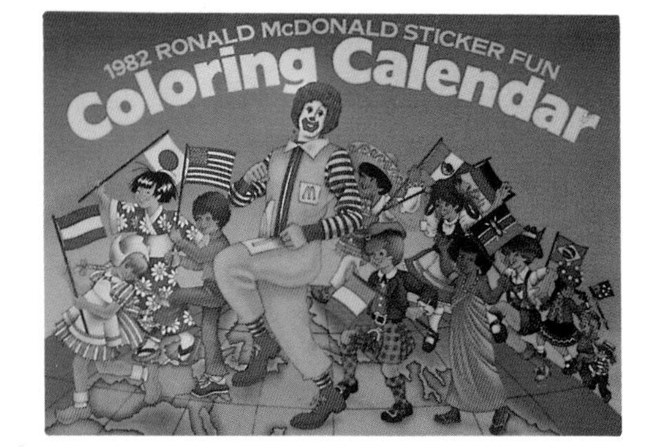

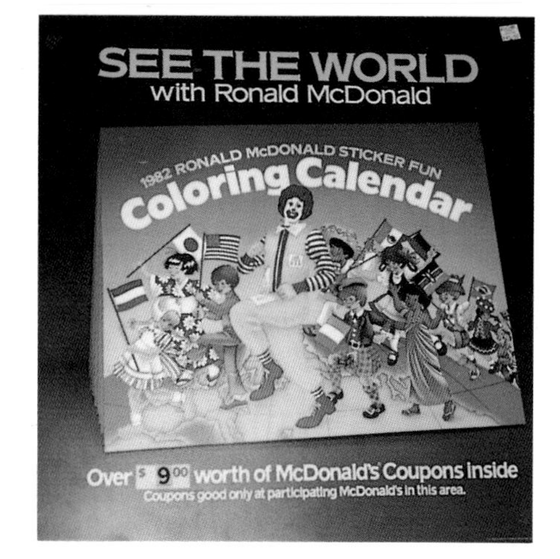

Right: Ge8244

Left: Ge8245

As8315

1983

- **"McDonald's & You" (repeated)**
- **Chicken McNuggets added to menu**
- **McNugget Buddies joins the cast of characters**
- **Hamburger University opens**

Astrosniks I Happy Meal, 1983

Boxes:

❑ ❑ USA As8315 **Hm Box - Astrosnik Rover with one wheel**, 1983. $100.00-150.00

❑ ❑ USA As8316 **Hm Box - Round Shape Space Ship/Volcano**, 1983. $100.00-150.00

❑ ❑ USA As8317 **Hm Box - Spaceship with 2 wings & front vents/Dinosaur/Astralia,** 1983. $100.00-150.00

❑ ❑ USA As8318 **Hm Box - Spaceship like Airplane/Moon Golf Course**, 1983. $100.00-150.00

As8316 (front)

As8316 (back)

As8317

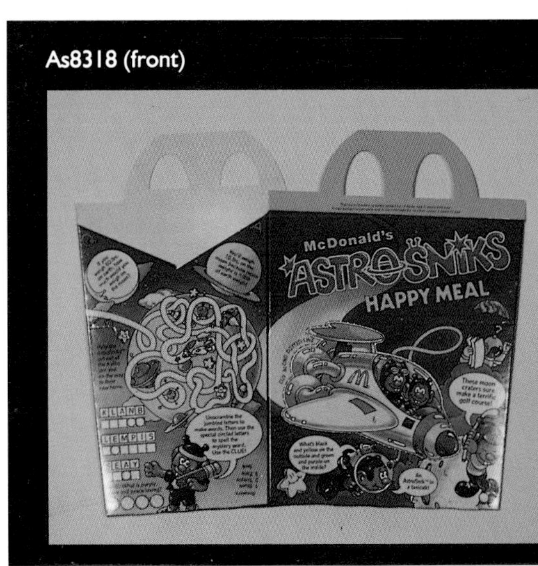

As8318 (front)

Premiums:

- ❑ ❑ USA As8301 **Astralia: Girl Holding Cone with Sticker MIP.** $30.00-40.00
- ❑ ❑ USA As8301 **Astralia: Girl Holding Cone without Sticker/Loose.** $5.00-7.00
- ❑ ❑ USA As8302 **Laser: with Gun with Sticker MIP.** $30.00-40.00
- ❑ ❑ USA As8302 **Laser: with Gun without Sticker/Loose.** $5.00-7.00
- ❑ ❑ USA As8303 **Robo-Robot in Gold with Sticker MIP.** $30.00-40.00
- ❑ ❑ USA As8303 **Robo-Robot in Gold without Sticker/Loose.** $10.00-12.00
- ❑ ❑ USA As8304 **Scout: Holding Flag with Sticker MIP.** $30.00-40.00
- ❑ ❑ USA As8304 **Scout: Holding Flag without Sticker/Loose.** $5.00-7.00
- ❑ ❑ USA As8305 **Skater: on Ice Skates with Sticker MIP.** $30.00-40.00
- ❑ ❑ USA As8305 **Skater: on Ice Skates without Sticker/Loose.** $5.00-7.00
- ❑ ❑ USA As8306 **Snikapotamus: Dinosaur with Sticker MIP.** $30.00-40.00
- ❑ ❑ USA As8306 **Snikapotamus: Dinosaur without Sticker/Loose.** $10.00-12.00
- ❑ ❑ USA As8307 **Sport: Holding Football with Sticker MIP.** $30.00-40.00
- ❑ ❑ USA As8307 **Sport: Holding Football without Sticker/Loose.** $5.00-7.00
- ❑ ❑ USA As8308 **Thirsty: Holding Drink with Sticker MIP.** $30.00-40.00
- ❑ ❑ USA As8308 **Thirsty: Holding Drink without Sticker/Loose.** $5.00-7.00

As8301 As8302 As8303 As8304

As8303

As8307 As8306 As8305 As8308

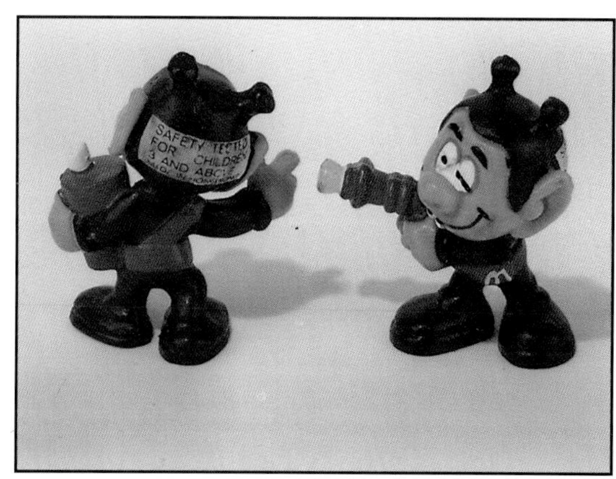

As8318 (back)

❑	❑	USA As8355 **Tray Liner**, 1983.	$15.00-20.00
❑	❑	USA As8356 **Table Tent**, 1983.	$15.00-25.00
❑	❑	USA As8365 **Translite/Lg**, 1983.	$65.00-90.00

Comments: Limited Regional Distribution: USA - 1983 in parts of New England. These Astrosnik figurines have a molded yellow "M" on their body and "McDonald's, Hong Kong, '83 Bully-Figuren" molded into the bottom of the feet. The figures were distributed MIP in a zip lock bag with a leaflet. Each came with a "Safety Tested" sticker attached to the back or side of the figurine. Stickers were easily removed with handling. Prices quoted are for Mint in the Zip Lock Packaging with perfect sticker attached and leaflet. Most figures available are loose figurines without the stickers. Loose Astrosniks sell for $5.00-12.00 each, the Gold Robo-Robo and Snikapotamus selling for $10.00-12.00 (loose Astrosnik figurines without the sticker).

As8355

As8356 (front)

As8356 (back)

Circus-3 Ring Circus Happy Meal, 1983

Boxes:

❑	❑	USA Ci8315 **Hm Box - Amazing Animal Acts**, 1983.	$25.00-40.00
❑	❑	USA Ci8316 **Hm Box - Circus Band**, 1983.	$25.00-40.00
❑	❑	USA Ci8317 **Hm Box - Clown Car,** 1983.	$25.00-40.00

Ci8315

Ci8316

Ci8317

❑ ❑ USA Ci8318 **Hm Box - High Wire Show**, 1983. $25.00-40.00

❑ ❑ USA Ci8319 **Hm Box - Monkey Cage**, 1983. $25.00-40.00

❑ ❑ USA Ci8320 **Hm Box - Tumblers & Jugglers**, 1983. $25.00-40.00

Premiums:

❑ ❑ USA Ci8301 **French Fry Faller,** 1982, Disk Rolls thru Ladder/Blu or Red or Yel. $20.00-25.00

❑ ❑ USA Ci8302 **Fun House Mirror- Ronald,** 1983, Blue with 3 Visual Effects. $10.00-13.00

❑ ❑ USA Ci8303 **Fun House Mirror - Hamburglar,** 1983, Blue with 3 Visual Effects. $10.00-13.00

❑ ❑ USA Ci8302 **Fun House Mirror- Ronald,** 1983, Red with 3 Visual Effects. $13.00-15.00

❑ ❑ USA Ci8303 **Fun House Mirror - Hamburglar,** 1983, Red with 3 Visual Effects. $13.00-15.00

❑ ❑ USA Ci8302 **Fun House Mirror - Ronald,** 1983, Orange/Yellowish with 3 Visual Effects. $15.00-20.00

❑ ❑ USA Ci8303 **Fun House Mirror - Hamburglar,** 1983, Orange/Yellowish with 3 Visual Effects. $15.00-20.00

Ci8320

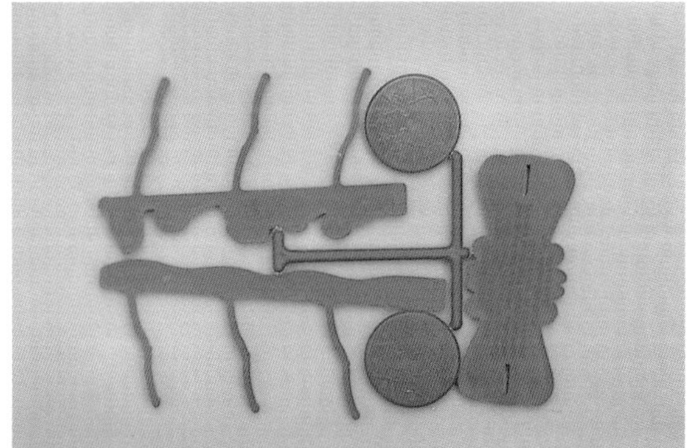
Ci8301

Ci8318

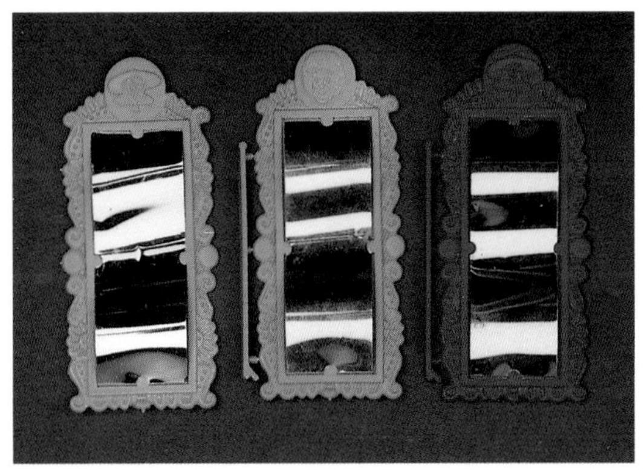
Ci8303 Ci8302 Ci8303

Ci8319

Ci8303

❑ ❑ USA Ci8304 **Grimace Strong Gong,** 1982, Mallet Knocks Grimace Off/Blu or Yel or Purp or Grn. $20.00-25.00

❑ ❑ USA Ci8305 **Ronald Acrobat,** 1980, Ronald Hangs from Bar/Yel or Blu or Red or Org. $25.00-35.00

❑ ❑ USA Ci8306 **Punch Outs Midway 1 - Puppet Show/Arcade/Professor in Car**, 1983. $45.00-60.00

❑ ❑ USA Ci8307 **Punch Outs Midway 2 - Fun House/How Strong/Elephant/Ronald**, 1983. $45.00-60.00

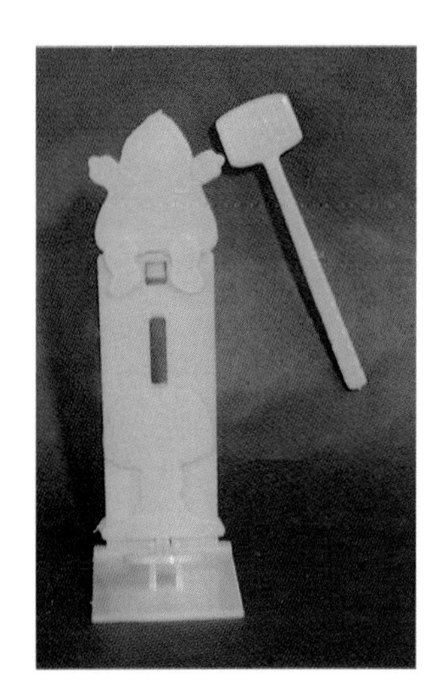

Ci8304

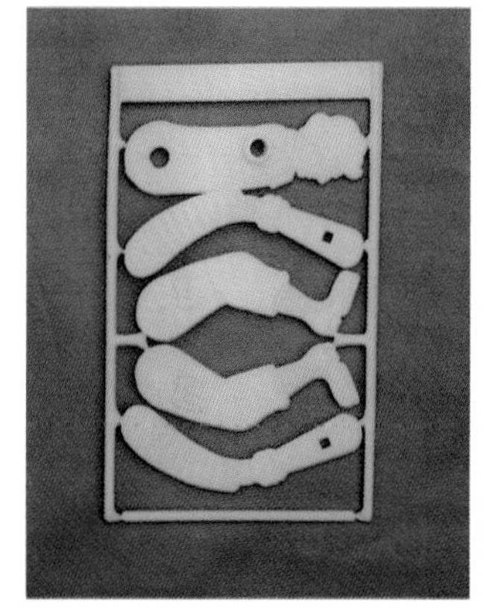

Below and right: Ci8305

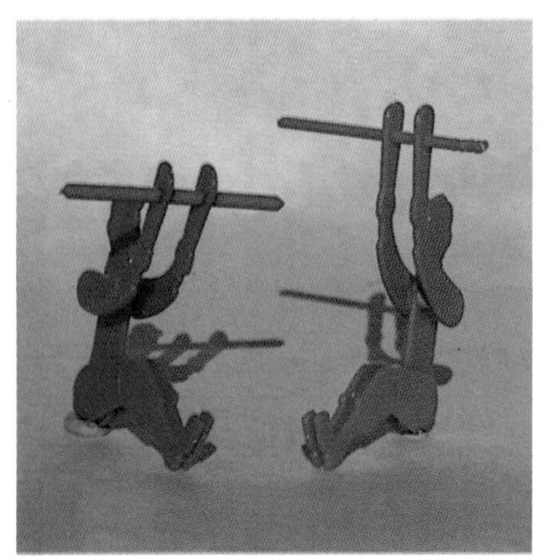

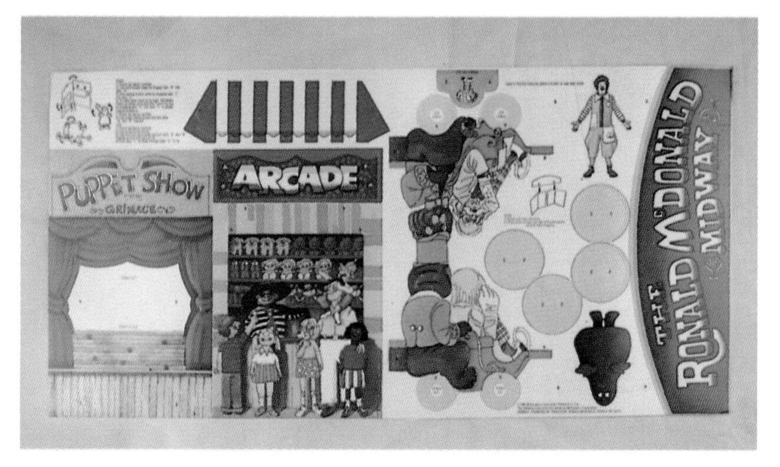

Ci8306

Ci8307

❑ ❑ USA Ci8308 **Punch Outs Left Tent - Lion/Tiger/Hamburglar**, 1983 $45.00-60.00

❑ ❑ USA Ci8309 **Punch Outs Right Tent - Birdie with Horse/Circus Banner on Right**, 1983. $75.00-100.00

❑ ❑ USA Ci8361 **MC Cardboard,** 1983. $25.00-40.00

❑ ❑ USA Ci8362 **Header Card**, 1983. $25.00-35.00

❑ ❑ USA Ci8365 **Translite/Lg**, 1983. $25.00-40.00

Comments: National Distribution: USA - September 30-November 23, 1983.

Ci8308

Ci8309

3 Ring Circus blue book.

Ci8365

Going Places/Hot Wheels Promotion, 1983

Premiums/Hot Wheel Cars:

Cars Distributed throughout the USA:

❑ ❑ USA Hw8330 **No. 9241 - Corvette Stingray.** $10.00-15.00

❑ ❑ USA Hw8331 **No. 3259 - Jeep CJ-7 Car.** $10.00-15.00

❑ ❑ USA Hw8335 **No. 2022 - Baja Breaker.** $10.00-15.00

❑ ❑ USA Hw8340 **No. 3250 - Firebird Funny Car.** $10.00-15.00

❑ ❑ USA Hw8342 **No. 3260 - Land Lord.** $10.00-15.00

❑ ❑ USA Hw8348 **No. 2019 - Sheriff Patrol/Blk/Wht.** $10.00-15.00

❑ ❑ USA Hw8352 **No. 1126 - Stutz Blackhawk.** $10.00-15.00

Cars Primarily Distributed on the East Coast:

❑ ❑ USA Hw8333 **No. 4341 - 56 Hi-Tail Hauler.** $10.00-15.00

❑ ❑ USA Hw8334 **No. 2013 - 57 T-Bird.** $10.00-15.00

❑ ❑ USA Hw8336 **No. 1698 - Cadillac Seville.** $10.00-15.00

❑ ❑ USA Hw8338 **No. 3255 - Datsun 200 SX.** $10.00-15.00

❑ ❑ USA Hw8339 **No. 3364 - Dixie Challenger.** $10.00-15.00

❑ ❑ USA Hw8341 **No. 3257 - Front Running Fairmont.** $10.00-15.00

❑ ❑ USA Hw8347 **No. 2021 - Race Bait 308.** $10.00-15.00

❑ ❑ USA Hw8350 **No. 1130 - Tricar X8.** $10.00-15.00

Hw8321

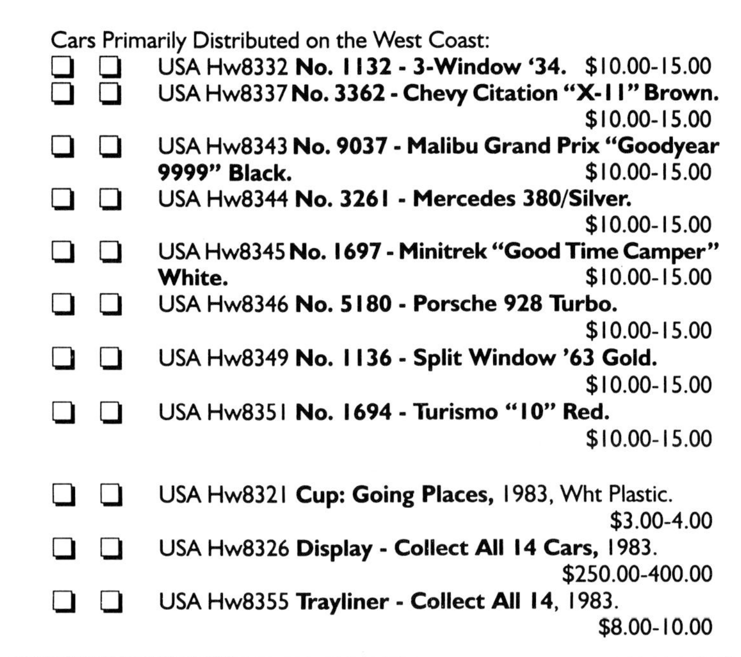

Cars Primarily Distributed on the West Coast:

- ❑ ❑ USA Hw8332 **No. 1132 - 3-Window '34.** $10.00-15.00
- ❑ ❑ USA Hw8337 **No. 3362 - Chevy Citation "X-11" Brown.** $10.00-15.00
- ❑ ❑ USA Hw8343 **No. 9037 - Malibu Grand Prix "Goodyear 9999" Black.** $10.00-15.00
- ❑ ❑ USA Hw8344 **No. 3261 - Mercedes 380/Silver.** $10.00-15.00
- ❑ ❑ USA Hw8345 **No. 1697 - Minitrek "Good Time Camper" White.** $10.00-15.00
- ❑ ❑ USA Hw8346 **No. 5180 - Porsche 928 Turbo.** $10.00-15.00
- ❑ ❑ USA Hw8349 **No. 1136 - Split Window '63 Gold.** $10.00-15.00
- ❑ ❑ USA Hw8351 **No. 1694 - Turismo "10" Red.** $10.00-15.00

- ❑ ❑ USA Hw8321 **Cup: Going Places,** 1983, Wht Plastic. $3.00-4.00
- ❑ ❑ USA Hw8326 **Display - Collect All 14 Cars,** 1983. $250.00-400.00
- ❑ ❑ USA Hw8355 **Trayliner - Collect All 14**, 1983. $8.00-10.00

Hw8355

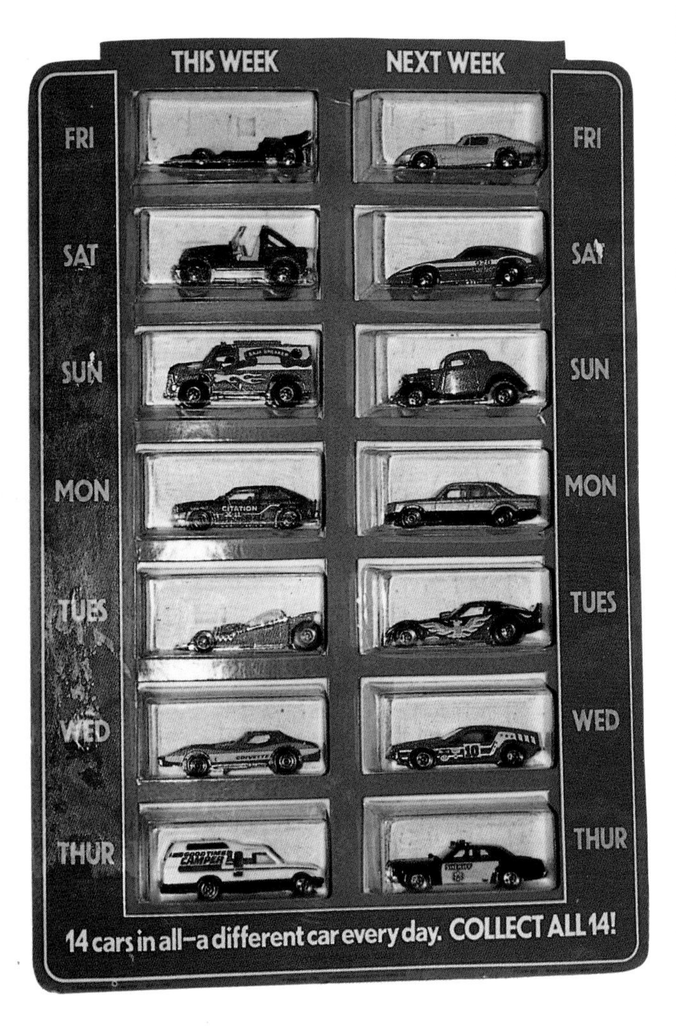

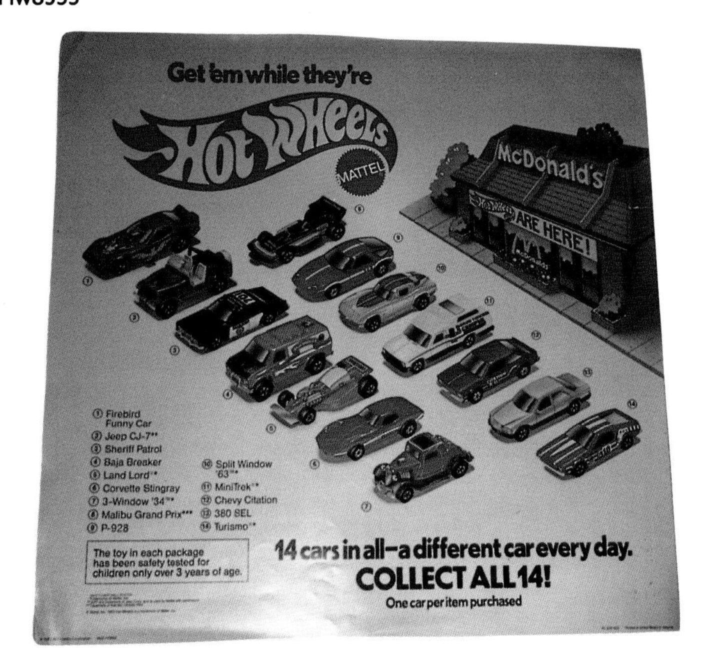

❑ ❑ USA Hw8361 **Hot Wheels Hub Cap Case/Round**, 1983, Grey/HW Logo. $75.00-100.00

❑ ❑ USA Hw8365 **Translite/Lg**, 1983. $35.00-40.00

❑ ❑ USA Hw8366 **Translite/Lg/Cardboard**, 1983. $35.00-40.00

Comments: National Optional Distribution: USA - 1983/89. Hot Wheels were used as a self-liquidating promotion with the "Going Places" boxes. Different cars were given out and/or sold in different regions. Cars could be purchased for $.59 — "Collect All 14. A Different Car Every Day." No under 3 (U-3) premiums were furnished. This was a national option in 1983. Dates stamped on cars do not always match the MIP package date; USA Go8330-31 carry 1981 dates, all others carry 1982 dates. USA Go8352 Stutz Black Hawk No. 1126 could have been substituted for USA Go8333 High Tail Hauler No. 4341 and/or USA Go8345 Minitrek Good Time Camper No. 1697. This promotion was run regionally as a self-liquidator in 1983, again in 1988, and in Texas, January 27-February 23, 1989. Cars were packaged on blister pack cards with no McDonald's logo. Hw8361 Hot Wheels Hub Case (7 1/4" round) was a crew advertising gift. The Hot Wheels that Mattel supplied McDonald's for the National Promotion in 1983 were test marketed in Chattanooga, Tennessee for a four week period in 1982.

Hw8361

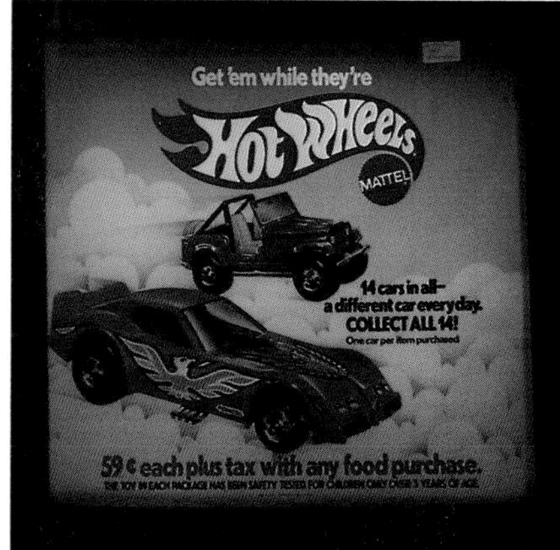

Hw8365

Hp8350 Hp8351 Hp8352

Happy Pail I Happy Meal, 1983

Premiums/Sand Pails:

❑ ❑ USA Hp8350 **Sand Pail: Ronald & Mayor under Umbrella,** 1983, Pnk Pail/Shovel/Lid. $20.00-35.00

❑ ❑ USA Hp8351 **Sand Pail: Ronald in an Inner Tube,** 1983, Wht Pail/Shovel/Lid. $20.00-35.00

❑ ❑ USA Hp8352 **Sand Pail: Airplane Pulling Banner,** 1983, Yel Pail/Shovel/Lid. $20.00-35.00

❑ ❑ USA Hp8365 **Translite/Lg**, 1983. $35.00-50.00

Comments: Regional Distribution: USA - 1983 in upper New York State and parts of New England. Lids had four holes in each and shovels/lids/pail colors matched.

Hp8365

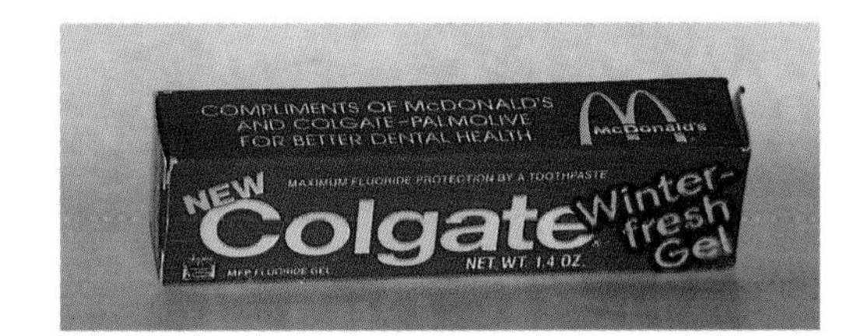

Ht8321

Ht8365

Happy Teeth Happy Meal, 1983

Premiums:

❑ ❑ USA Ht8320 **Toothbrush,** 1983, Reach Toothbrush/No Logo/Red or Yel or Blu or Org or Grn. $8.00-12.00

❑ ❑ USA Ht8321 **Toothpaste,** 1983, Colgate/1.4 oz./Red Box with McDonald's Logo. $15.00-20.00

❑ ❑ USA Ht8365 **Translite/Lg**, 1983, Get a Reach Youth-Size Toothbrush with Every Hm. $——

Comments: Regional Distribution: USA - 1983 in New England. Happy Meal promoted in conjunction with Dental Health month. Expiration date on toothpaste: 05/84. Toothbrushes came in five colors.

Lego Building Sets I Test Market Happy Meal, 1983

Boxes:

❑ ❑ USA Le8310 **Hm Box - Master the Maze,** 1983. $40.00-65.00

❑ ❑ USA Le8311 **Hm Box - What's Wrong?**, 1983. $40.00-65.00

U-3 Premiums:

❑ ❑ USA Le8305 **U-3 Wk #1 Blu Pkg,** 1983, Duplo. $25.00-40.00

❑ ❑ USA Le8306 **U-3 Wk #2 Yel Pkg,** 1983, Bird/Duplo/3 Sm Grn/1 Lg Grn/1 Sm Yel. $25.00-40.00

❑ ❑ USA Le8307 **U-3 Wk #3 Red Pkg,** 1983, Duplo. $25.00-40.00

❑ ❑ USA Le8308 **U-3 Wk #4 Grn Pkg,** 1983, Duplo. $25.00-40.00

Premiums:

❑ ❑ USA Le8302 **Set 1 Ship,** 1983, Lego/Yellow Package 27p. $20.00-25.00

❑ ❑ USA Le8304 **Set 2 Airplane,** 1983, Lego/Grn Package 18p. $20.00-25.00

❑ ❑ USA Le8303 **Set 3 Helicopter,** 1983, Lego/Blue Package 19p. $20.00-25.00

❑ ❑ USA Le8301 **Set 4 Truck,** 1983, Lego/Red Package 17p. $20.00-25.00s

❑ ❑ USA Le8326 **Display** - Collect All Lego Building Sets, 1983. $500.00-750.00

❑ ❑ USA Le8365 **Translite/Lg**, 1983. $65.00-90.00

Comments: Regional Distribution: USA - June 1983. Test Market in Salt Lake City, Utah - Summer 1983.

Le8306

Le8302 (front)

Le8302 (back)

McDonaldland Junction Happy Meal, 1983

Boxes:

❑ ❑ USA Ju8310 **Hm Box - Engine Barn**, 1983. $10.00-15.00
❑ ❑ USA Ju8311 **Hm Box - Post Office,** 1983. $10.00-15.00
❑ ❑ USA Ju8312 **Hm Box - Signal Tower**, 1983. $10.00-15.00
❑ ❑ USA Ju8313 **Hm Box - Station**, 1983. $10.00-15.00
❑ ❑ USA Ju8314 **Hm Box - Town Hall,** 1983. $10.00-15.00
❑ ❑ USA Ju8315 **Hm Box - Train Tunnel,** 1983. $10.00-15.00

Ju8310

Ju8311

Ju8312

Ju8313

Ju8314

Ju8315

Ju8300 Ju8305 Ju8302 Ju8303 Ju8307
Ju8304
Ju8301 Ju8306

Premiums: Nationally Distributed:

- ❑ ❑ USA Ju8300 **#1 Ronald Train Engine - red,** 1982, Red. $8.00-10.00
- ❑ ❑ USA Ju8302 **#2 Flat Car with 4 Fry Kids - green,** 1982, Grn. $8.00-10.00
- ❑ ❑ USA Ju8304 **#3 Birdie Parlor Car - yellow,** 1982, Yel. $8.00-10.00
- ❑ ❑ USA Ju8306 **#4 Grimace Caboose - purple,** 1982, Pur. $8.00-10.00

Premiums: Regionally Distributed:

- ❑ ❑ USA Ju8301 **#1 Ronald Train Engine - blue,** 1982 Blue. $35.00-50.00
- ❑ ❑ USA Ju8303 **#2 Flat Car with 4 Fry Kids - white,** 1982, Wht. $35.00-50.00
- ❑ ❑ USA Ju8305 **#3 Birdie Parlor Car - pink,** 1982, Pink. $35.00-50.00
- ❑ ❑ USA Ju8307 **#4 Grimace Caboose - orange,** 1982, Orange. $35.00-50.00

- ❑ ❑ USA Ju8324 **Amtrak Pamphlet with Stickers**, 1983. $20.00-25.00
- ❑ ❑ USA Ju8361 **MC Insert/Cardboard**, 1983. $25.00-30.00
- ❑ ❑ USA Ju8362 **Header Card**, 1983. $20.00-25.00

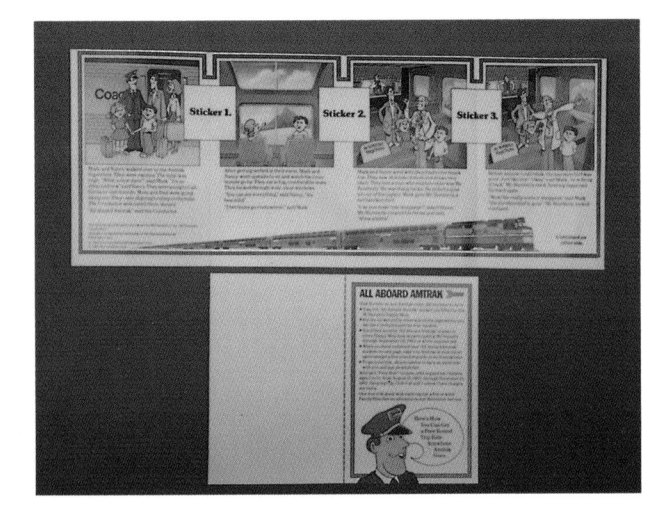

Ju8324

Ju8361

Ju8362

❑ ❑ USA Ju8363 **Menu Board Lug on**, 1983. $20.00-25.00
❑ ❑ USA Ju8365 **Translite/Lg**, 1983. $25.00-30.00
❑ ❑ USA Ju8366 **Translite/Lg,** 1983, Get Today's Toy. $25.00-30.00

Comments: National Distribution: USA - January 17-March 27, 1983. The Amtrak "All Aboard" sticker sheet with pamphlet was distributed with the Happy Meal in the northeast. USA Ju8300/02/04/06 were nationally distributed. USA Ju8301/03/05/07 were regionally distributed.

Mystery Happy Meal, 1983

Boxes:
❑ ❑ USA My8310 **Hm Box - Dog-Gone Mystery**, 1983. $20.00-25.00
❑ ❑ USA My8311 **Hm Box - Golden Key Case,** 1983. $20.00-25.00
❑ ❑ USA My8312 **Hm Box - Mysterious Map**, 1983. $20.00-25.00

Ju8365

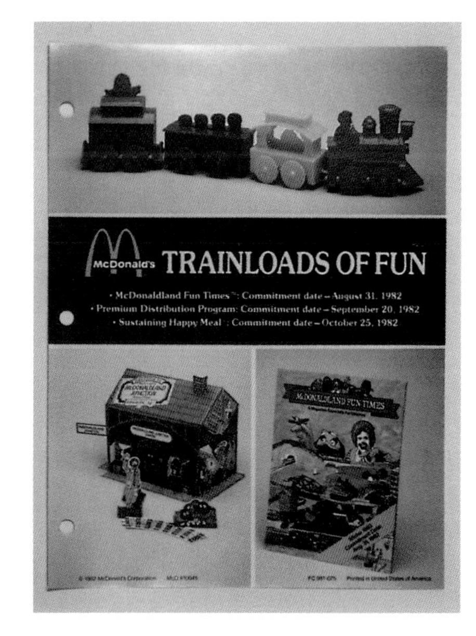

My8310

My8311

My8312

My8313

My8314

❑ ❑ USA My8313 **Hm Box - Ocean's Away!,** 1983. $20.00-25.00

❑ ❑ USA My8314 **Hm Box - Thump Blam Bump Mystery**, 1983. $20.00-25.00

Premiums:

❑ ❑ USA My8300 **Detective Kit - Box with Tweezers**, 1982, Blu or Grn or Org or Red/2p. $15.00-20.00

❑ ❑ USA My8301 **Crystal Ball (McDonaldland),** 1982, Flat Paper forms a Cube/Clear Ball with Ronald Pic/3p. $75.00-100.00

My8301

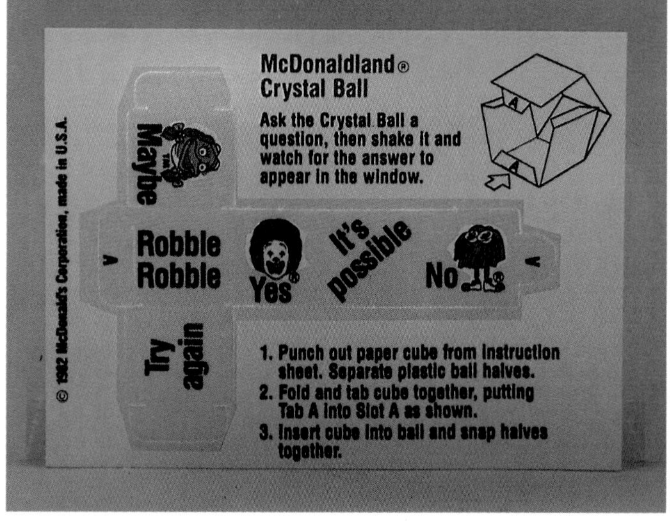

My8301

My8300

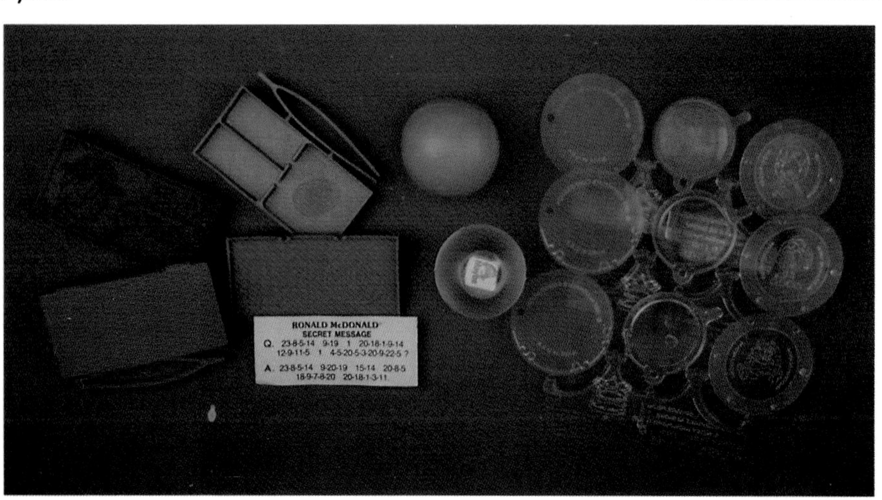

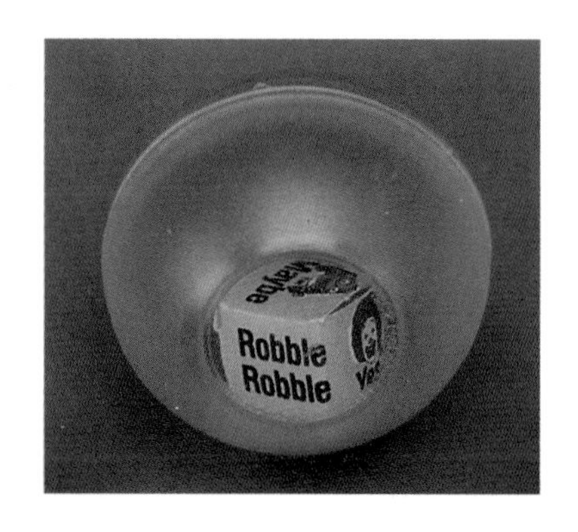

❑ ❑ USA My8302 **Magni-Finder Glass - Ronald,** 1982, Clear. $35.00-40.00
❑ ❑ USA My8304 **Magni-Finder Glass - Fry Kid,** 1982, Clear. $35.00-40.00
❑ ❑ USA My8305 **Magni-Finder Glass - Birdie,** 1982, Clear. $35.00-40.00

❑ ❑ USA My8303 **Unpredict-A-Ball,** 1983, Ronald's Face on ball/Wobbles/Blu or Red or Yel/2p. $20.00-25.00

❑ ❑ USA My8361 **MC/Cardboard**, 1983. $35.00-45.00
❑ ❑ USA My8362 **Header Card**, 1983. $20.00-30.00
❑ ❑ USA My8363 **Menu Board/Lug on**, 1983. $20.00-30.00
❑ ❑ USA My8365 **Translite/Lg**, 1983. $30.00-35.00

Comments: National Distribution: USA - March 28-June 5, 1983. USA My8301 was recalled prior to national distribution.

My8305 My8304 My8302

Play-Doh I '83 Happy Meal, 1983

Premiums: Play-Doh Cans:
❑ ❑ USA Pl8301 **Blue Play-Doh,** 1981, **Cardboard Can with Tin Bottom.** $20.00-25.00
❑ ❑ USA Pl8302 **Red Play-Doh,** 1981, **Cardboard Can with Tin Bottom.** $20.00-25.00
❑ ❑ USA Pl8303 **White Play-Doh,** 1981, **Cardboard Can with Tin Bottom.** $20.00-25.00
❑ ❑ USA Pl8304 **Yellow Play-Doh,** 1981, **Cardboard Can with Tin Bottom.** $20.00-25.00

❑ ❑ USA Pl8365 **Translite/Lg**, 1983. $25.00-40.00

Comments: Regional Distribution: USA - May 1983 in the Boston, Massachusetts area. The 2 ounce cans of Play-Doh by Kenner were made of cardboard with tin bottoms; they had no McDonald's markings. In 1984, the Wichita, Kansas (March 23-April 22) and Nebraska (March 2-April 1) areas ran this Play-Doh promotion. In 1985, Play-Doh was offered again with two additional cans (See USA Pl8529-30, Play-Doh Happy Meal, 1985).

Pl8301 Pl8302 Pl8303 Pl8304

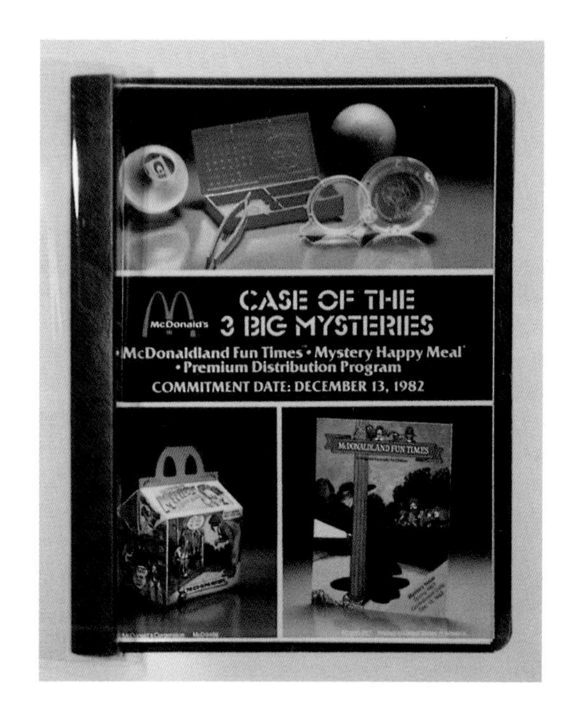

Mystery blue book.

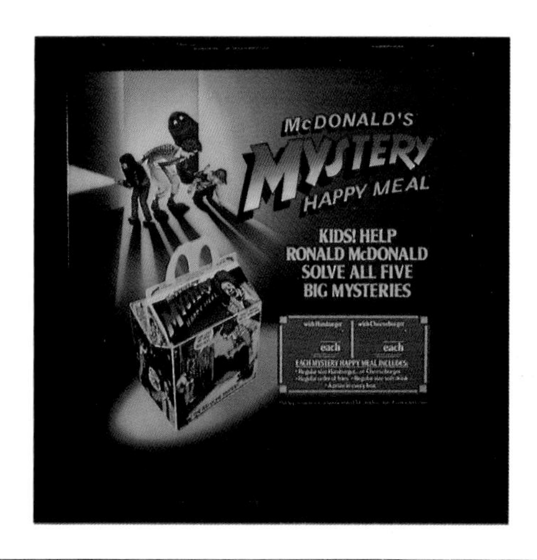

My8365

Pl8365

Sh8303 Sh8304 Sh8302 Sh8301

Ship Shape I '83 Happy Meal, 1983

Premiums: Vacuum Formed Containers:

❑ ❑ USA Sh8301 **Splash Dasher - Hamburglar,** 1983, Wht Top/Org Bottom. $10.00-15.00
❑ ❑ USA Sh8302 **Tubby Tugger - Grimace,** 1983, Pnk Top/Blu Bottom. $10.00-15.00
❑ ❑ USA Sh8303 **Rub-A-Dub Sub - Captain Crook,** 1983, Grn Top/Grn Bottom/Submarine. $10.00-15.00
❑ ❑ USA Sh8304 **River Boat - Ronald,** 1983, Yel Top/Red Bottom. $10.00-15.00

❑ ❑ USA Sh8305 **Sticker Sheet - Splash Dasher,** 1983. $4.00-5.00
❑ ❑ USA Sh8306 **Sticker Sheet - Tubby Tugger,** 1983. $4.00-5.00
❑ ❑ USA Sh8307 **Sticker Sheet- Rub-A-Dub-Sub,** 1983. $4.00-5.00
❑ ❑ USA Sh8308 **Sticker Sheet - River Boat,** 1983. $4.00-5.00

❑ ❑ USA Sh8326 **Display,** 1983. $150.00-200.00
❑ ❑ USA Sh8361 **MC/Cardboard**, 1983. $35.00-45.00
❑ ❑ USA Sh8365 **Translite/Lg**, 1983. $40.00-50.00

Comments: National Distribution: USA June 6-July 18, 1983. Decals and containers carried 1983 dates. The "Splash Dasher" issued in 1985 had a larger decal. In 1983, Captain Crook was called "Captain Crook" on the sticker sheet. In 1984, Captain Crook's name was changed to "The Captain." The 1985 sticker sheet for Rub-A-Dub-Sub was named, "The Captain."

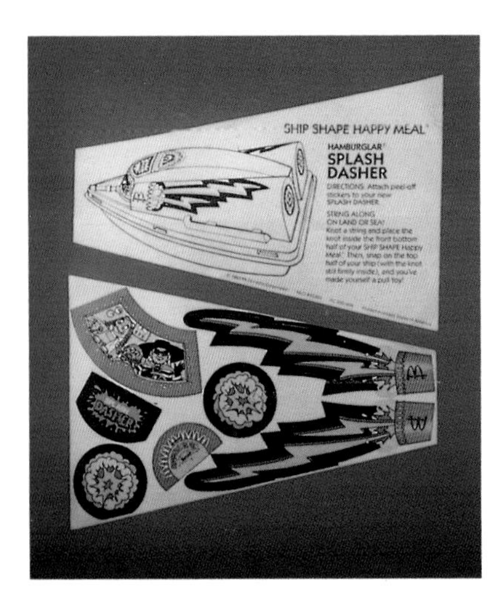

Sh8305

Sh8305, 1983

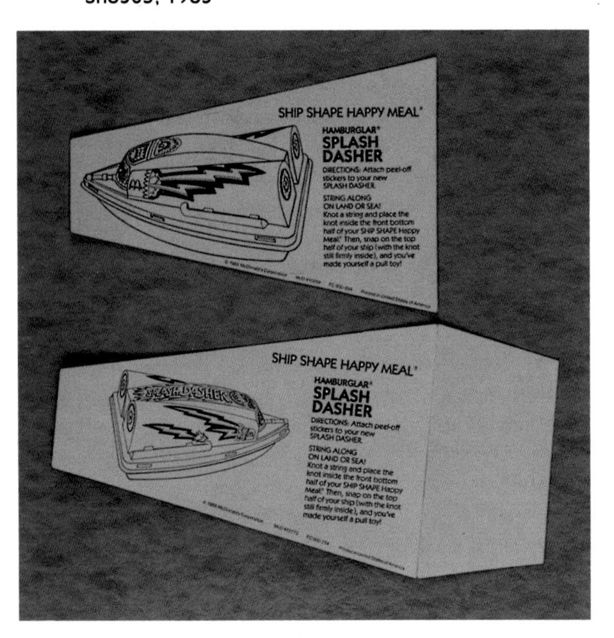

Sh8505, 1985

Sh8326

Sh8365

Winter Worlds Happy Meal, 1983

Boxes:

❑ ❑ USA Wi8310 **Hm Box - Birds of Ice and Snow**, 1983. $5.00-8.00

❑ ❑ USA Wi8311 **Hm Box - Lands of Ice and Snow**, 1983. $5.00-8.00

❑ ❑ USA Wi8312 **Hm Box - Lands of the Midnight Sun**, 1983. $5.00-8.00

❑ ❑ USA Wi8313 **Hm Box - Mammals on the Icy Shores**, 1983. $5.00-8.00

❑ ❑ USA Wi8314 **Hm Box - People of the Frosty Frontier**, 1983. $5.00-8.00

Wi8310

Wi8311

Wi8312

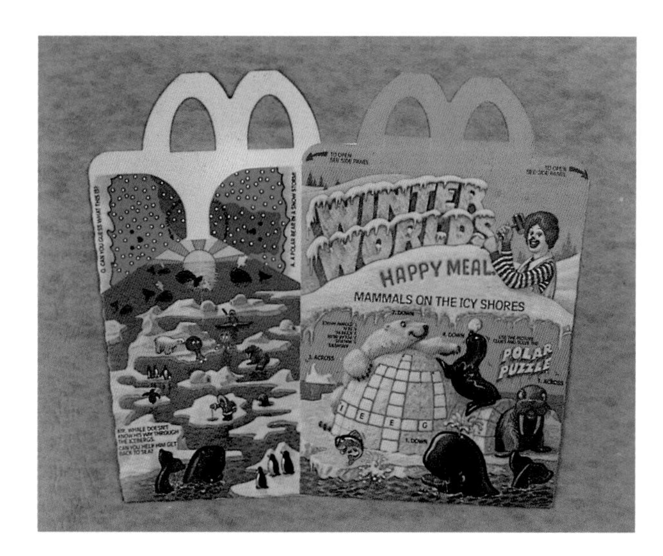

Wi8313

Wi8314

Wi8301 Wi8302 Wi8303

Wi8304 Wi8305

Premiums:

❑ ❑ USA Wi8301 **Ornament - Birdie, 1983,** Yel/Pnk Vinyl 4" with Looped Cord. $5.00-8.00
❑ ❑ USA Wi8302 **Ornament - Grimace, 1983,** Pur Vinyl 4" with Looped Cord. $5.00-8.00
❑ ❑ USA Wi8303 **Ornament - Hamburglar, 1983,** Blk/Wht Stripes/Vinyl 4" with Looped Cord. $5.00-8.00
❑ ❑ USA Wi8304 **Ornament - Mayor, 1983,** Yel/Pnk/Pur Vinyl 4" with Looped Cord. $8.00-10.00
❑ ❑ USA Wi8305 **Ornament - Ronald, 1983,** Red/Yel Vinyl 4" with Looped Cord. $5.00-8.00

❑ ❑ USA Wi8361 **MC/Cardboard**, 1983. $20.00-25.00
❑ ❑ USA Wi8362 **Header Card**, 1983. $20.00-25.00
❑ ❑ USA Wi8363 **Menu Board/Lug on**, 1983. $20.00-25.00
❑ ❑ USA Wi8365 **Translite/Lg**, 1983. $15.00-20.00

Comments: Limited National Distribution: USA November 28, 1983- February 5, 1984. Grimace/Ronald/Hamburglar vinyl ornaments were reissued Christmas 1984, dated 1984 (not 1983). The Winter Worlds Happy Meal distributed ornaments were dated 1983 only.

USA Generic Promotions, 1983

❑ ❑ USA Ge8301 **Spinner Baseball,** 1983, Grn/5p/Ronald/Grimace/Hamburglar/Fry Kid. $4.00-5.00
❑ ❑ USA Ge8302 **Fun Ruler,** 1983, Inches/Centimeters. $2.00-3.00
❑ ❑ USA Ge8303 **Ronald Styro-glider with red nose weight,** red/white, 1983. $7.00-10.00

Wi8363

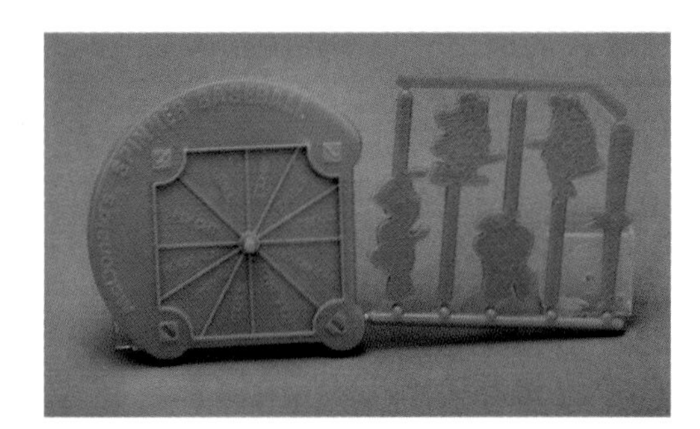

Ge8301

Ge8302

Wi8365

❑ ❑ USA Ge8304 **Ronald McDonald large adult size Hand Puppet,** Bubble Red/Yellow print, full color/full figure, no number on puppet, TM under foot and name. Larger size called The Big Ones because they were designed big enough for "Mom and Dad" hands too. $7.00-10.00

❑ ❑ USA Ge8305 **Ronald McDonald smaller size Hand Puppet,** bubble Red/Yellow print, full color/full figure, no number on puppet, TM under foot and name, redesigned to fit child's hand. $1.00-2.00

❑ ❑ USA Ge8306 **White Plastic Cup: 1983 Ronald McDonald's Happy Cup,** 5" WHITE cup. Ronald on roller skates holding banner, 5" WHITE. $4.00-5.00

❑ ❑ USA Ge8307 **Record: Merry McBirthday from Ronald McDonald,** 78 RPM with jacket. Made by Kid Stuff Records. Sold in retail stores. $7.00-10.00

❑ ❑ USA Ge8308 **Fun Times Magazine: Vol. 4 No. 3 Mystery Issue, 1983.** $4.00-5.00

❑ ❑ USA Ge8309 **Fun Times Magazine: Vol. 4 No. 4 Transportation Issue, 1983.** $4.00-5.00

❑ ❑ USA Ge8310 **Fun Times Magazine: Vol. 5 No. 1 Circus Issue, 1983.** Vol. 5 No. 1 Circus Fun Times magazine issue was distributed at the same time as Season Premier issues in regional parts of the USA where ABC-TV was not highlighted. Vol. 5 No. 2 Holiday Fun was only distributed in Canada. After Vol. 5 No. 2 was distributed in Canada in 1983, Fun McDonalds closed out Vol. 5. Each successive volume began in January instead of the Fall. $4.00-5.00

❑ ❑ USA Ge8311 **Fun Times Magazine: Season Premier ABC WJZ-TV 13, 1983.** In 1983 Season Premier ABC-TV issues featured all the new cartoon characters. Fun Times Magazine issues were regionally distributed based on television markets. $5.00-8.00

❑ ❑ USA Ge8312 **Fun Times Magazine: Season Premier ABC WPVI-TV 6, 1983.** $5.00-8.00

❑ ❑ USA Ge8313 **Fun Times Magazine: Season Premier ABC WPVI-TV 13, 1983.** $5.00-8.00

❑ ❑ USA Ge8314 **Fun Times Magazine: Season Premier ABC WABC-TV 7, 1983.** $5.00-8.00

❑ ❑ USA Ge8315 **Fun Times Magazine: Season Premier ABC WABC-TV 13, 1983.** $5.00-8.00

❑ ❑ USA Ge8316 **Fun Times Magazine: Season Premier ABC WJLA-TV 7, 1983.** $5.00-8.00

❑ ❑ USA Ge8317 **Fun Times Magazine: Season Premier ABC KGO-TV 7, KOVR-TV 13, and KNTV-TV 11, 1983** $5.00-8.00

❑ ❑ USA Ge8318 **Fun Times Magazine: Season Premier ABC KGO-TV 13, 1983.** $5.00-8.00

❑ ❑ USA Ge8319 **Fun Times Magazine: Season Premier ABC KABC-TV 7, 1983.** $5.00-8.00

Ge8304 Ge8305

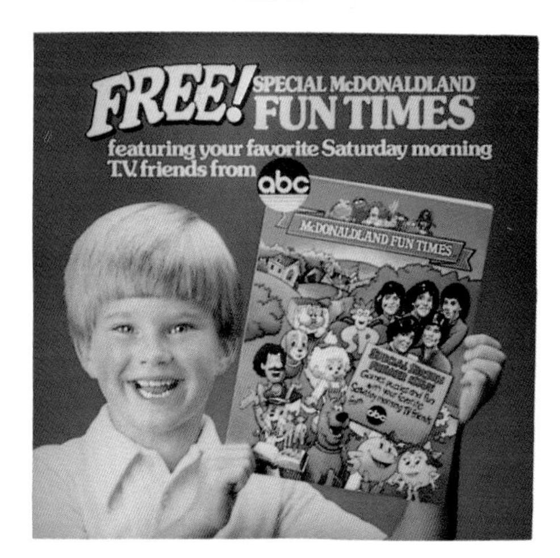

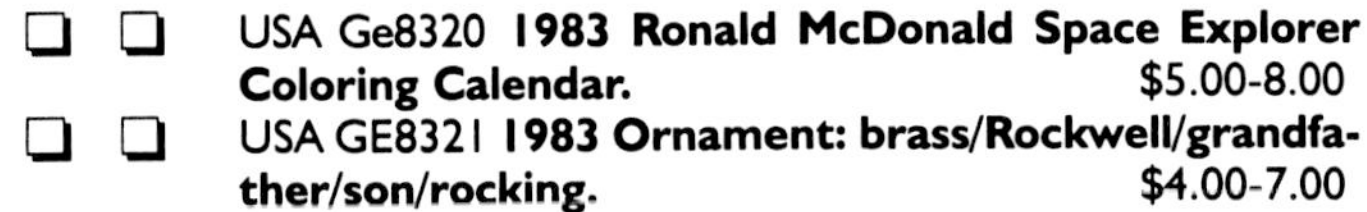

❑ ❑ USA Ge8320 **1983 Ronald McDonald Space Explorer Coloring Calendar.** $5.00-8.00

❑ ❑ USA GE8321 **1983 Ornament: brass/Rockwell/grandfather/son/rocking.** $4.00-7.00

Comments: Regional Distribution: USA - 1983 during Clean-Up weeks. This Is a sampling of generic premiums given away and/or sold in retail stores during this period. **Chicken McNuggets were introduced along with McNugget Buddies characters in 1983.** McDonald's restaurants were now in thirty-one countries; the 7,000th restaurant opened in Falls Church, Virginia and they served the 45th billion hamburger. A Christmas ornament was given free with the purchase of Gift Certificates.

Ge8320

Ge8321

1984

Astrosniks II Happy Meal, 1984
Fast Macs I Promotion, 1984
Good Sports Happy Meal, 1984
Happy Holidays Happy Meal, 1984
Happy Pail II Olympic Theme Happy Meal, 1984
Lego Building Sets II '84 Happy Meal, 1984
Olympic Beach Ball Promotion, 1984
Olympic Sports [Zip Action] HM (canceled), 1984
Olympic Sports I Happy Meal, 1984
Popoids Crazy Creatures I Test Market HM, 1984
School Days Happy Meal, 1984
USA Generic Promotions, 1984

- **"It's a Good Time for the Great Taste of McDonald's" jingle**
- **"When the USA Wins You Win" ad campaign**
- **Ray A. Kroc, founder of McDonald's dies (1902-1984)**
- **10th Anniversary of Ronald McDonald House**
- **8th National O/O Convention**
- **"One of a Kind" - O/O advertising theme**
- **World Boy Scout Jamboree sponsored by McDonald's on TV**

Astrosniks II Happy Meal, 1984

Boxes:

❑ ❑ USA As8450 **Hm Box - McDonald's Store/Racer**, 1984. $75.00-100.00

❑ ❑ USA As8451 **Hm Box - Snik Station Earth/Perfido**, 1984. $75.00-100.00

Premiums:

❑ ❑ USA As8431 **Copter - with Helicopter Blades on Head/MIP**, 1984. $20.00-25.00

❑ ❑ USA As8431 **Copter - with Helicopter Blades on Head/Loose**, 1984. $4.00-5.00

❑ ❑ USA As8432 **Racing - on Sled/MIP**, 1984. $20.00-25.00

❑ ❑ USA As8432 **Racing - on Sled/Loose**, 1984. $4.00-5.00

❑ ❑ USA As8433 **Ski - with Skis/Goggles/MIP**, 1984. $20.00-25.00

❑ ❑ USA As8433 **Ski - with Skis/Goggles/Loose**, 1984. $4.00-5.00

❑ ❑ USA As8434 **Commander - with Black Mask/MIP**, 1984. $20.00-25.00

❑ ❑ USA As8434 **Commander - with Black Mask/Loose**, 1984. $5.00-7.00

❑ ❑ USA As8435 **Drill - with Drill in Hands/MIP**, 1984. $20.00-25.00

❑ ❑ USA As8435 **Drill - with Drill in Hands/Loose**, 1984. $4.00-5.00

❑ ❑ USA As8436 **Perfido - with Red Cape/MIP**, 1984. $20.00-25.00

❑ ❑ USA As8436 **Perfido - with Red Cape/Loose**, 1984. $5.00-7.00

As8450 As8315 As8451

As8431

As8431 As8432 As8433 As8434 As8435 As8436

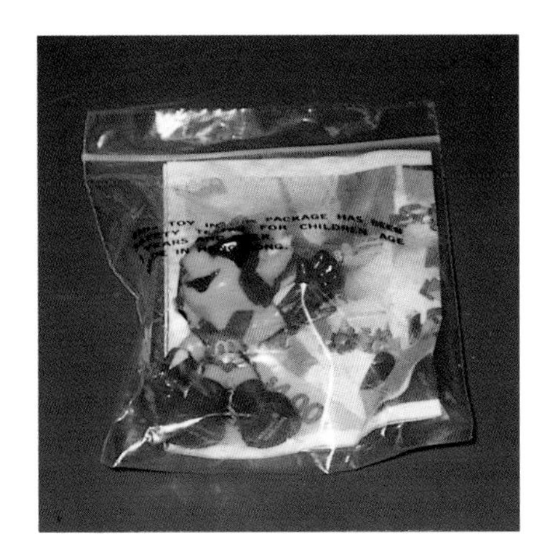

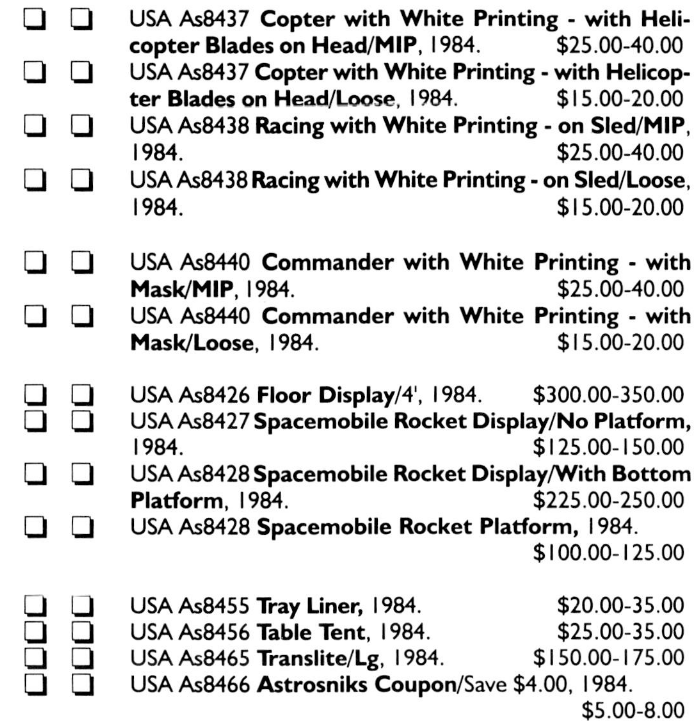

❑ ❑ USA As8437 **Copter with White Printing - with Helicopter Blades on Head/MIP**, 1984. $25.00-40.00

❑ ❑ USA As8437 **Copter with White Printing - with Helicopter Blades on Head/Loose**, 1984. $15.00-20.00

❑ ❑ USA As8438 **Racing with White Printing - on Sled/MIP**, 1984. $25.00-40.00

❑ ❑ USA As8438 **Racing with White Printing - on Sled/Loose**, 1984. $15.00-20.00

❑ ❑ USA As8440 **Commander with White Printing - with Mask/MIP**, 1984. $25.00-40.00

❑ ❑ USA As8440 **Commander with White Printing - with Mask/Loose**, 1984. $15.00-20.00

❑ ❑ USA As8426 **Floor Display**/4', 1984. $300.00-350.00

❑ ❑ USA As8427 **Spacemobile Rocket Display/No Platform**, 1984. $125.00-150.00

❑ ❑ USA As8428 **Spacemobile Rocket Display/With Bottom Platform**, 1984. $225.00-250.00

❑ ❑ USA As8428 **Spacemobile Rocket Platform,** 1984. $100.00-125.00

❑ ❑ USA As8455 **Tray Liner,** 1984. $20.00-35.00

❑ ❑ USA As8456 **Table Tent**, 1984. $25.00-35.00

❑ ❑ USA As8465 **Translite/Lg**, 1984. $150.00-175.00

❑ ❑ USA As8466 **Astrosniks Coupon**/Save $4.00, 1984. $5.00-8.00

As8440

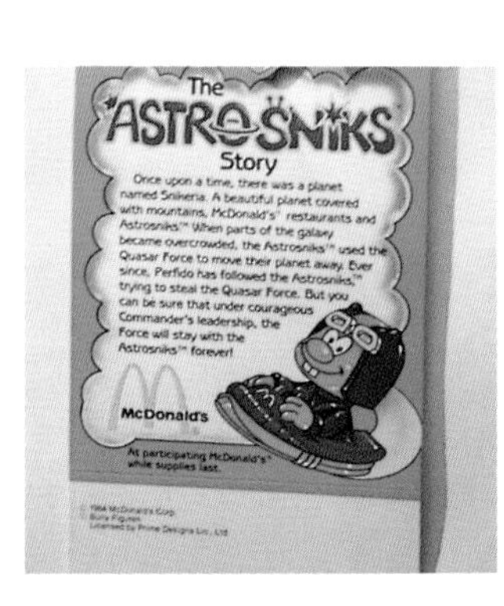

As8456

As8427

As8466

Comments: Limited Regional Distribution: USA - March/April and August 10-September 23, 1984 in St. Louis, Missouri area. All mint astrosnik figurines were enclosed in a ziplock bag with a $4 off coupon for the Spacemobile display/sold in retail stores. Bags said, "This toy in this package has been safety tested for children age 3 years and over. Made in Hong Kong." USA As8431-36 figurines have a molded yellow logo "M" on figurine and trademark data '84 Bully-Figuren TM Astrosnik McDonald's Hong Kong printed onto the bottom of the figurine. USA As8437-42 were limited regional distribution: USA - 1984 in New England areas; the trademark data was printed in white on the bottom of each figurine and the yellow "M" was painted on. **Prices quoted are for Mint in the Package figurines. Loose figurines without the white printing sell for $5.00 - 10.00. Loose figurines with the white printing sell for $15.00 - 20.00. The majority of the figurines on the market are loose without the white printing.** After further research, it appears the following figurines do not exist: USA As8439 Ski/Wht Printing, 1984, with Skis/Goggles; USA As8441 Drill/Wht Printing, 1984, with Drill in Hands; and USA As8442 Perfido/Wht Printing, 1984, with Red Cape. It appears only three figurines were given out with the white printing on the bottom. The platform was a separate boxed piece available from toy stores.

Fast Macs I Promotion, 1984

Premiums:

❑ ❑ USA Fa8401 **Big Mac - in Wht Squad/Police Car**, 1985. $4.00-5.00

❑ ❑ USA Fa8402 **Hamburglar - in Red Sports Car**, 1985. $4.00-5.00

❑ ❑ USA Fa8403 **Mayor - in Pink Sun Cruiser/Wht Arches Windshield**, 1985. $20.00-25.00

❑ ❑ USA Fa8404 **Ronald - in Yellow Jeep with Wht Rectangle Windshield.** $35.00-50.00

Comments: Limited Regional Distribution: USA - 1984. Blister pack cards/cars carry 1984 date. Fast Mac cars were sold for 59 cents on a blister pack.. USA Fa8503-04/Mayor/Ronald were redesigned for national distribution in 1985, becoming USA Fa8506/08.

Fa8404

Fa8402

Fa8403

Fa8401

Fa8404 Fa8403

Good Sports Happy Meal, 1984

Boxes:

❑ ❑ USA Go8415 **Hm Box - Skiing**, 1984. $4.00-5.00

Go8415

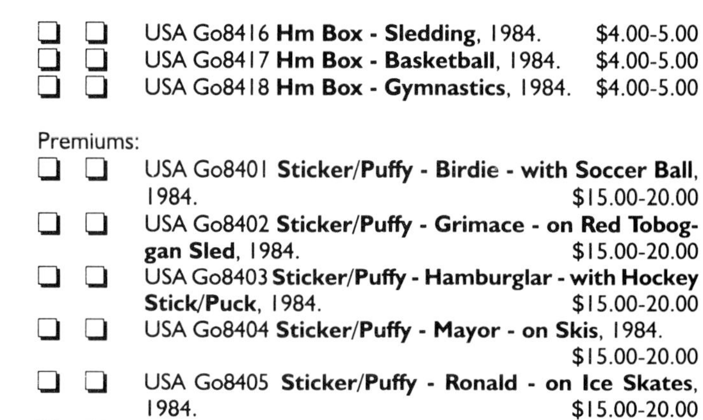

❑ ❑ USA Go8416 **Hm Box - Sledding**, 1984. $4.00-5.00
❑ ❑ USA Go8417 **Hm Box - Basketball**, 1984. $4.00-5.00
❑ ❑ USA Go8418 **Hm Box - Gymnastics**, 1984. $4.00-5.00

Premiums:
❑ ❑ USA Go8401 **Sticker/Puffy - Birdie - with Soccer Ball**, 1984. $15.00-20.00
❑ ❑ USA Go8402 **Sticker/Puffy - Grimace - on Red Toboggan Sled**, 1984. $15.00-20.00
❑ ❑ USA Go8403 **Sticker/Puffy - Hamburglar - with Hockey Stick/Puck**, 1984. $15.00-20.00
❑ ❑ USA Go8404 **Sticker/Puffy - Mayor - on Skis**, 1984. $15.00-20.00
❑ ❑ USA Go8405 **Sticker/Puffy - Ronald - on Ice Skates**, 1984. $15.00-20.00
❑ ❑ USA Go8406 **Sticker/Puffy - Sam the Olympic Eagle - with Basketball**, 1984. $20.00-25.00

Go8416

Go8401 Go8402 Go8403

Go8417

Go8404 Go8405 Go8406

Go8418

Go8403

❑ ❑ USA Go8461 **MC Insert/Cardboard**, 1983. $20.00-25.00
❑ ❑ USA Go8462 **Header Card**, 1983. $15.00-20.00
❑ ❑ USA Go8465 **Translite/Lg**, 1983. $15.00-25.00

Comments: National Distribution: USA - February 5-April 15, 1984. Sam the Olympic Eagle puffy sticker is dated 1984 and 1981/ front and back.

Happy Holidays Happy Meal, 1984

Boxes:
❑ ❑ USA Hh8405 **Hm Box - Red Box/Ronald with Sleigh,** 1984. $8.00-10.00
❑ ❑ USA Hh8406 **Hm Box - Grn Box/Gingerbread House**, 1984. $8.00-10.00

Premiums:
❑ ❑ USA Hh8400 **Card/Stickers - Gingerbread House with Stickers**, 1984. $20.00 -25.00
❑ ❑ USA Hh8401 **Card/Stickers - Train with Stickers**, 1984. $20.00 -25.00

Go8462

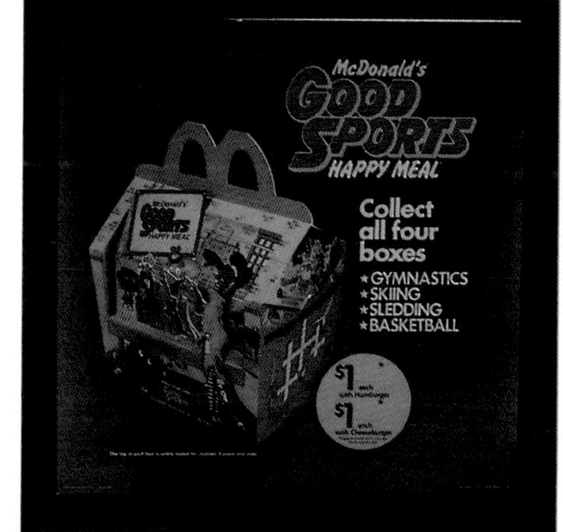

Go8465

Hh8405

Hh8406

Hh8400

Hh8401

Happy Holidays blue book.

❑ ❑ USA Hh8461 **MC Insert/Cardboard**, 1984. $15.00-25.00
❑ ❑ USA Hh8462 **Header Card,** 1984. $15.00-20.00
❑ ❑ USA Hh8463 **Menu Board Lug-On**, 1984. $15.00-20.00
❑ ❑ USA Hh8465 **Translite/Lg**, 1984. $15.00-25.00

Comments: Limited National Distribution: USA - November 23-December 24, 1984.

Happy Pail II Olympic Theme Happy Meal, 1984

Premiums: Happy Meal Pails
❑ ❑ USA Hp8470 **Hm Pail - Athletics,** 1983, Beige Lid/Pail with Yel Shovel. $8.00-10.00
❑ ❑ USA Hp8471 **Hm Pail - Cycling,** 1983, Yel Pail/Lid with Yel Shovel. $8.00-10.00
❑ ❑ USA Hp8472 **Hm Pail - Olympic Games,** 1983, Wht Pail/Lid with Yel Shovel. $8.00-10.00
❑ ❑ USA Hp8473 **Hm Pail - Swimming,** 1983, Blue Pail/Lid with Yel Shovel. $8.00-10.00

❑ ❑ USA Hp8426 **Display with 1 Pail,** 1984. $35.00-50.00
❑ ❑ USA Hp8441 **Dangler Display with 4 Pails,** 1984. $45.00-50.00
❑ ❑ USA Hp8461 **MC Insert,** 1984. $15.00-25.00
❑ ❑ USA Hp8465 **Translite/Lg**, 1984. $15.00-25.00

Comments: National Distribution: USA - May 18-June 17, 1984. Lids had four open vent holes. Yellow shovel had the warning "Safety Tested for Children 3 Years and Over" molded on them.

Hp8470 Hp8471

Hp8441

Hp8472 Hp8473

Hp8426

Hp8465

Lego Building Sets II Happy Meal, 1984

Boxes:

❑ ❑ USA Le8435 **Hm Box - Find the Fry Guy,** 1984. $5.00-8.00

❑ ❑ USA Le8436 **Hm Box - Master the Maze**, 1984. $5.00-8.00

❑ ❑ USA Le8437 **Hm Box - Ship Shape**, 1984. $5.00-8.00

❑ ❑ USA Le8438 **Hm Box - What's Wrong?** 1984.$5.00-8.00

U-3 Premiums:

❑ ❑ USA Le8405 **U-3 Boat with Sailor,** 1984, Duplo/Blu Pkg "Ages 1-4"/5p. $3.00-4.00

❑ ❑ USA Le8406 **U-3 Bird with Eye,** 1984, Duplo/Red Pkg "Ages 1-4"/5p. $3.00-4.00

Le8437

Le8435

Le8438

Le8436

Le8406

Le8401 Le8402 Le8403 Le8404 Le8405 Le8406

Premiums:

- ❑ ❑ USA Le8401 **Set 1 Truck,** 1984, Lego/Red Pkg/17p. $3.00-4.00
- ❑ ❑ USA Le8402 **Set 2 Ship,** 1984, Lego/Blu Pkg/27p. $3.00-4.00
- ❑ ❑ USA Le8403 **Set 3 Helicopter,** 1984, Lego/Yel Pkg/19p. $3.00-4.00
- ❑ ❑ USA Le8404 **Set 4 Airplane,** 1984, Lego/Grn Pkg/18p. $3.00-4.00

- ❑ ❑ USA Le8426 **Display/Ronald,** 1984, Lego with Stickers. $500.00-650.00
- ❑ ❑ USA Le8441 **Dangler,** 1984. Each $10.00-15.00
- ❑ ❑ USA Le8455 **Trayliner,** 1984. $5.00-8.00
- ❑ ❑ USA Le8461 **MC Insert/Cardboard,** 1984. $35.00-50.00
- ❑ ❑ USA Le8465 **Translite/Lg,** 1984. $15.00-25.00

Comments: National Distribution: USA - October 26-November 25, 1984. Auction prices have increased the price for the Lego Ronald display. There are at least two different versions of the stickers on the Lego Man. One version has the LEGO sticker on the menu board and another version has the LEGO sticker on the square in his right hand. One version has no "M" decals on any of the pockets and another has the decals on each pocket. The Lego Man without the "M" decals may be from a pre-promotional kit sent to the store managers, prior to their receipt of the POP Kit with the decaled Lego Man.

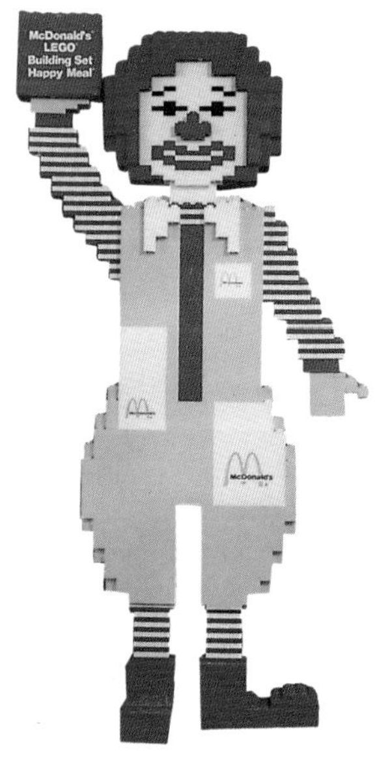

Le8426

Le8441

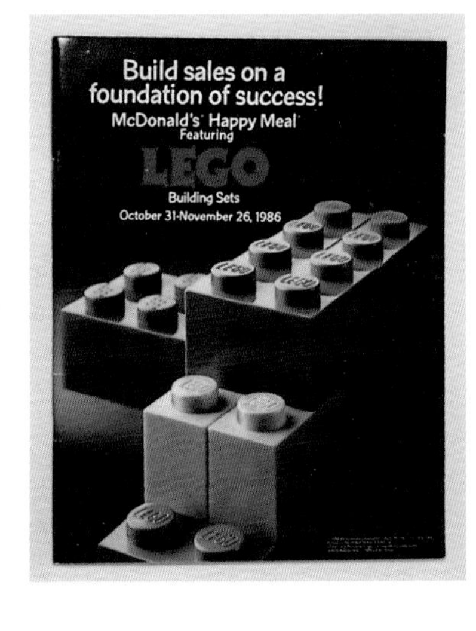

Lego Building Set blue book.

Olympic Beach Ball Promotion, 1984

Premiums: Beach Balls

❑ ❑ USA Be8400 **Beach Ball: Birdie in a Sailboat** - with Olympic Logo on Sail/Blue, 1984. $20.00-25.00

❑ ❑ USA Be8401 **Beach Ball: Grimace in a Kayak** - with Olympic Logo on Sail/Grn, 1984. $20.00-25.00

❑ ❑ USA Be8402 **Beach Ball: Ronald in Olympic Event** - with Olympic Logo on Banner/Red, 1984. $20.00-25.00

Comments: Regional Distribution: USA - 1984. MIP came polybagged with scotch tape closure. This promotion could have been a regional promotion and not a Happy Meal.

Be8400 Be8401 Be8402

Olympic Sports [Zip Action] Happy Meal (canceled), 1984

Premiums: Zip Action Pull Toys:

❑ ❑ USA O18430 **Ronald Riding a Bicycle,** 1984, red/3p. $50.00-75.00

❑ ❑ USA O18431 **Birdie Roller Skating,** 1984, pink/3p. $50.00-75.00

❑ ❑ USA O18432 **Grimace, Captain & Birdie Rowing,** 1984, blue/3p. $50.00-75.00

❑ ❑ USA O18433 **Grimace Running,** 1984, purple/3p. $50.00-75.00

❑ ❑ USA O18434 **Hamburglar playing Soccer,** 1984, burnt orange/3p. $50.00-75.00

Comments: Projected for USA Distribution: June 18-August 20, 1984 (canceled). The above five toys were produced in prototype form. They did not pass McDonald's safety testing; consequently they were canceled. The Happy Meal time slot was replaced with Olympic Sports Happy Meal Guess N' Glow Puzzles. Each prototype toy came on a "Mint on Tree" plastic form. The two halves snapped together with a wheel in the center. A Zip Action Strip was pulled through the center of the toy, propelling the toy into action. Auction prices for the toys have exceeded book price.

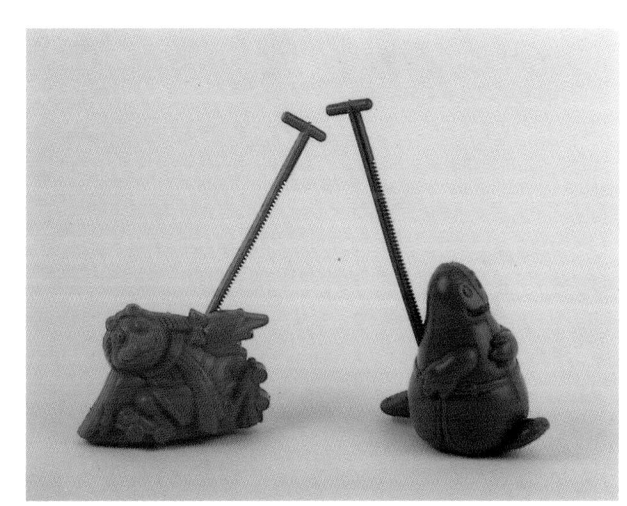

O18431 O18433

O18410

Olympic Sports I Happy Meal, 1984

Boxes:

❑ ❑ USA O18410 **Hm Box - Boats Afloat**, 1984. $5.00-8.00
❑ ❑ USA O18411 **Hm Box - In the Swim**, 1984. $5.00-8.00
❑ ❑ USA O18412 **Hm Box - Just for Kicks**, 1984. $5.00-8.00
❑ ❑ USA O18413 **Hm Box - Making Tracks**, 1984. $5.00-8.00
❑ ❑ USA O18414 **Hm Box - Pedal Power,** 1984. $5.00-8.00

O18411

O18412

O18413

O18414

O18400 O18403 O18401

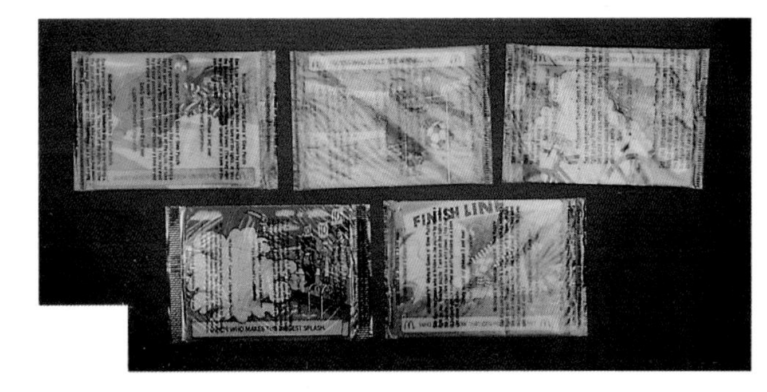

O18402 O18404

Premiums: Puzzles:

❑ ❑ USA O18400 **Puzzle - Guess Which Guy Comes under Wire,** 1984, Fry Guy Glows. $35.00-50.00

❑ ❑ USA O18401 **Puzzle - Guess Who Finished Smiles Ahead,** 1984, Birdie Glows. $35.00-50.00

❑ ❑ USA O18402 **Puzzle - Guess Who Makes Biggest Splash,** 1984, Grimace Glows. $35.00-50.00

❑ ❑ USA O18403 **Puzzle - Guess Who Stole the Winning Goal,** 1984, Hamburglar Glows. $35.00-50.00

❑ ❑ USA O18404 **Puzzle - Who Do You Know Can Help Them Row?** 1984, Ronald Glows. $35.00-50.00

❑ ❑ USA O18461 **MC Insert/Cardboard**, 1984. $25.00-35.00

❑ ❑ USA O18462 **Header Card,** 1984. $15.00-20.00

O18462

O18463

O18465

❑ ❑ USA O18463 **Menu Board/Lug on**, 1984. $15.00-20.00
❑ ❑ USA O18465 **Translite/Lg,** 1984. $15.00-25.00

Comments: National Distribution: USA June 18-August 20, 1984. Puzzles were 4 3/4" x 3 3/4". Other premiums, characters with push handles, were originally slated to be distributed. Due to production problems, these were canceled and puzzles substituted.

Popoids/Crazy Creatures I Test Market Happy Meal, 1984

Boxes:
❑ ❑ USA Po8410 **Hm Box - Elephoid/Jungle,** 1984. $40.00-75.00
❑ ❑ USA Po8411 **Hm Box - Octopoid/Undersea**, 1984. $40.00-75.00

Premiums: Popoids:
❑ ❑ USA Po8400 **Popoids #1,** 1984, 2 Bellows/Blu/Dk Blu & 1 Wht Ball. $40.00-50.00
❑ ❑ USA Po8401 **Popoids #2,** 1984, 2 Bellows/Blue/White & 1 White Cube With 6 Holes. $40.00-50.00
❑ ❑ USA Po8402 **Popoids #3,** 1984, 2 Bellows/Blu/Dk Blu & 1 Cube With 6 Holes. $40.00-50.00
❑ ❑ USA Po8403 **Popoids #4,** 1984, 2 Bellows/Red/Yel & 1 Pentahedron/Org. $40.00-50.00
❑ ❑ USA Po8404 **Popoids #5,** 1984, 2 Bellows/Red/Yel & 1 Column/Org. $40.00-50.00
❑ ❑ USA Po8405 **Popoids #6,** 1984, 3 Bellows/Blu/Dk Blue/Yel. $40.00-50.00
❑ ❑ USA Po8455 **Trayliner**, 1984. $8.00-10.00

Po8410

Po8455

Po8411

❑ ❑ USA Po8456 **Table Tent**, 1984. $20.00-25.00

Comments: Regional Distribution: USA - March 30-May 6, 1984 in St. Louis, Missouri. Pullum Pushum Twistum Bendum Popoid sets came in clear polybag with "Popoids" name printed in red letters across the bag. No McDonald's markings on the MIP. The table tent and trayliner refer to four sets, with the other two sets probably being U-3 premiums. Each bag included a Popoid paper page showing sets which could be purchased in local retail stores. The packages containing the "ball" and "cube" were probably the U-3 intended sets.

Po8456

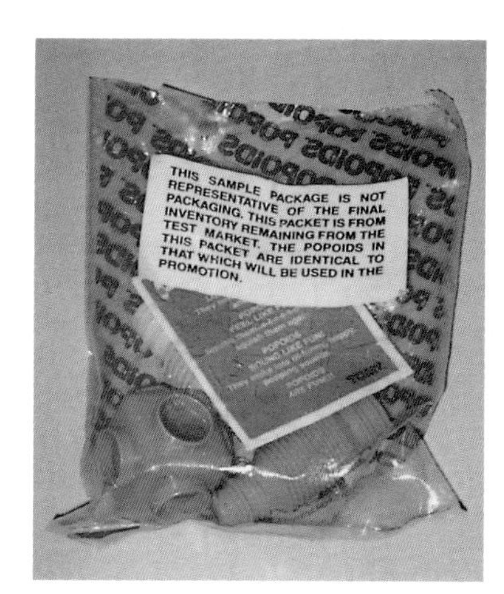

School Days Happy Meal, 1984

Boxes:

❑ ❑ USA Sc8420 **Hm Box - 123's,** 1984. $5.00-8.00
❑ ❑ USA Sc8421 **Hm Box - ABC's,** 1984. $5.00-8.00
❑ ❑ USA Sc8422 **Hm Box - History**, 1984. $5.00-8.00
❑ ❑ USA Sc8423 **Hm Box - Science,** 1984. $5.00-8.00

Sc8420

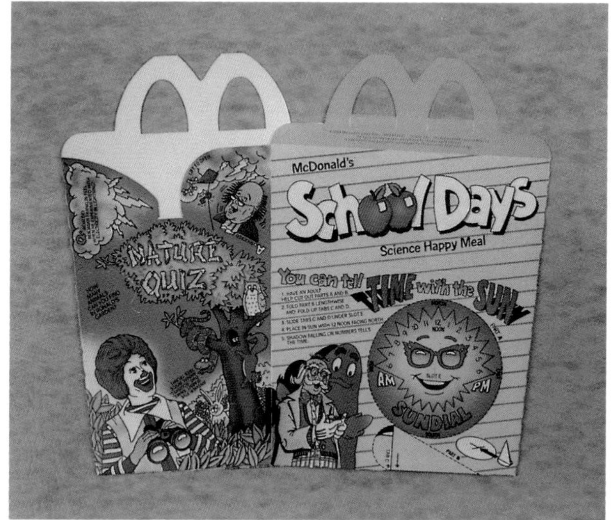

Sc8422

Sc8421

Sc8423

Sc8401 Sc8403 Sc8400
Sc8409 Sc8404 Sc8402 Sc8410

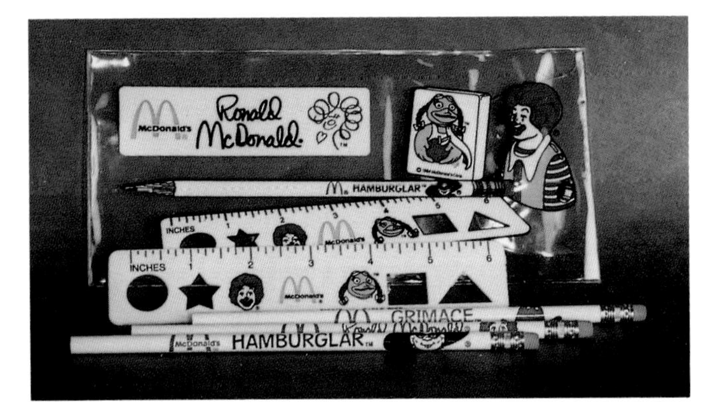

Sc8408 Sc8400 Sc8404
Sc8411
Sc8405
Sc8407

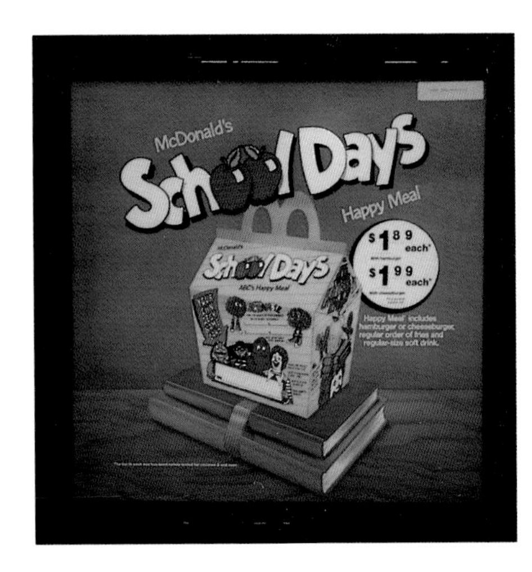

Sc8465

Premiums:

❑ ❑ USA Sc8400 **Eraser - Birdie,** 1984, Holding an Apple. $5.00-7.00

❑ ❑ USA Sc8401 **Eraser - Captain,** 1984, Holding a Ruler and Parrot. $5.00-7.00

❑ ❑ USA Sc8402 **Eraser - Grimace,** 1984, with Pencil. $5.00-7.00

❑ ❑ USA Sc8403 **Eraser - Hamburglar,** 1984, with Lunch Bag. $5.00-7.00

❑ ❑ USA Sc8404 **Eraser - Ronald,** 1984, with History Book. $5.00-7.00

❑ ❑ USA Sc8405 **Pencil - Grimace,** Nd, Wht with Pur Grimace. $7.00-10.00

❑ ❑ USA Sc8406 **Pencil - Hamburglar,** Nd, Wht Org Logo Blk Hamb. $7.00-10.00

❑ ❑ USA Sc8407 **Pencil - Ronald,** Nd, Scrip Wht/Red Logo. $7.00-10.00

❑ ❑ USA Sc8408 **Pencil Case - Ron/Birdie,** 1984, Thin Clear Vinyl. $3.00-4.00

❑ ❑ USA Sc8409 **Pencil Sharpener - Grimace Bust,** 1984. $5.00-8.00

❑ ❑ USA Sc8410 **Pencil Sharpener - Ronald Bust,** 1984. $4.00-5.00

❑ ❑ USA Sc8411 **Ruler/No Metric Scale, 1984,** Ron Birdie(front)/Hamb Capt(back). $2.00-4.00

❑ ❑ USA Sc8434 **Food Item/Lug on**, 1984. $8.00-12.00
❑ ❑ USA Sc8461 **MC Insert/Cardboard**, 1984. $15.00-20.00
❑ ❑ USA Sc8462 **Header Card,** 1984. $15.00-20.00
❑ ❑ USA Sc8463 **Menu Board/Lug on,** 1984. $15.00-20.00
❑ ❑ USA Sc8465 **Translite/Lg**, 1984. $15.00-25.00

Comments: National Distribution: USA - August 20-October 25, 1984. The ruler can be confused with 1985 versions with metric scale included.

USA Generic Promotions, 1984

❑ ❑ USA Ge8401 **Spinner Bike Race - Ronald/Hamburglar,** 1984, Pink or Turq/4p. $3.00-5.00

❑ ❑ USA Ge8402 **Coloring book: Learn the ABC's.** $4.00-5.00

❑ ❑ USA Ge8403 **McDonaldland Cookies/Small packets:** Ronald. $1.00-1.50

❑ ❑ USA Ge8404 **McDonaldland Cookies/Small packets:** Fry Kid. $1.00-1.50

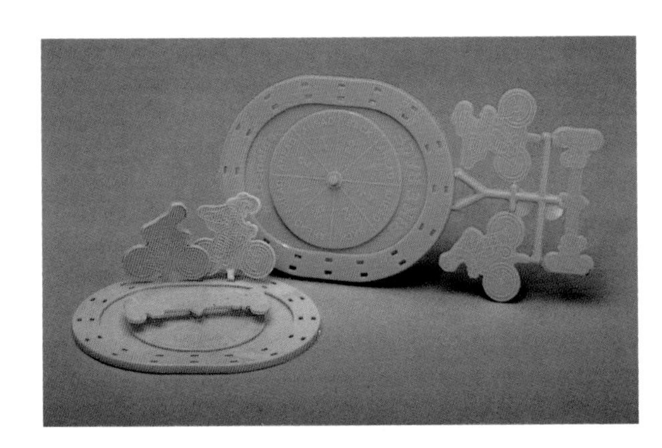

Ge8401

❑ ❑ USA Ge8405 **1984 Set of 4 character cookie boxes: Ronald McDonald (full face)/green box.** $3.00-3.50

❑ ❑ USA Ge8406 **1984 Set of 4 character cookie boxes: Birdie, The Early Bird (full face)/blue box.** $3.00-3.50

❑ ❑ USA Ge8407 **1984 Set of 4 character cookie boxes: Grimace (full face)/purple box.** $3.00-3.50

❑ ❑ USA Ge8408 **1984 Set of 4 character cookie boxes: Fry Guy (full face)/orange box.** $3.00-3.50

❑ ❑ USA Ge8409 **1984 - Chocolaty Chip/white box.** $2.00-3.00

Ge8408 Ge8407 Ge8405 Ge8406

Comments: Starting with Vol. 6 No. 1, the Fun Times Magazine ceased being a quarterly publication. Starting in 1984 it was issued six times a year, beginning with the January/February time frame. Vol. 6 No. 1 offered a mail order premium on the back cover -- a McDonaldland Fun Times T-shirt. The Vol. 6 No. 2 mail order premium was a plastic blow-up Grimace doll. Vol. 6 No. 3 featured Sam the Eagle (Los Angeles Olympic logo mascot) with the McDonaldland characters. Vol. 6 No. 3 mail order premium was a Fun Times McDonaldland backpack. Note: Vol. 6 No. 3 June/July 1984 Canadian version FTM did not have Sam the Eagle anywhere in the issue nor was there any mail-in premium offer. Canada's Quebec version FTM carries a 5 cent price on the front cover. USA Vol. 6 No. 4 offered a McDoodle desk and school starter kit. Vol. 6 No. 5 offered a McDonaldland Cup and Bowl set, with the cup featuring a special snap-on yellow sipper lid. Vol. 6 No. 6 illustrated a calendar with no mail-in premium offered. Apparently the response level was not enough to continue the offers in future issues.

❑ ❑ USA Ge8410 **Fun Times Magazine: Vol. 6 No. 1 Feb/Mar, 1984.** $8.00-10.00

❑ ❑ USA Ge8411 **Fun Times Magazine: Vol. 6 No. 2 Apr/May, 1984.** $8.00-10.00

❑ ❑ USA Ge8412 **Fun Times Magazine: Vol. 6 No. 3 Jun/Jul, 1984.** $8.00-10.00

❑ ❑ USA Ge8413 **Fun Times Magazine: Vol. 6 No. 4 Aug/Sept, 1984.** $8.00-10.00

❑ ❑ USA Ge8414 **Fun Times Magazine: Vol. 6 No. 5 Oct/Nov, 1984.** $8.00-10.00

❑ ❑ USA Ge8415 **Fun Times Magazine: Vol. 6 No. 6 Dec/Jan, 1984.** $8.00-10.00

❑ ❑ USA Ge8416 **Fun Times Magazine T-shirt, 1984.** $8.00-10.00

❑ ❑ USA Ge8417 **Fun Times Magazine Blow-Up Grimace - 30" poly-vinyl, 1984.** $10.00-15.00

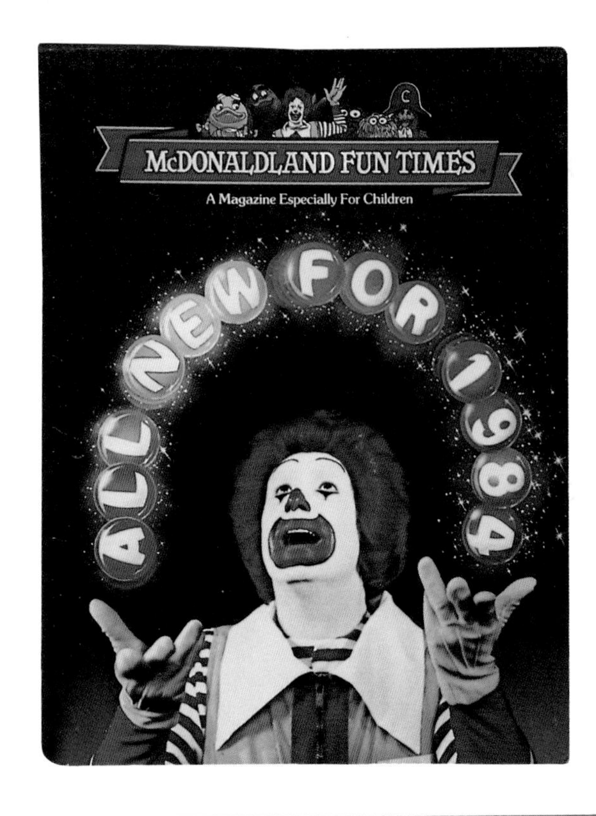

Ge8416

Ge8417

Ge8418

Below: Ge8419

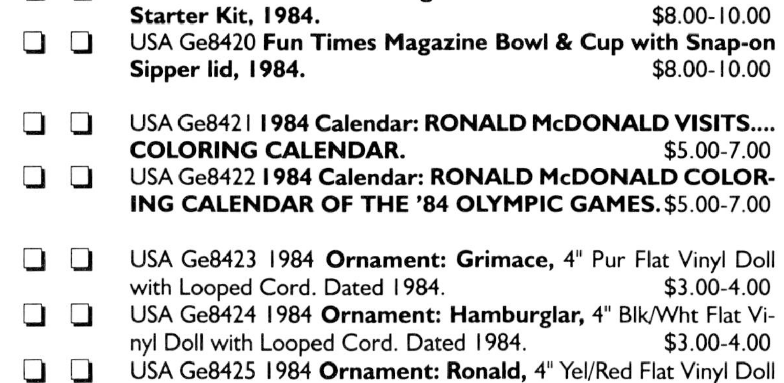

❑ ❑ USA Ge8418 **Fun Times Magazine McDonaldland Fun Times Backpack, 1984.** $15.00-20.00

❑ ❑ USA Ge8419 **Fun Times Magazine McDoodle Desk & School Starter Kit, 1984.** $8.00-10.00

❑ ❑ USA Ge8420 **Fun Times Magazine Bowl & Cup with Snap-on Sipper lid, 1984.** $8.00-10.00

❑ ❑ USA Ge8421 **1984 Calendar: RONALD McDONALD VISITS.... COLORING CALENDAR.** $5.00-7.00

❑ ❑ USA Ge8422 **1984 Calendar: RONALD McDONALD COLORING CALENDAR OF THE '84 OLYMPIC GAMES.** $5.00-7.00

❑ ❑ USA Ge8423 1984 **Ornament: Grimace,** 4" Pur Flat Vinyl Doll with Looped Cord. Dated 1984. $3.00-4.00

❑ ❑ USA Ge8424 1984 **Ornament: Hamburglar,** 4" Blk/Wht Flat Vinyl Doll with Looped Cord. Dated 1984. $3.00-4.00

❑ ❑ USA Ge8425 1984 **Ornament: Ronald,** 4" Yel/Red Flat Vinyl Doll with Looped Cord. Dated 1984. $3.00-4.00

Comments: Regional Distribution: USA - 1984 during Clean-Up weeks and/or with Fun Times Magazine promotions. Learn the ABC's with Ronald McDonald Coloring Book showcases the introduction of the Chicken McNuggets. This is a sampling of generic premiums given away during this period.

"It's A Good Time For The Great Taste of McDonald's" (June 1984) slogan captures the voice of American youth who either visit McDonald's for "Food, Folks and Fun" in the summer months or who seek out entry level employment with their local franchisee. Either way it is a winning combination. Their youth and money is being spent at McDonald's. By 1984, McDonald's becomes a community meeting site for all age groups.

The 10th Anniversary of the Ronald McDonald House is celebrated with a national fund-raiser. McDonald's commitment to children's charities and the Ronald McDonald House is a presence in many communities. Over five million dollars was raised and seventy-three Ronald McDonald Houses are providing services to families in need. Giving back to the communities is a focal point of Ray Kroc's ventures.

Ray Kroc dies on January 14, 1984, but his American dream lives on. Kroc is eulogized as one who epitomizes excellence in American business, from the beginning to his untimely death. The Ronald McDonald Children's Charities is established in Ray Kroc's memory.

The 8th National O/O Convention is held in San Francisco, California with the theme being "One of a Kind."

McDonald's salutes scouting by sponsoring the World Boy Scout Jamboree on ABC-TV.

Ge8420

2 center figures: Ge8424-Ge8425

As8571 As8572 As8573 As8574 As8575 As8576 As8577

As8578

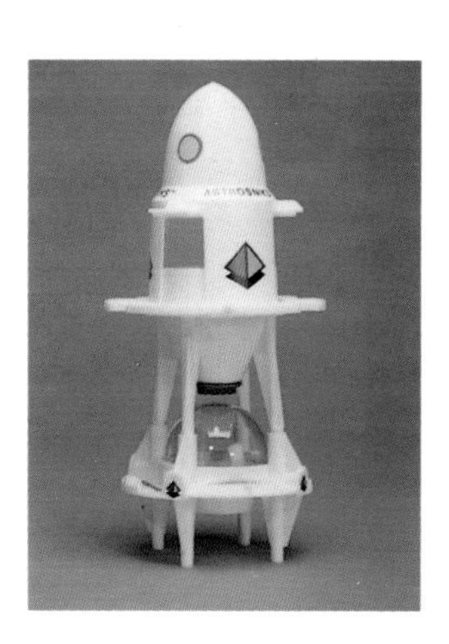

As8595

1985

Astrosniks III Happy Meal, 1985
Beach Ball Characters Promotion, 1985
Commandrons Test Market Happy Meal, 1985
Crazy Creatures with Popoids II Happy Meal, 1985
Day & Night Happy Meal, 1985
E.T. Happy Meal, 1985
Fast Macs II Promotion, 1985
Feeling Good Happy Meal, 1985
Florida Beach Ball/Olympic Beach Ball Promotion, 1985
Halloween '85 Happy Meal, 1985
Hobby Box Happy Meal, 1985
Little Travelers with Lego Building Sets/Super Travelers HM, 1985
Magic Show Happy Meal, 1985
Music Happy Meal, 1985
On the Go I Happy Meal, 1985
Picture Perfect Happy Meal, 1985
Play-Doh II Happy Meal, 1985
Santa Claus the Movie Happy Meal, 1985
Ship Shape II Happy Meal, 1985
Sticker Club Happy Meal, 1985
Stomper Mini 4 x 4 I (Collect All 6 Push-Alongs)/Test Market HM, 1985
Toothbrush Happy Meal, 1985
Transformers/My Little Pony Happy Meal, 1985
USA Generic Promotions, 1985

- **"Large Fries for Small Fries" (August) ad campaign**
- **"The Hot Stays Hot and the Cool Stays Cool" (November 4) jingle**
- **30th year celebration**
- **McBlimp advertises on a large scale**

Astrosniks III Happy Meal, 1985

Premiums: Astrosniks Mint (Loose prices are 50% off lowest price):

❑ ❑ USA As8571 **Banner - Scout Holding Yel Flag**, "Astrosniks" on Flag, No "M", 1983. $20.00-25.00
❑ ❑ USA As8572 **C.B. - Guy with Headphones with Radio**, No "M", 1983. $20.00-25.00
❑ ❑ USA As8573 **Commander - with Black Mask**, No Yel "M" on Belt, 1983. $20.00-25.00
❑ ❑ USA As8574 **Jet - on Rocket**, No "M", 1983. $20.00-25.00
❑ ❑ USA As8575 **Junior - Holding Ice Cream Cone**, No "M", 1983. $20.00-25.00
❑ ❑ USA As8576 **Laser - with Gold Gun**, No "M" on Belt, 1983. $20.00-25.00
❑ ❑ USA As8577 **Perfido - with Red Cape**, Letter "P" on Back, 1983. $20.00-25.00
❑ ❑ USA As8578 **Pyramido - Pyramid Shape with Left Hand Raised**, No "M", 1983. $20.00-25.00
❑ ❑ USA As8579 **Robo-Robot - Gold colored Astrosnik**, No "M" on Front, 1983. $20.00-25.00
❑ ❑ USA As8580 **Snikapotamus - Dinosaur**, No "M" on Saddle, 1983. $20.00-25.00
❑ ❑ USA As8581 **Astralia - Girl Not Holding Cone**, No "M", 1983. $20.00-25.00

❑ ❑ USA As8595 **Counter Card**, 1983, Features 11 Astrosniks/ Cardboard. $25.00-40.00

Comments: Limited Regional Distribution: USA - November 1985 in Oklahoma. Astrosnik figurines do not have an "M" or "McDonald's" markings on them. Mint figures came wrapped in ziplock bag which said, "This toy in this package has been safety tested for children Age 3 Years and over. Made in Hong Kong." "83 Bully-Figuren '83 Schaper - Astrosniks TM - Made in H.K" Was molded into the bottom of figurine. These same figurines were sold in retail stores. **Prices quoted Are for Mint in the Package (MIP). Loose figurines are selling for $10.00 - 15.00 each. Figurines Do not have an "M", not marked McDonald's.**

Beach Ball Characters Promotion, 1985

Premiums: Beach Balls

❑ ❑ USA Be8505 **Beach Ball: Birdie - In Sailboat**, No Logo/ Blue, 1985. $8.00-10.00
❑ ❑ USA Be8506 **Beach Ball: Grimace - with Kayak**, No Logo/ Grn, 1985. $8.00-10.00
❑ ❑ USA Be8507 **Beach Ball: Ronald - With Beach Ball**, No Logo/Red. $8.00-10.00

Comments: Regional Distribution: USA - 1985. These beach balls are similar to 1984 Olympic beach balls (USA Be8400-02), but dated 1985 without McDonald's logo.

Commandrons Test Market Happy Meal, 1985

Premiums: Commandron Changeables in Blister Pack:

❑ ❑ USA Co8500 **Commander Magna,** 1985, with "Airborne!" Comic/Red/Blu/Wht Tomy/**Blister Pack.** $10.00-15.00

❑ ❑ USA Co8501 **Motron,** 1985, with "Robo-Mania" Comic/Red/Blu/Wht Tomy/**Blister Pack.** $10.00-15.00

❑ ❑ USA Co8502 **Solardyn,** 1985, with "The Copy-Bats!" Comic/Red/Blu/Wht Tomy/**Blister Pack.** $10.00-15.00

❑ ❑ USA Co8503 **Velocitor,** 1985, with "Dawn of the Commandrons!" Comic/Red/Blu/Wht Tomy/**Blister Pack.** $10.00-15.00

Premiums: Commandron Changeables in Light Blue Box:

❑ ❑ USA Co8504 **Commander Magna,** 1985, with "Airborne!" Comic/Red/Blu/Wht Tomy/**Light Blue Box.** $10.00-12.00

❑ ❑ USA Co8505 **Motron,** 1985, with "Robo-Mania" Comic/Red/Blu/Wht Tomy/**Light Blue Box.** $10.00-12.00

❑ ❑ USA Co8506 **Solardyn,** 1985, with "The Copy-Bats!" Comic/Red/Blu/Wht Tomy/**Light Blue Box.** $10.00-12.00

❑ ❑ USA Co8507 **Velocitor,** 1985, with "Daw/Commandrons!" Com/Red/Blu/Wht Tomy/**Light Blu Box.** $10.00-12.00

Comments: Regional Distribution: USA - August 22 - September 21, 1985. Blister packs were priced at 99 cents each with any food purchase. One set of Commandrons came blister packaged with a mini-comic book, a serialized story about that robot/vehicle. No McDonald's logo on the vehicles. McDonald's logo is on each comic book in the blister pack. A second set exists packaged in a light blue box with a comic like "Airborne!" with Robo Strux offer, expires Nov. 30, 1985. The Commandrons/loose were sold in retail stores; note the shape of the decals. **Prices Reflect MIP, not loose.**

Co8500 Co8501 Co8502 Co8503

Co8504

Crazy Creatures with Popoids II Happy Meal, 1985

Boxes:

❑ ❑ USA Cr8510 **Hm Box - Elephoids,** 1985. $10.00-15.00

❑ ❑ USA Cr8511 **Hm Box - Dragonoids,** 1985. $10.00-15.00

❑ ❑ USA Cr8512 **Hm Box - Octopoid,** 1985. $10.00-15.00

❑ ❑ USA Cr8513 **Hm Box - Scorpoid,** 1985. $10.00-15.00

Cr8510

Cr8511

Cr8512

Cr8513

Premiums: Popoids:

❑ ❑ USA Cr8550 **Wk 1: 2 Poppin'/1 Red/1 Blu/ 1 Column 10 Holes,** 1985. $12.00-15.00

❑ ❑ USA Cr8551 **Wk 2: 2 Poppin'/1 Yel/1 Red/ 1 Cube 6 Holes,** 1985. $12.00-15.00

❑ ❑ USA Cr8552 **Wk 3: 2 Poppin'/1 Yel/1 Blu/ 1 Ball 6 Holes,** 1985. $12.00-15.00

❑ ❑ USA Cr8553 **Wk 4: 2 Poppin'/1 Red/1 Yel/ 1 Five Sided Connector,** 1985. $12.00-15.00

Cr8551 Cr8552 Cr8550 Cr8553

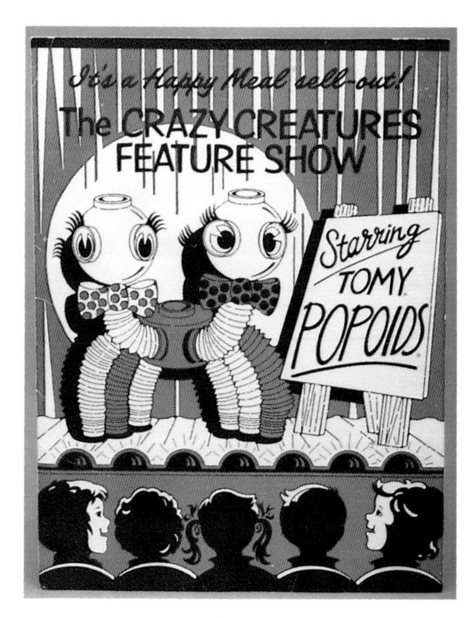

Cr8543

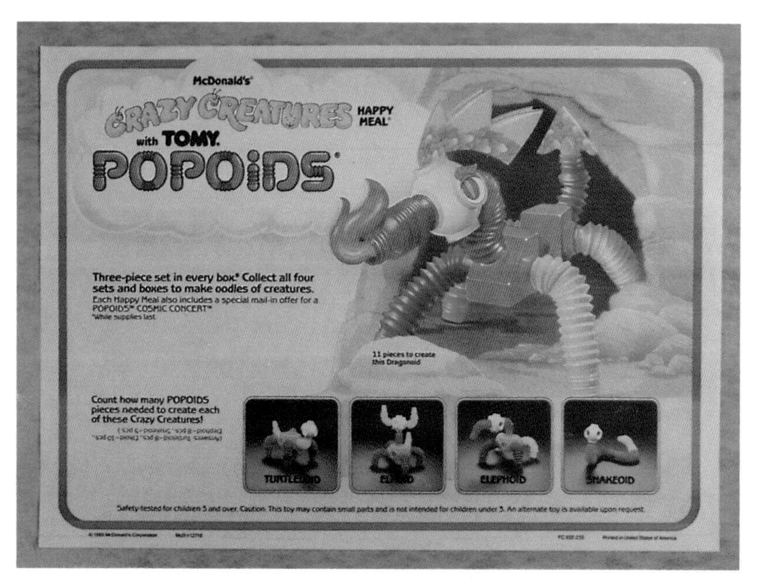

Cr8555

Crazy Creature Birth Certificate
For McDONALD's® Crazy Creatures Happy Meal® Featuring TOMY® Popoids®

Name of Creature ______
Date of Birth ______
(THE FIRST TIME YOU PUT THE CREATURE TOGETHER)
Number of Popoids® body parts ______
What is the Creature's favorite color? ______
Name of Owner ______
Age ____ Birthday ______
Names of your other pets:
Cats ______
Dogs ______
Fish ______
Birds ______
Other ______

McDonald's Crazy Creatures Happy Meal with TOMY POPOIDS

©1985 McDonald's Corporation McD #13399 FC #832-262 Printed in United States of America

Cr8562

❑ ❑ USA Cr8526 **Display,** 1985. $150.00-200.00
❑ ❑ USA Cr8543 **Manager's Guide,** 1985. $10.00-15.00
❑ ❑ USA Cr8555 **Trayliner,** 1985, Crazy Creatures/Popoids. $8.00-10.00
❑ ❑ USA Cr8561 **MC Insert/Cardboard,** 1985. $15.00-25.00
❑ ❑ USA Cr8562 **Birth Certificate.** $.50-1.00
❑ ❑ USA Cr8564 **Translite/Sm,** 1985. $10.00-15.00
❑ ❑ USA Cr8565 **Translite/Lg,** 1985. $15.00-25.00

Comments: National Distribution: USA August 2-September 2, 1985. Previously test marketed in the St. Louis region with two boxes and six premiums - March/May 1984; USA Po8400-05. Some stores gave away Birth Certificates, a forerunner to the Cabbage Patch Kids Birth Certificates.

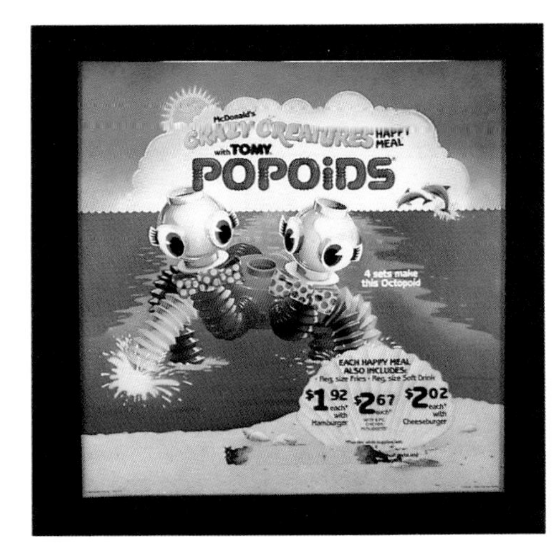

Cr8565

Da8525

Day & Night Happy Meal, 1985

Boxes:
❑ ❑ USA Da8525 **Hm Box - All Star Sunday,** 1985. $4.00-5.00
❑ ❑ USA Da8526 **Hm Box - Who's Afraid of the Dark,** 1985. $4.00-5.00

Comments: Regional Distribution: USA - 1985 during Clean-Up week Happy Meal; generic premiums were given.

Da8526

E.T. Happy Meal, 1985

Boxes:

❑ ❑ USA Et8510 **Hm Box - ET Makes Friends,** 1985. $4.00-5.00

❑ ❑ USA Et8511 **Hm Box - The Great Adventure,** 1985. $4.00-5.00

Premiums: Posters

❑ ❑ USA Et8500 **Poster: ET Boy with Basket/Bike,** 1985, Poster. $5.00-10.00

❑ ❑ USA Et8501 **Poster: ET Boy & Girl with Spaceship,** 1985, Poster/Touching Fingers. $5.00-10.00

❑ ❑ USA Et8502 **Poster: ET with Glowing Finger,** 1985, Poster. $5.00-10.00

❑ ❑ USA Et8503 **Poster: ET with Radio Device,** 1985, Poster. $5.00-10.00

Et8510

Et8500

Et8501

Et8511

Et8502

Et8503

Et8523

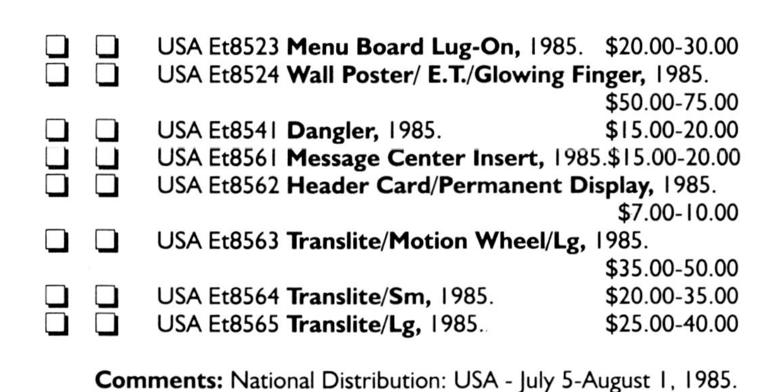

- ❑ ❑ USA Et8523 **Menu Board Lug-On,** 1985. $20.00-30.00
- ❑ ❑ USA Et8524 **Wall Poster/ E.T./Glowing Finger,** 1985. $50.00-75.00
- ❑ ❑ USA Et8541 **Dangler,** 1985. $15.00-20.00
- ❑ ❑ USA Et8561 **Message Center Insert,** 1985.$15.00-20.00
- ❑ ❑ USA Et8562 **Header Card/Permanent Display,** 1985. $7.00-10.00
- ❑ ❑ USA Et8563 **Translite/Motion Wheel/Lg,** 1985. $35.00-50.00
- ❑ ❑ USA Et8564 **Translite/Sm,** 1985. $20.00-35.00
- ❑ ❑ USA Et8565 **Translite/Lg,** 1985. $25.00-40.00

Comments: National Distribution: USA - July 5-August 1, 1985.

Fast Macs II Promotion, 1985

Premiums:

- ❑ ❑ USA Fa8505 **Big Mac - In White Squad**, Police Car, 1985. $3.00-4.00
- ❑ ❑ USA Fa8506 **Birdie - In Pink Sun Cruiser**, 1985. $3.00-4.00
- ❑ ❑ USA Fa8507 **Hamburglar - In Red Sports Car**, 1985. $3.00-4.00
- ❑ ❑ USA Fa8508 **Ronald - In Yellow Jeep**, 1985. $3.00-4.00
- ❑ ❑ USA Fa8515 **Promo Sign - Turn Here for Fast Macs,** 1985. $7.00-10.00
- ❑ ❑ USA Fa8516 **Promo Sign - Stop Here for Fast Macs,** 1985. $7.00-10.00

Et8541

Et8565

Fa8516

Fa8508 Fa8505 Fa8506 Fa8507

Fa8515

Fa8526

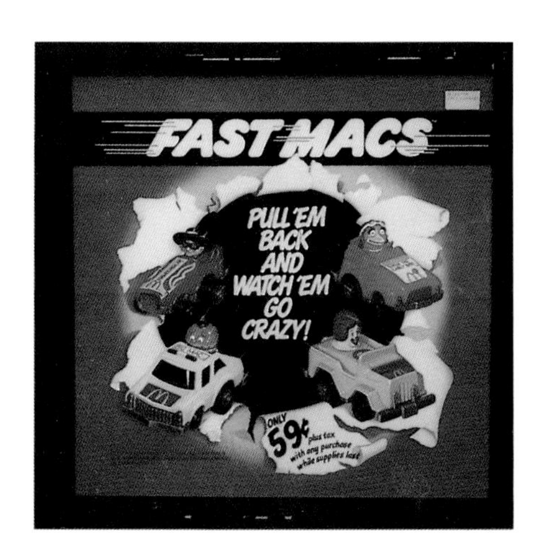

Fa8565

❑ ❑ USA Fa8526 **Display,** 1985. $50.00-75.00
❑ ❑ USA Fa8565 **Translite/Lg,** 1985. $15.00-25.00

Comments: Regional Distribution: USA - 1985. Fast Mac cars were sold for 59 cents on a blister pack and/or used as Happy Meal premium.

Feeling Good Happy Meal, 1985

Boxes:

❑ ❑ USA Fe8515 **Hm Box - Child in Mirror/Guess Who,** 1985. $4.00-5.00
❑ ❑ USA Fe8516 **Hm Box - Hidden Toothbrushes/Find the Pairs,** 1985. $4.00-5.00
❑ ❑ USA Fe8517 **Hm Box - Characters Warm up Exercises/ Snooze Blues,** 1985. $4.00-5.00

Fe8515

Fe8516

Fe8517

Fe8518

❑ ❑ USA Fe8518 **Hm Box - Reverse Message/The Birdie Path,** 1985. $4.00-5.00

U-3 Premiums:

❑ ❑ USA Sh8554 **U-3 Grimace in a Tub,** 1984/5, Floater/Yel or Grn. $5.00-7.00

❑ ❑ USA Sh8555 **U-3 Fry Guy on Duck,** 1984/5, Floater/Yel or Grn. $5.00-7.00

❑ ❑ USA Sh8556 **U-3 Grimace in a Tub,** 1984/5, Floater/Bright Pink. $8.00-10.00

❑ ❑ USA Sh8557 **U-3 Fry Guy on Duck,** 1984/5, Floater/Bright Pink. $8.00-10.00

Premiums:

❑ ❑ USA Fe8501 **Toothbrush - Ronald Bust,** 1985, Wht-Red/Polybagged. $4.00-5.00

❑ ❑ USA Fe8502 **Toothbrush - Hamburglar Bust,** 1985, Wht-Blk/Polybagged. $4.00-5.00

❑ ❑ USA Fe8503 **Soap Dish,** 1985, Grimace with Spread Arms/Pur. $4.00-5.00

❑ ❑ USA Fe8504 **Sponge - Fry Guy/Shower,** 1985, Grn. $2.00-3.00

❑ ❑ USA Fe8505 **Mirror - Birdie,** 1985, Yel Rectangle Mirror/Polybagged. $4.00-5.00

❑ ❑ USA Fe8506 **Comb - Captain,** 1985, Red. $1.00-2.00

❑ ❑ USA Fe8561 **MC Insert/Cardboard,** 1985. $15.00-20.00
❑ ❑ USA Fe8563 **Menu Board Lug-On,** 1985. $15.00-20.00
❑ ❑ USA Fe8564 **Translite/Sm,** 1985. $10.00-15.00
❑ ❑ USA Fe8565 **Translite/Lg,** 1985. $15.00-20.00

Comments: National Distribution: USA - December 26, 1985-March 9, 1986. Other character combs may have been distributed with this Happy Meal.

Fe8503 Fe8504 Fe8505 Fe8506

Fe8565

Florida Beach Ball/Olympic Beach Ball Promotion, 1985

Premiums: Beach Balls:

❑ ❑ USA Fl8550 **Beach Ball: Birdie - In Sailboat with Palm Trees**, Florida/Blue, 1985. $15.00-20.00

❑ ❑ USA Fl8551 **Beach Ball: Grimace - Kayak with Palm Trees**, Florida/Grn, 1985. $15.00-20.00

❑ ❑ USA Fl8552 **Beach Ball: Ronald - with Beach Ball with Palm Trees**, Florida/Red, 1985. $15.00-20.00

Comments: Regional Distribution: USA - 1985 in Florida. This could have been a regional promotion.

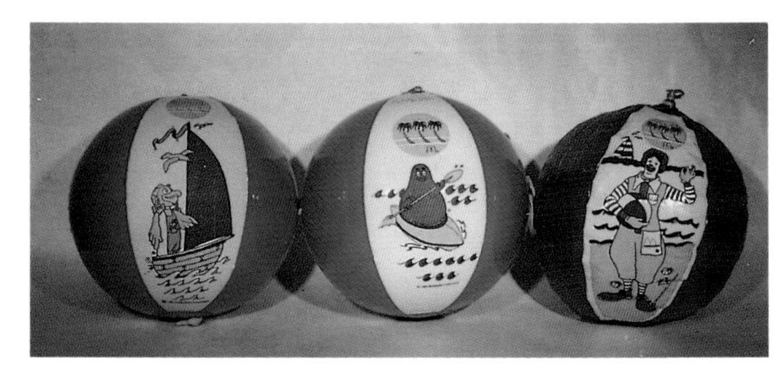

Fl8550 Fl8551 Fl8552

Halloween '85 Happy Meal, 1985

Premiums: Halloween Pumpkins with Lids:

❑ ❑ USA Ha8500 **McPunky with 1985 date,** Org Pail with McPunk'n Face. $10.00-15.00

❑ ❑ USA Ha8501 **McPunk'n with 1985 date,** Org Pail.$7.00-10.00

❑ ❑ USA Ha8502 **McGoblin with 1985 date,** Org Pail.$7.00-10.00

❑ ❑ USA Ha8503 **McJack with 1985 date,** Org Pail with McGoblin Face. $10.00-15.00

❑ ❑ USA Ha8504 **McBoo with 1985 date,** Org Pail. $10.00-15.00

❑ ❑ USA Ha8565 **Translite/Lg,** 1985. $15.00-20.00

Comments: Regional Distribution: USA - October 11-31, 1985. Pails were test marketed in New England **with 1985 date**. Lids had four 1/2" openings with twelve hole grid mesh. Assorted lids were distributed with these 1985 dated pails. Lids with finger entrapment type holes were distributed and changed to mesh type holes.

Ha8500 Ha8501 Ha8502 Ha8504

Ha8565

Hobby Box Happy Meal, 1985

Premiums: Lunch Boxes:

❑ ❑ USA Lu8500 **Lunch Box - Light Green,** 7 1/2 x 6 x 2 1/2" Rectangle with Rect Handle/Lg M Logo. $10.00-15.00

❑ ❑ USA Lu8501 **Lunch Box - Yellow,** 7 1/2 x 6 x 2 1/2" Rectangle with Rect Handle/Lg M Logo. $10.00-15.00

❑ ❑ USA Lu8502 **Lunch Box - Red,** 7 1/2 x 6 x 2 1/2" Rectangle with Rect Handle/Lg M Logo. $10.00-15.00

❑ ❑ USA Lu8503 **Lunch Box - Blue,** 7 1/2 x 6 x 2 1/2" Rectangle with Rect Handle/Lg M Logo. $10.00-15.00

❑ ❑ USA Lu8565 **Translite/Lg,** 1985. $25.00-40.00

Comments: Regional Distribution: USA - 1985 in the South. McDonald's logo on one side with "Whirley Industries, Inc., Warren, Pa U.S.A." on back.

Lu8500 Lu8501 Lu8502 Lu8503

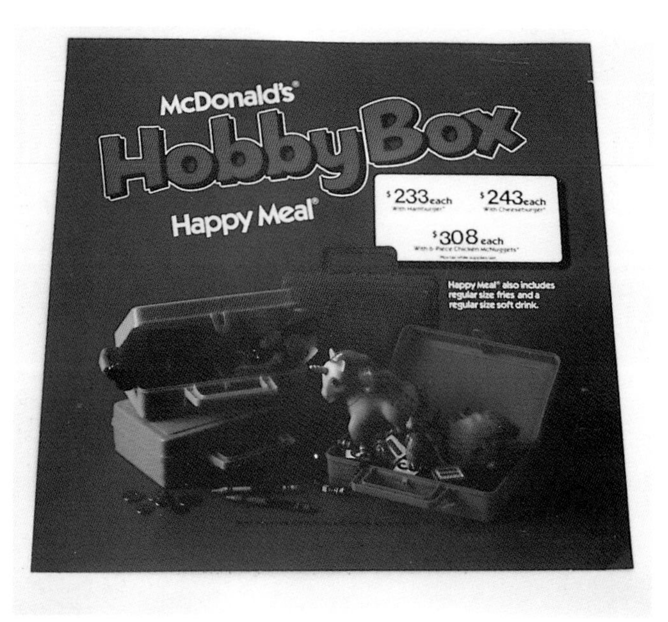

Lu8565

Le8515

Little Travelers with Lego Building Sets/Super Travelers Happy Meal, 1985

Boxes:

❑ ❑ USA Le8515 **Hm Box - Capt Crook/Fry Guy/Which Came First,** 1985. $40.00-65.00

❑ ❑ USA Le8516 **Hm Box - Grimace/Vacation/Match a Patch,** 1985. $40.00-65.00

❑ ❑ USA Le8517 **Hm Box - Grimace/San Francisco/Animal Power,** 1985. $40.00-65.00

❑ ❑ USA Le8518 **Hm Box - Ronald/Record Trips Around the Globe,** 1985. $40.00-65.00

Premiums:

❑ ❑ USA Le8504 **Set 1 Helicopter,** 1984, Lego/Packaged 36p. $35.00-50.00

❑ ❑ USA Le8505 **Set 2 Airplane with Pilot,** 1984, Lego/Packaged 26p. $35.00-50.00

❑ ❑ USA Le8506 **Set 3 Tanker/Boat,** 1984, Lego/Packaged 38p. $35.00-50.00

❑ ❑ USA Le8507 **Set 4 Roadster/Race Auto,** 1984, Lego/Packaged 19p. $35.00-50.00

Le8516

Le8505 Le8504

Le8517

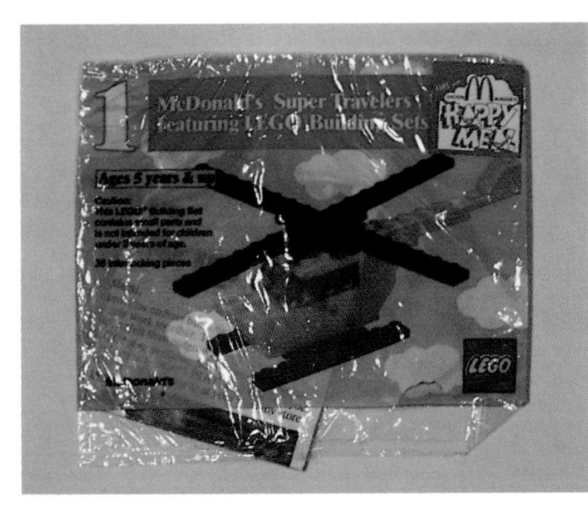

Le8504

Le8518

Le8506 Le8507

❑ ❑ USA Le8526 **Display with Premiums,** 1985.$450.00-600.00
❑ ❑ USA Le8565 **Translite/Little Travelers/Lg,** 1985. $65.00-90.00
❑ ❑ USA Le8566 **Translite/Super Travelers/Lg,** 1985. $50.00-75.00
❑ ❑ USA Le8567 **Translite/Lego Toy Days/Lg,** 1985. $25.00-40.00

Comments: Limited Regional Distribution: USA - Oklahoma 1985. Packaging is like Lego Building Sets Happy Meal, except blue color.

Magic Show Happy Meal, 1985

Boxes:
❑ ❑ USA Ma8510 **Hm Box - Ghost Writer/Crying Quarter,** 1985. $10.00-15.00
❑ ❑ USA Ma8511 **Hm Box - Eggs with Legs/Nickel Trick,** 1985. $10.00-15.00
❑ ❑ USA Ma8512 **Hm Box - Sticky Card Trick,** 1985. $10.00-15.00
❑ ❑ USA Ma8513 **Hm Box - Tug-O-War/Moving Checkers,** 1985. $10.00-15.00

Ma8510

Ma8512

Ma8511

Ma8513

Premiums:
- ❑ ❑ USA Ma8500 **Magic Egg Trick - Disappearing Hamburglar,** 1985, Egg Shaped/Red or Blu/3p. $5.00-8.00
- ❑ ❑ USA Ma8501 **Magic String Trick - Birdie,** 1985, Org or Grn. $1.00-2.00
- ❑ ❑ USA Ma8502 **Magic Tablet,** 1985, Paper/Ron Holding Slate. $15.00-20.00

- ❑ ❑ USA Ma8503 **Magic Picture - Make Ronald Appear in Color**, 1985. $15.00-20.00
- ❑ ❑ USA Ma8504 **Magic Picture - Make Grimace Appear in Color**, 1985. $15.00-20.00

- ❑ ❑ USA Ma8561 **MC Insert/Cardboard,** 1985. $15.00-25.00
- ❑ ❑ USA Ma8562 **Header Card,** 1985. $15.00-25.00
- ❑ ❑ USA Ma8564 **Translite/Sm,** 1985. $15.00-20.00
- ❑ ❑ USA Ma8565 **Translite/Lg,** 1985. $25.00-40.00

Comments: National Distribution: USA - April/May 1985. The U-3 toys were USA Sh8554-55.

Music Happy Meal, 1985

Boxes:
- ❑ ❑ USA Mu8510 **Hm Box - Audience Clapping,** 1985. $25.00-40.00
- ❑ ❑ USA Mu8511 **Hm Box - Can You Find,** 1985. $25.00-40.00
- ❑ ❑ USA Mu8512 **Hm Box - Jam Session,** 1985. $25.00-40.00
- ❑ ❑ USA Mu8513 **Hm Box - Ronald Directing,** 1985. $25.00-40.00

Premiums: Records:
- ❑ ❑ USA Mu8500 **Record: If You're Happy/Ronald One Man Band,** 1985, Record/Fp/Blu Pkg. $4.00-5.00
- ❑ ❑ USA Mu8501 **Record: Coming Round the Mt/Obj Is Music,** 1985, Record/Fp/Grn Pkg. $4.00-5.00
- ❑ ❑ USA Mu8502 **Record: Great to Be Crazy/Music Machine,** 1985, Record/Fp/Pnk Pkg. $4.00-5.00
- ❑ ❑ USA Mu8503 **Record: Hokey Pokey/Ronald Orchestra,** 1985, Record/Fp/Yel Pkg. $4.00-5.00

- ❑ ❑ USA Mu8565 **Translite/Lg,** 1985. $50.00-65.00
- ❑ ❑ USA Mu8561 **MC Insert/Cardboard,** 1985. $40.00-50.00

Comments: Regional Distribution: USA - 1985 in St. Louis, Missouri. Records are Fisher-Price 33 1/3 RPM phonograph records.

On the Go I Happy Meal, 1985

Boxes:
- ❑ ❑ USA On8510 **Hm Box - Bridge,** 1985. $4.00-5.00
- ❑ ❑ USA On8511 **Hm Box - Drive thru,** 1985. $4.00-5.00
- ❑ ❑ USA On8512 **Hm Box - Garage,** 1985. $4.00-5.00
- ❑ ❑ USA On8513 **Hm Box - Tunnel,** 1985. $4.00-5.00

Mu8512

Mu8500 Mu8501 Mu8503 Mu8502

Mu8565

On8512 On8513

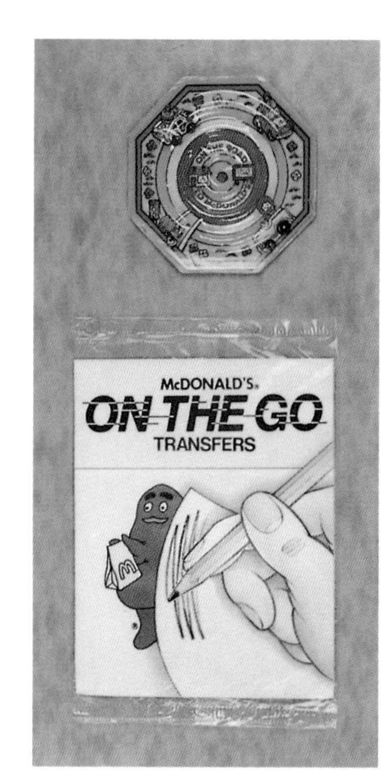

On8501

On8505

Premiums:

❑ ❑ USA On8501 **Bead Game: On the Road to McDonald's** - Octagon Shaped, 1985. $35.00-40.00

❑ ❑ USA On8504 **Bead Game: Stop & Go** - Hamburglar & Ronald at Traffic Light/Rectangle shaped, 1985. $35.00-40.00

❑ ❑ USA On8502 **Magic Slate Board,** 1985, Hamb/Lift Pad/Red Race Car. $15.00-20.00

❑ ❑ USA On8503 **Magic Slate Board,** 1985, Ronald/Lift Pad/Yellow Car. $15.00-20.00

❑ ❑ USA On8505 **On the Go Transfers: Grimace - Decals,** 1985. $8.00-10.00

❑ ❑ USA On8564 **Translite/Sm,** 1985. $15.00-20.00

❑ ❑ USA On8565 **Translite/Lg,** 1985. $25.00-40.00

Comments: Regional Distribution: USA - 1985. Auction prices have increased the price of On the Go I Happy Meal premiums.

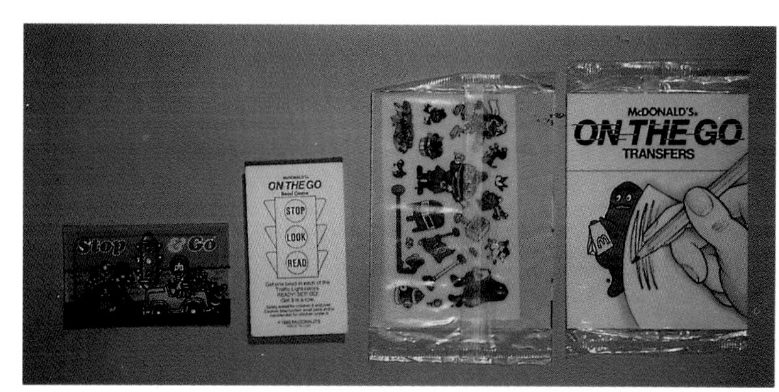

On8501 On8502 On8503 On8504 On8505

Picture Perfect Happy Meal, 1985

Boxes:

❑ ❑ USA Pi8510 **Hm Box - Birdie,** 1984. $4.00-5.00

❑ ❑ USA Pi8511 **Hm Box - Fry Guys,** 1984. $4.00-5.00

❑ ❑ USA Pi8512 **Hm Box - Grimace,** 1984. $4.00-5.00

❑ ❑ USA Pi8513 **Hm Box - Ronald McDonald,** 1984. $4.00-5.00

Pi8511

Pi8510

Pi8512

Pi8513

Pi8565

Premiums: Markers and Crayons:

❑ ❑ USA Pi8501 **Markers/Coloring - Blue or Red/Thin/5 5/8".** $8.00-10.00

❑ ❑ USA Pi8502 **Markers/Drawing - Orange or Green/Thick/5".** $8.00-10.00

❑ ❑ USA Pi8503 **Crayons,** 1985, 3/Yel/Red/Blu in Yel/Grn Binney/Smith/Promo Pkg. $5.00-8.00

❑ ❑ USA Pi8504 **Crayons,** 1985, 6/Blk/Blu/Brn/Grn/Red/Yel/3 5/8" x 5/16". $5.00-8.00

❑ ❑ USA Pi8561 **MC Insert/Cardboard,** 1984. $15.00-20.00

❑ ❑ USA Pi8562 **Header Card,** 1984. $15.00-20.00

❑ ❑ USA Pi8563 **Menu Board/Lug on,** 1984. $15.00-20.00

❑ ❑ USA Pi8565 **Translite/Lg,** 1984. $15.00-25.00

Comments: National Optional Distribution: USA - December 28, 1984-January 25, 1985.

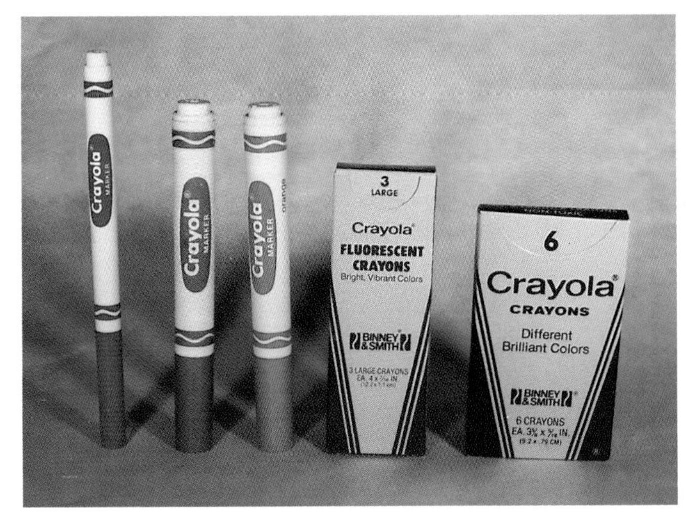

Pi8501 Pi8502 Pi8503 Pi8504

Play-Doh II Happy Meal, 1985

Box:

❑ ❑ USA PI8535 **Hm Box - Play-Doh Place,** 1985. $15.00-20.00

PI8535

Premiums: Play-Doh Containers:

- ❑ ❑ USA PI8529 **Container with Pink dough,** 1981, **Cardboard with Tin Bottom**, 1981. $20.00-25.00
- ❑ ❑ USA PI8530 **Container with Green dough,** 1981, **Cardboard with Tin Bottom**, 1981. $20.00-25.00

- ❑ ❑ USA PI8556 **Table Tent,** 1985. $15.00-20.00
- ❑ ❑ USA PI8561 **MC Insert/Cardboard,** 1985. $15.00-25.00
- ❑ ❑ USA PI8565 **Translite/Lg,** 1985. $15.00-25.00

Comments: Regional Distribution: USA - February 15-March 29, 1985 in Kansas, Missouri, Illinois, Tennessee, Arkansas, Oklahoma, Alabama, Texas, and Indiana. USA PI8301-04 were again distributed with USA PI8529-30 for a total of six cans of Play-Doh with tin bottoms.

PI8529

PI8530

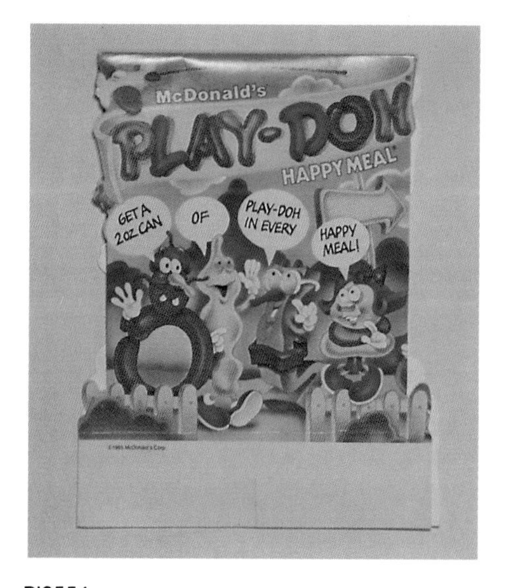

PI8556

PI8556

PI8565

Santa Claus the Movie Happy Meal, 1985

Boxes:

- ❑ ❑ USA Sa8510 **Hm Box - Santa's Cottage,** 1985. $4.00-5.00
- ❑ ❑ USA Sa8511 **Hm Box - Workshop,** 1985. $4.00-5.00

Sa8511

Sa8510

Sa8500 Sa8501

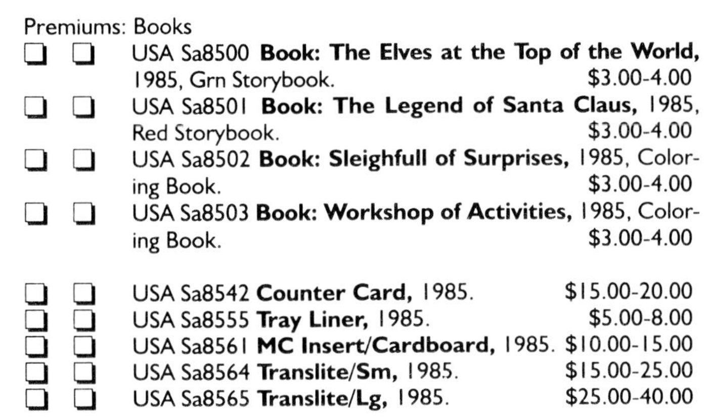

Premiums: Books

- ❑ ❑ USA Sa8500 **Book: The Elves at the Top of the World,** 1985, Grn Storybook. $3.00-4.00
- ❑ ❑ USA Sa8501 **Book: The Legend of Santa Claus,** 1985, Red Storybook. $3.00-4.00
- ❑ ❑ USA Sa8502 **Book: Sleighfull of Surprises,** 1985, Coloring Book. $3.00-4.00
- ❑ ❑ USA Sa8503 **Book: Workshop of Activities,** 1985, Coloring Book. $3.00-4.00

- ❑ ❑ USA Sa8542 **Counter Card,** 1985. $15.00-20.00
- ❑ ❑ USA Sa8555 **Tray Liner,** 1985. $5.00-8.00
- ❑ ❑ USA Sa8561 **MC Insert/Cardboard,** 1985. $10.00-15.00
- ❑ ❑ USA Sa8564 **Translite/Sm,** 1985. $15.00-25.00
- ❑ ❑ USA Sa8565 **Translite/Lg,** 1985. $25.00-40.00

Comments: National Distribution: USA - November 22-December 24, 1985.

Sa8500

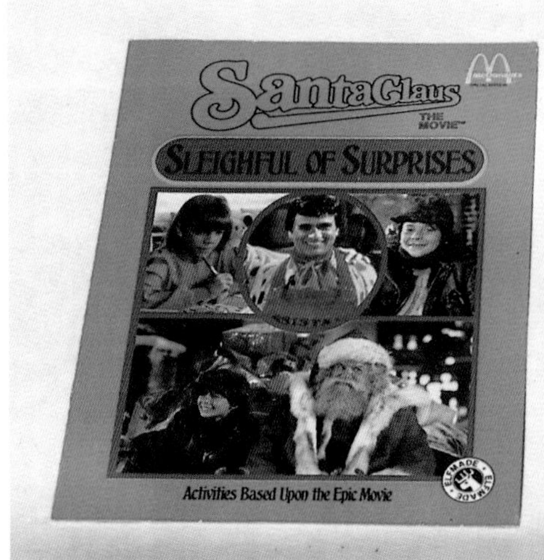

Sa8502 Sa8503

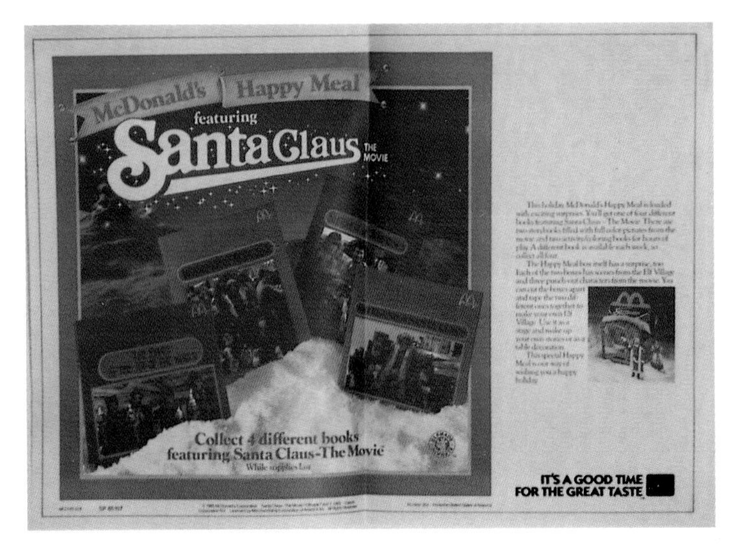

Sa8555

Sa8565

Ship Shape II Happy Meal, 1985

U-3 Premiums:
❑ ❑ USA Sh8554 **U-3 Grimace in Tub,** 1985, Yel Dated '85/ Grn Dated '84. $5.00-7.00
❑ ❑ USA Sh8555 **U-3 Fry Kid on Duck,** 1985, Grn Dated '85/ Yel Dated '84. $5.00-7.00

Premiums: Vacuum Formed Containers:
❑ ❑ USA Sh8301 **Hamburglar Splash Dasher,** 1983, Wht Top/ Org Bottom. $8.00-10.00
❑ ❑ USA Sh8302 **Grimace Tubby Tugger,** 1983, Pnk Top/Blu Bottom. $8.00-10.00
❑ ❑ USA Sh8303 **Capt Rub-A-Dub Sub,** 1983, Grn Top/Grn Bottom. $8.00-10.00
❑ ❑ USA Sh8304 **Ronald River Boat,** 1983, Yel Top/Red Bottom. $8.00-10.00
❑ ❑ USA Sh8505 **Sticker Sheet - Splash Dasher,** 1985. $3.00-4.00
❑ ❑ USA Sh8506 **Sticker Sheet - Tubby Tugger,** 1985. $3.00-4.00
❑ ❑ USA Sh8507 **Sticker Sheet - Rub-A-Dub-Sub - The Captain,** 1985. $5.00-8.00
❑ ❑ USA Sh8508 **Sticker Sheet - River Boat,** 1985. $3.00-4.00

❑ ❑ USA Sh8526 **Display/Motion/1 Premium,** 1985. $150.00-200.00

Sh8555 Sh8554

Sh8303 Sh8302
Sh8301

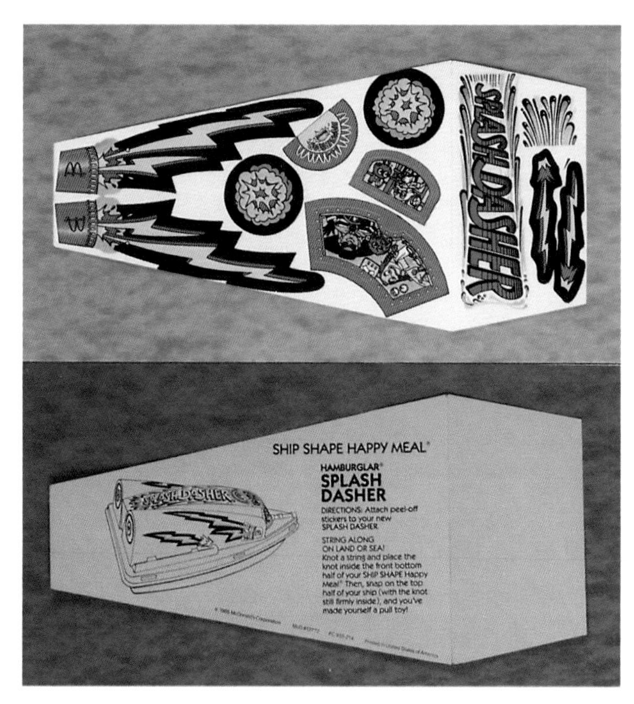

Sh8505

Sh8304

Sh8506

Sh8555

- ❑ ❑ USA Sh8555 **Trayliner.** $2.00-3.00
- ❑ ❑ USA Sh8561 **MC Insert/Cardboard,** 1985. $15.00-25.00
- ❑ ❑ USA Sh8563 **Menu Board/Lug-On/3 Premiums,** 1985. $25.00-40.00
- ❑ ❑ USA Sh8564 **Translite/Sm,** 1985. $10.00-15.00
- ❑ ❑ USA Sh8565 **Translite/Lg,** 1985. $15.00-25.00

Comments: National Distribution: USA - May 31-Jun 30, 1985. USA Sh8301-04 were redistributed in 1985 with new decals. Note that Splash Dasher had a large decal running the length of the boat vs. the 1983 smaller decal. All Ship Shape '85 decals were dated 1985. In 1985, Captain Crook was called "The Captain."

Sh8565

Sticker Club Happy Meal, 1985

Boxes:

- ❑ ❑ USA St8510 **Hm Box - Club House Meeting,** 1984. $5.00-8.00
- ❑ ❑ USA St8511 **Hm Box - Sticker Club Party,** 1984. $5.00-8.00
- ❑ ❑ USA St8512 **Hm Box - Sticker Club Picnic,** 1984. $5.00-8.00

St8511

St8510

St8512

❑ ❑ USA St8513 **Hm Box - Trading Days,** 1984. $5.00-8.00

Premiums: Sticker Sheets:

❑ ❑ USA St8501 **Sticker - Paper,** 1985, 12 Stickers/McDonaldland Characters. $5.00-8.00

❑ ❑ USA St8502 **Sticker - Scented/Scratch & Sniff,** 1985, 5 Stickers/Hamburglar/Birdie/Ronald/Fry Guy/Grimace. $5.00-10.00

❑ ❑ USA St8503 **Sticker - Motion (3-D),** 1985, 4 Stickers/Ron/Prof/Hamb/Birdie. $5.00-7.00

❑ ❑ USA St8504 **Sticker - Shiny/Prismatic,** 1985, 6 Stickers/Ron/Grim/Arches/Birdie/McCheese/Fries. $5.00-7.00

❑ ❑ USA St8505 **Sticker - Puffy,** 1985, 2 Stickers/Ronald & Grimace. $5.00-8.00

❑ ❑ USA St8518 **Sticker Album - Hawaii**, 1984. $25.00-35.00

❑ ❑ USA St8519 **Stickers: Hawaii.** 6 different varieties. Each $7.00-10.00

St8513

St8501

St8502

St8504

St8505

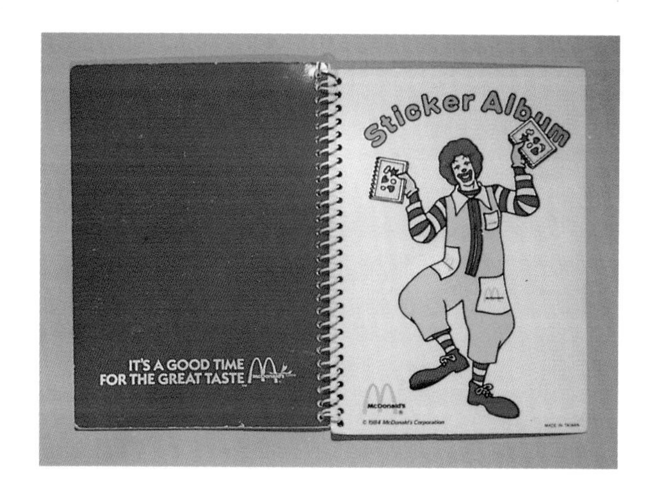

St8518

St8503

St8519

St8565

❑ ❑ USA St8561 **MC Insert/Cardboard**, 1985. $15.00-20.00
❑ ❑ USA St8562 **Header Card/Permanent Display**, 1985. $15.00-20.00
❑ ❑ USA St8563 **Menu Board Lug-On**, 1985. $15.00-20.00
❑ ❑ USA St8565 **Translite/Lg**, 1985. $15.00-25.00

Comments: National Distribution: USA - March 11 - May 19, 1985. Hawaii distributed an entirely different set of six stickers. These were given out, just like the Continental USA stickers, except some stores gave away Sticker Albums, to store and save stickers. Sticker collecting was "the fad" in Hawaii in 1985.

Pu8515

Stomper Mini 4 x 4 I (Collect All 6 Push-Alongs)/Test Market Happy Meal, 1985

Box:
❑ ❑ USA Pu8515 **Hm Box - Stompers with Desert,** 1985. $40.00 - 50.00

U-3 Premiums:
❑ ❑ USA Pu8507 **U-3 Chevy Van,** 1986, Yel with Plastic Tires/Black Wheels (NO white centers). $10.00-15.00
❑ ❑ USA Pu8508 **U-3 Jeep Renegade,** 1986, Org with Plastic Tires/Black Wheels (NO white centers). $10.00-15.00

Premiums: Cars:
❑ ❑ USA Pu8501 **Speedy Chevy S-10 Pick-Up,** 1986, Blk-Sil with Rubber Tires/Wht Wheels. $7.00-10.00
❑ ❑ USA Pu8502 **Speedy Chevy S-10 Pick-Up,** 1986, Yel-Pur with Rubber Tires/Wht Wheels. $7.00-10.00

❑ ❑ USA Pu8503 **Sporty Chevy Van,** 1986, Red-Yel with Rubber Tires/Wht Wheels. $7.00-10.00

❑ ❑ USA Pu8504 **Free-Wheelin' Dodge Rampage,** 1986, Wht-Blu with Rubber Tires/Wht Wheels. $7.00-10.00
❑ ❑ USA Pu8505 **Free-Wheelin' Dodge Rampage,** 1986, Blu-Mar with Rubber Tires/Wht Wheels. $7.00-10.00

❑ ❑ USA Pu8506 **Jeep Renegade 78,** 1986, Maroon-White with Rubber Tires/Wht Wheels. $7.00-10.00

❑ ❑ USA Pu8509 **Limited Edition Stomper 4 x 4 (Chev),** 1986, Wht/Grn with Arches/Send Away Premium. $12.00-15.00

❑ ❑ USA Pu8555 **Trayliner,** 1986. $10.00-15.00

Pu8509

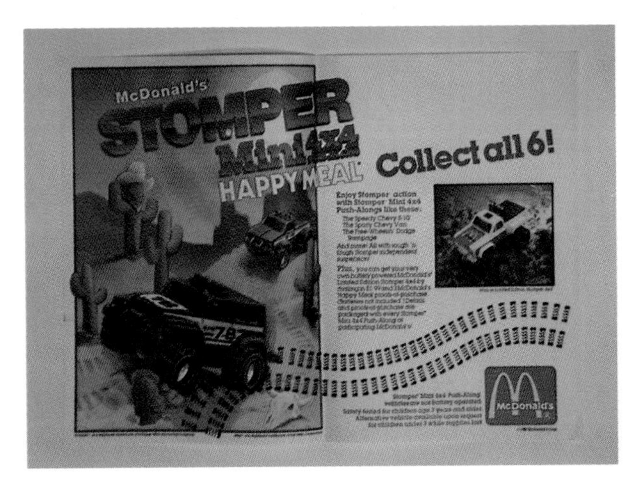

Pu8555

Pu8556

❑ ❑ USA Pu8556 **Table Tent,** 1986. $15.00-20.00
❑ ❑ USA Pu8564 **Translite/Sm,** 1986. $25.00-40.00
❑ ❑ USA Pu8565 **Translite/Lg,** 1986, Collect All 6. $60.00-75.00

Comments: Limited Regional Distribution: USA - September 6-October 17, 1985. The cars are the same as offered in the Stomper Mini II 4 x 4's Happy Meal, 1986. **Prices quoted are for Mint in the Package (MIP) only. Loose Stomper cars sell for $3.00 - 4.00 with the U-3 selling loose for $4.00 - 7.00.**

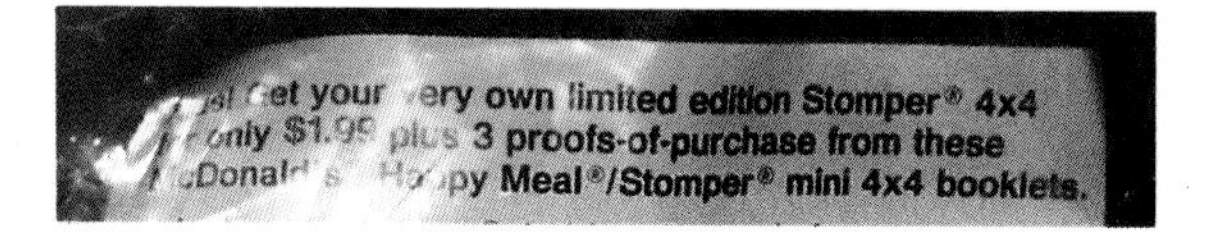

Pu8565

Toothbrush Happy Meal, 1985

Premium: Toothbrushes:
❑ ❑ USA To8501 **Toothbrush: Ronald/Full Figure,** 1984, Yellow or Red. $40.00-50.00
❑ ❑ USA To8501 **Toothbrush: Ronald/Full Figure,** 1984, Blue. $75.00-100.00

❑ ❑ USA To8565 **Translite/Lg,** 1985. $50.00-75.00

Comments: Limited Regional Distribution: USA - 1985 in New England. Distribution was only for a two week period.

To8565

To8501

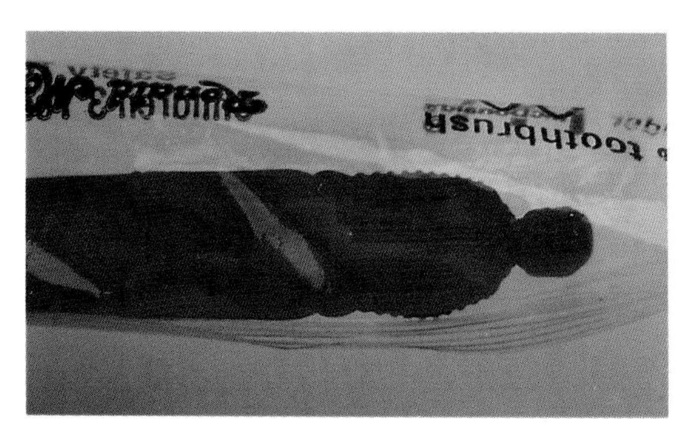

Detail of above top toothbrush.

Transformers/My Little Pony Happy Meal, 1985

Box:

❑ ❑ USA Tr8520 **Hm Box - My Little Pony/Transformers,** 1985. $75.00-100.00

Premiums for the Boys: Transformers (MIP: $85.00-100.00; Loose: $45.00-50.00):

❑ ❑ USA Tr8501 **Brawn,** 1984, Arms Out: Grn-Blu/Red-Blu/Blu-Yel/Red-Grn/Red-Yel/Grn-Yel/Loose. $45.00-50.00

❑ ❑ USA Tr8502 **Bumblebee,** 1984, Small Window: Blk-Red/Wine-Blk/Yel-Blk/Blk-Blu/Blk-Grn/Vio-Blu/Teal-Blk/Loose. $45.00-50.00

❑ ❑ USA Tr8503 **Cliffjumper,** 1984, Large Window: Blk-Red/Wine-Blk/Yel-Blk/Blk-Blu/Blk-Grn/Vio-Blu/Teal-Blk/Loose. $45.00-50.00

❑ ❑ USA Tr8504 **Gears,** 1984, Arms at Side: Grn-Blu/Red-Blu/Blu-Yel/Red-Grn/Red-Yel/Grn-Yel/Loose. $45.00-50.00

Premiums for the Girls: Charms (MIP: $85.00-100.00; Loose: $15.00-25.00 depending on sticker imprint appearance):

❑ ❑ USA Tr8510 **Charm - Blossom,** 1984, Purp with Flower Imprint/Hasbro/Loose. $15.00-25.00

❑ ❑ USA Tr8511 **Charm - Blue Belle,** 1984, Blu/Gry with Star Imprint/Hasbro/Loose. $15.00-25.00

❑ ❑ USA Tr8512 **Charm - Butterscotch,** 1984, Butterscotch with Butterfly Imprint/Hasbro/Loose. $15.00-25.00

❑ ❑ USA Tr8513 **Charm - Cotton Candy,** 1984, Pnk with Paw Print Imprint/Hasbro/Loose. $15.00-25.00

❑ ❑ USA Tr8514 **Charm - Minty,** 1984, Grn with Three Leaf Imprint/Hasbro/Loose. $15.00-25.00

❑ ❑ USA Tr8515 **Charm - Snuzzle,** 1984, Grey with Heart Imprint/Hasbro/Loose. $15.00-25.00

❑ ❑ USA Tr8505 **Car: Overdrive, Red Transformer.** Hasbro send away premium. $——

Tr8520

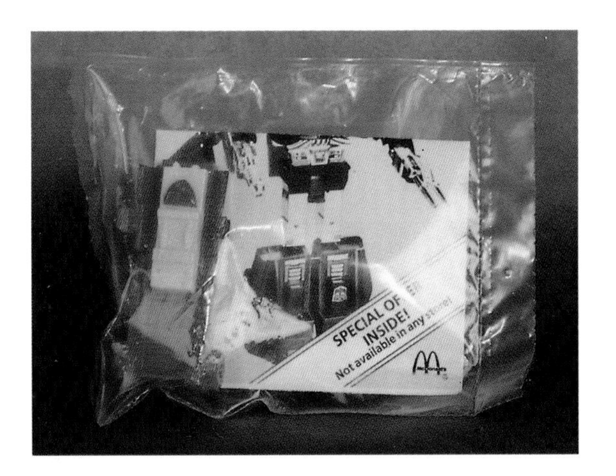

Tr8501

Tr8501 Tr8502 Tr8503 Tr8504

Tr8505

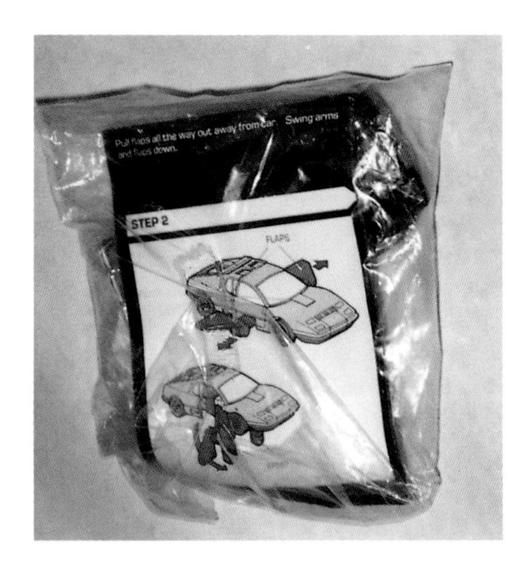

Tr8505, see previous page.

Tr8510 Tr8511 Tr8512 Tr8513 Tr8514 Tr8515

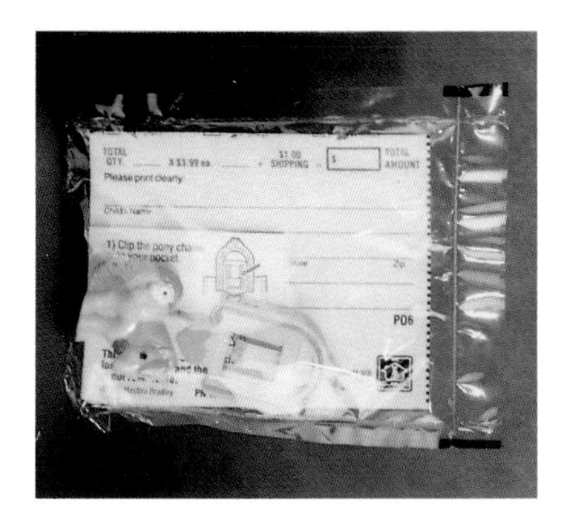

Tr8513

Save this valuable Proof of Purchase COUPON

COUPON

YES! I want to help the Autobot fleet! Please send me ______ OVERDRIVE(S). I've enclosed 3 Proof of Purchase coupons and $5.99 for each OVERDRIVE I order, plus $1.00 for shipping and handling.

TOTAL QTY. ____ X $5.99 ea. = ____ + $1.00 SHIPPING = $ ____ TOTAL AMOUNT

Please print clearly:

Child's Name

Address

City State Zip

Mo /Day /Year Birthday

Parent's Name

T04

Mail to: **Hasbro — Transformers Offer**
P.O. Box 7050
N. Hollywood, CA 91609

This special offer and pricing is good until January 31, 1986. Please allow 6–8 weeks for delivery. Offer good only in the United States except where prohibited by law.

Part Number: 169633.00A © 1985 Hasbro Bradley. All rights reserved.

THE TRANSFORMERS®
MORE THAN MEETS THE EYE!

BULLETIN:
The Transformers have landed on Earth. Help the fleet by sending for OVERDRIVE today!

SPECIAL OFFER INSIDE!
Not available in any store!

Tr8514 Tr8510

Their struggle began on a faraway planet light years from Earth. Heroic Autobots vs. Evil Decepticons. Robot against Robot. Now as the battle continues, the tide begins to turn against the Autobots.

But *you* can help by sending for their powerful and dependable reinforcement — OVERDRIVE.

OVERDRIVE is ready to take on the Decepticons wherever they may be hiding... Just pull back, swing open, fold down, and in a snap you've created OVERDRIVE, a new, heroic Autobot warrior.

Here's how to send OVERDRIVE racing to *your* house.

Every time you buy a McDonald's Happy Meal you'll find a Proof of Purchase coupon enclosed. Clip and save them. As soon as you've collected 3 coupons, you'll be ready to send OVERDRIVE hurtling through space to your house.

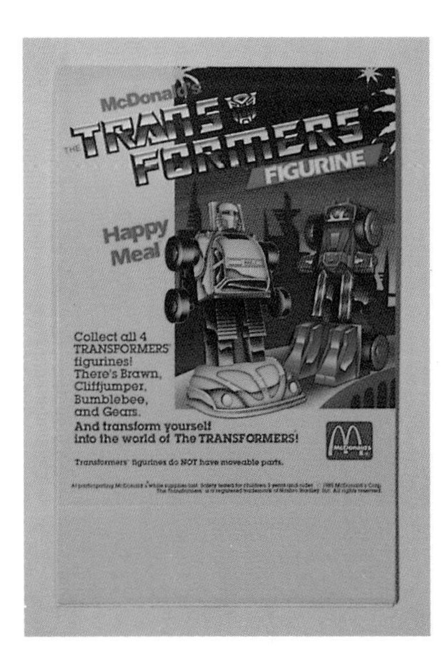

Tr8556

- ❑ ❑ USA Tr8555 **Tray Liner,** 1985. $7.00-10.00
- ❑ ❑ USA Tr8556 **Table Tent,** 1985. $15.00-20.00
- ❑ ❑ USA Tr8565 **Translite/Lg,** 1985. $40.00-50.00

Comments: Limited Regional Distribution: USA - 1985 in St. Louis, Missouri in the summer. Premium markings: "1985 Hasbro." With the addition of the teal-black transformers, the total numbers are: 26 transformers/6 My Little Pony charms. **Prices quoted are for loose and Mint in the Package. Loose transformers are selling for $45.00 - 50.00 with those at the higher end being the transformer with his arms extended. Loose charms are selling for $15.00 - 25.00 each, depending on condition of the imprint on the hindside of the My Little Pony charm.** For three Proof of Purchase coupons from Happy Meal boxes, plus $5.99 and postage, one could receive the Special Offer OVERDRIVE Transformer. Offer ended January 31, 1986.

Tr8555

Ge8501

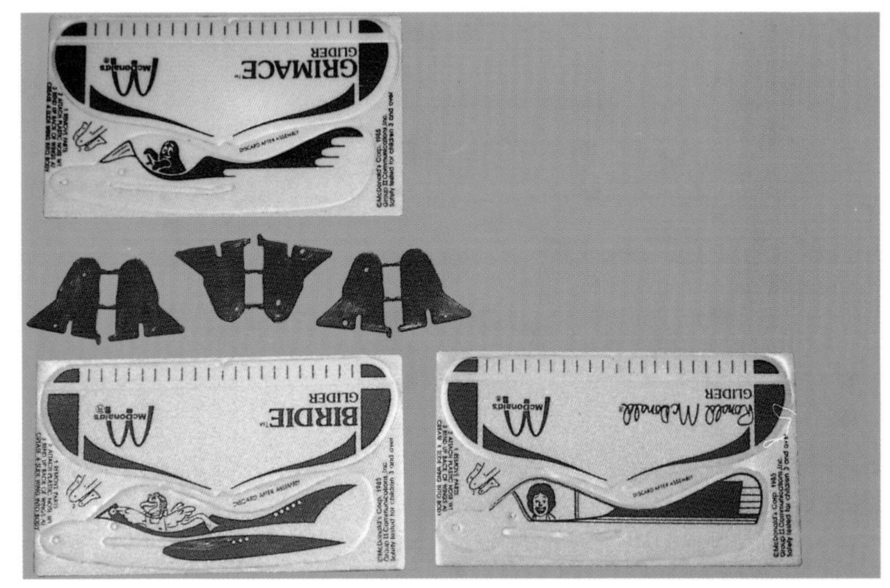

Ge8519

Ge8520

USA Generic Promotions, 1985

- ❑ ❑ USA Ge8501 **Grimace Glider,** purple/white, 1985. $5.00-8.00
- ❑ ❑ USA Ge8519 **Birdie Glider,** pink/white, 1985. $5.00-8.00
- ❑ ❑ USA Ge8520 **Ronald Glider,** red/white, 1985. $5.00-8.00
- ❑ ❑ USA Ge8521 **Hamburglar Glider,** black/white, 1985. $5.00-8.00

❑ ❑ USA Ge8502 **Photo Card (Large): Squire Fridell as Ronald McDonald**, 1985. Squire Fridell was Ronald McDonald for a short time only. $10.00-15.00

Vol. 7 No. 1 Fun Times Magazine repeated the offer of a McDonaldland T-shirt. From then on, the offers were discontinued.

❑ ❑ USA Ge8503 **Fun Times Magazine: Vol. 7 No. 1 Feb/Mar, 1985.** $2.00-3.00

❑ ❑ USA Ge8504 **Fun Times Magazine: Vol. 7 No. 2 Apr/May, 1985.** $2.00-3.00

❑ ❑ USA Ge8505 **Fun Times Magazine: Vol. 7 No. 3 June/July, 1985.** $2.00-3.00

❑ ❑ USA Ge8506 **Fun Times Magazine: Vol. 7 No. 4 September, 1985.** $2.00-3.00

❑ ❑ USA Ge8507 **Fun Times Magazine: Vol. 7 No. 5 Oct/Nov, 1985.** $2.00-3.00

❑ ❑ USA Ge8508 **Fun Times Magazine: Vol. 7 No. 6 Dec/Jan, 1985.** $2.00-3.00

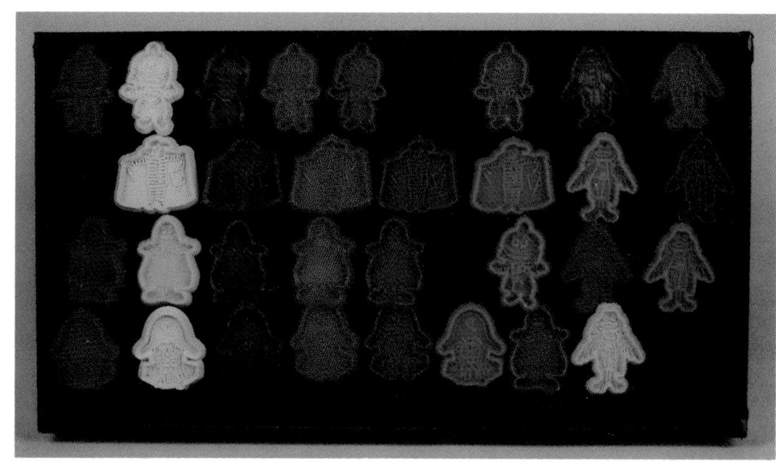

Ge8509-13

❑ ❑ USA Ge8509 **Ring: Play-Doh Cut Out Grimace:** red or yellow or green or blue or orange or purple. $6.00-10.00

❑ ❑ USA Ge8910 **Ring: Play-Doh Cut Out Hamburglar:** red or yellow or green or blue or orange or purple. $6.00-10.00

❑ ❑ USA Ge8511 **Ring: Play-Doh Cut Out Captain Crook:** red or yellow or green or blue or orange or purple. $6.00-10.00

❑ ❑ USA Ge8512 **Ring: Play-Doh Cut Out Big Mac:** red or yellow or green or blue or orange or purple. $6.00-10.00

❑ ❑ USA Ge8513 **Ring: Play-Doh Cut Out Birdie:** red or yellow or green or blue or orange or purple. $6.00-10.00

❑ ❑ USA Ge8514 **1985 Calendar: Ronald McDonald Secret Solver Coloring Calendar.** $5.00-7.00

❑ ❑ USA Ge8515 **Plate: Ronald McDonald with Birdie on Moon Next to Spaceship,** 9". $10.00-20.00

❑ ❑ USA Ge8516 **Plate: Ronald McDonald with Fry Guys in Biplane,** 9". $10.00-20.00

❑ ❑ USA Ge8517 **Plate: Ronald McDonald with Grimace in McDonald Land Express Train,** 9". $10.00-20.00

❑ ❑ USA Ge8518 **Plate: Ronald McDonald with Grimace in Sailboat,** 9". $10.00-20.00

❑ ❑ USA Ge8519 **Ornament: Reindeer - stuffed Rudolph,** 3 1/2" $4.00-6.00

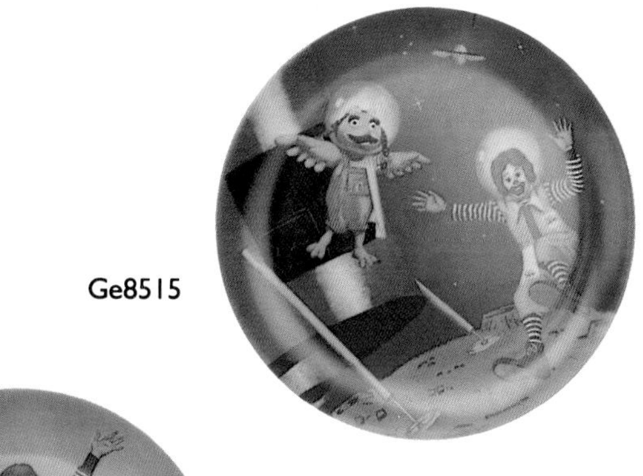

Ge8515

Ge8516

Ge8517

Ge8518

Ge8522

Squire Fridell as Ronald McDonald.

Comments: Regional Distribution: USA - 1985. This is a sampling of the generic premiums given away in 1985. The Play-Doh rings came in light green and dark green and light blue and dark blue. Shading color occurred during the different distribution cycles. The Stuffed Reindeer was given out free with the purchase of Gift Certificates in the stores.

30th year of operation is celebrated in 1985. "Large Fries for Small Fries" slogan begins. McBlimp, a real blimp, floats over cities and sporting events catching the advertising eye of the consumer. "The Hot Stays Hot and The Cool Stays Cool" introduces the McDonald's L. T. sandwich. The first European Ronald McDonald House opens in Amsterdam, Holland.

Squire Fridell is selected as the National Ronald McDonald (1985-90). He was also called the Toyota Ronald because he did local Toyota commercials along with being the National Ronald McDonald.

1986

Airport Happy Meal, 1986
An American Tail Happy Meal, 1986
Beach Ball Happy Meal, 1986
Beachcomber Happy Meal, 1986
Berenstain Bears I Test Market Happy Meal, 1986
Colorforms Happy Meal, 1986
Construx Action Building System Test Market Happy Meal, 1986
Crayola/Crayon Magic I Test Market Happy Meal, 1986
Glo-Tron Spaceship Test Market Happy Meal, 1986
Halloween '86 Happy Meal, 1986
Happy Pail III '86 Happy Meal, 1986
High Flying Kite Happy Meal, 1986
Lego Building Sets III Happy Meal, 1986
Muppet Babies I Test Market Happy Meal, 1986
Old McDonald's Farm/Barnyard Happy Meal, 1986
Play-Doh III Happy Meal, 1986
Stomper Mini 4 x 4 II Happy Meal, 1986
Story of Texas Happy Meal, 1986
Tinosaurs Happy Meal, 1986
Young Astronauts I Happy Meal, 1986
USA Generic Promotions, 1986

- **Signs changed to "MORE THAN 60 BILLION SERVED."**
- **Regional Happy Meal promotions tested**
- **9th National O/O Convention**
- **"Back To Our Future" - O/O advertising theme**

Ai8675

Ai8676

Airport Happy Meal, 1986

Boxes:
❑ ❑ USA Ai8675 **Hm Box - Control Tower,** 1986. $4.00-5.00
❑ ❑ USA Ai8676 **Hm Box - Hangar,** 1986. $4.00-5.00

❑ ❑ USA Ai8677 **Hm Box - Terminal,** 1986. $4.00-5.00
❑ ❑ USA Ai8678 **Hm Box - Luggage Claim Area,** 1986. $4.00-5.00

Ai8677

U-3 Premiums:
❑ ❑ USA Ai8666 **U-3 Fry Guy Friendly Flyer,** 1986, Blu or Red Floater. $5.00-7.00
❑ ❑ USA Ai8667 **U-3 Grimace Smiling Shuttle,** 1986, Blu or Red Floater. $5.00-7.00

Premiums:
❑ ❑ USA Ai8651 **Big Mac Helicopter - green,** 1982, Grn. $8.00-10.00
❑ ❑ USA Ai8652 **Fry Guy Flyer Airplane - blue,** 1986, 3p Blu Airplane. $8.00-10.00
❑ ❑ USA Ai8653 **Ronald Seaplane - red,** 1986, 4p Red Airplane. $8.00-10.00
❑ ❑ USA Ai8654 **Grimace Ace Biplane - purple,** 1986, 3p Pur Biplane. $8.00-10.00
❑ ❑ USA Ai8655 **Birdie Bent Wing Brazer Airplane - pink,** 1986, 5p Pnk Birdie the Early Bird Bent Wing. $8.00-10.00

❑ ❑ USA Ai8661 **MC Insert/Cardboard,** 1986. $15.00-20.00
❑ ❑ USA Ai8663 **Menu Board Lug-On with 1 Premium,** 1986. $10.00-15.00
❑ ❑ USA Ai8664 **Translite/Sm,** 1986. $10.00-15.00
❑ ❑ USA Ai8665 **Translite/Lg,** 1986. $15.00-25.00

Comments: National Distribution: USA - Mar 10-May 18, 1986. **The red and blue Big Mac helicopters (selling loose for $6.00-10.00), like the green Big Mac helicopter, were a 1982 giveaway and were used again with Airport Happy Meal. Also distributed were the green and red Fry Guy copter/flyer, selling loose for $3.00-5.00.** These other planes/helicopters were used as generic fill-in premiums.

Ai8678

Ai8666 Ai8667

Ai8651 Ai8652 Ai8653 Ai8654

Ai8651 Ai8652

Ai8655

An8610

An8611

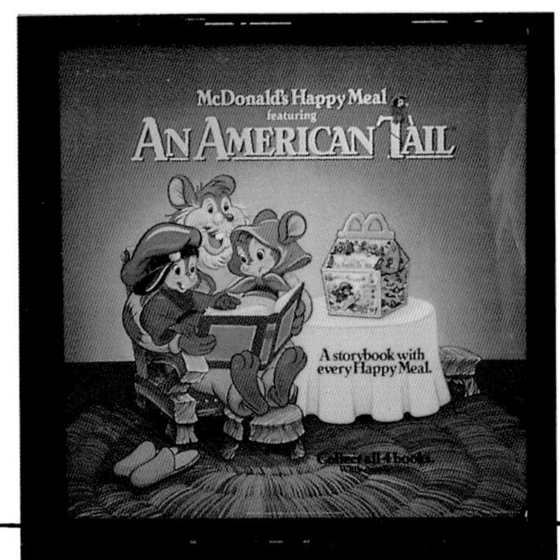

An8665

An American Tail Happy Meal, 1986

Boxes:
- ❑ ❑ USA An8610 **Hm Box - Mouse in the Moon,** 1986. $4.00-5.00
- ❑ ❑ USA An8611 **Hm Box - Slippery Solutions,** 1986. $4.00-5.00

Premiums: Books:
- ❑ ❑ USA An8601 **Book - Fievel's Boat Trip,** 1986, Pnk. $2.00-4.00
- ❑ ❑ USA An8602 **Book - Fievel's Friends,** 1986, Yel. $2.00-4.00
- ❑ ❑ USA An8603 **Book - Fievel and Tiger,** 1986, Purp. $2.00-4.00
- ❑ ❑ USA An8604 **Book - Tony and Fievel,** 1986, Blu/Grn. $2.00-4.00

- ❑ ❑ USA An8661 **MC Insert/Cardboard,** 1986. $15.00-20.00
- ❑ ❑ USA An8664 **Translite/Sm,** 1986. $10.00-15.00
- ❑ ❑ USA An8665 **Translite/Lg,** 1986. $15.00-25.00

Comments: National Distribution: USA - November 28-December 24, 1986.

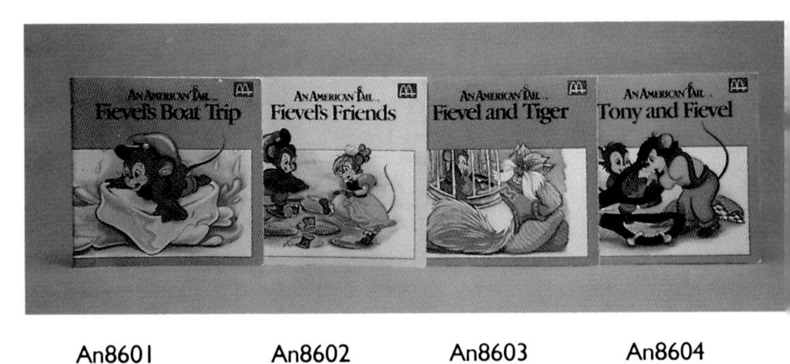

An8601 An8602 An8603 An8604

Beach Ball Happy Meal, 1986

Box:
- ❑ ❑ USA Be8607 **Hm Box - Having a Wonderful Time/By the Sea,** 1986. $7.00-10.00

Premiums: Beach Balls:
- ❑ ❑ USA Be8601 **Beach Ball: Birdie on Beach with Sand Castle**, Sailboat/Blue, 1986. $15.00-25.00
- ❑ ❑ USA Be8602 **Beach Ball: Grimace with Beach Umbrella with Bird/Fish**, Sailboat/Yel, 1986. $15.00-25.00
- ❑ ❑ USA Be8603 **Beach Ball: Ronald Waving on Beach with Pelican**, Seahorse/Sun/Red, 1986. $15.00-25.00

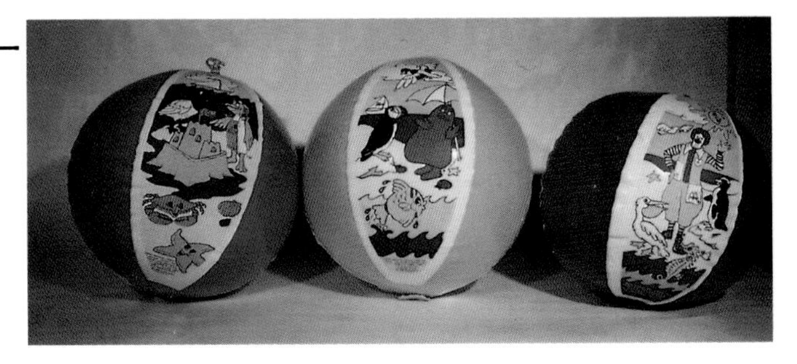

Be8601 Be8602 Be8603

Be8607

❑ ❑ USA Be8665 **Translite/Lg,** 1986. $20.00-25.00

Comments: Regional Distribution: USA - 1986 in Washington, New York, and Colorado. MIP package came polybagged with scotch tape enclosure.

Be8665

Beachcomber Happy Meal, 1986

Premiums: Sand Pails

❑ ❑ USA Bc8650 **Sand Pail - Grimace,** 1986, Wht Pail. $8.00-10.00

❑ ❑ USA Bc8651 **Sand Pail - Mayor,** 1986, Wht Pail. $8.00-10.00

❑ ❑ USA Bc8652 **Sand Pail - Ronald,** 1986, Wht Pail. $8.00-10.00

❑ ❑ USA Bc8665 **Translite/Lg,** 1986. $35.00-50.00

Comments: Regional Distribution: USA - South Carolina - 1986. All white pails had white lids with yellow shovels. Lids had four 1/2" holes in each.

Bc8651 Bc8652 Bc8650

Bc8665

Berenstain Bears I Test Market Happy Meal, 1986

Boxes:

❑ ❑ USA Bb8610 **Hm Box - Holly Wreath on Front Door/ Holiday Barn Dance,** 1986. $25.00-40.00

❑ ❑ USA Bb8611 **Hm Box - Bear Country General Store/Holly Wreath,** 1986. $25.00-40.00

Bb8610

Bb8611

Bb8612

- ❑ ❑ USA Bb8612 **Hm Box - Heave-Ho!/Wreath on Front Door,** 1986. $25.00-40.00
- ❑ ❑ USA Bb8613 **Hm Box - Bear Country School/Wreaths on Windows,** 1986. $25.00-40.00

Premiums:

- ❑ ❑ USA Bb8603 **Set 1 Sister with painted hands and feet/ not natural rubber color,** 1986, 2p Arms at Side on Red Sled. $50.00-75.00
- ❑ ❑ USA Bb8602 **Set 2 Mama with painted hands and feet/ not natural rubber color,** 1986, 2p Flocked Head with Pants with Yel Shopping Cart. $50.00-75.00
- ❑ ❑ USA Bb8601 **Set 3 Papa with painted hands and feet/ not natural rubber color,** 1986, 2p Flocked Head with Burnt Org Wheelbarrow. $50.00-75.00
- ❑ ❑ USA Bb8604 **Set 4 Brother with painted hands and feet/ not natural rubber color,** 1986, 2p Flocked Head on Yel Scooter/Grn Handle Bars. $50.00-75.00

- ❑ ❑ USA Bb8664 **Translite/Sm,** 1986. $15.00-25.00
- ❑ ❑ USA Bb8665 **Translite/Lg,** 1986. $25.00-40.00

Comments: Regional Distribution: USA - November 28-December 24, 1986 in Evansville, Indiana. The National Promotion followed in Oct-Nov 1987 with four redesigned boxes and two new Happy Meal bags. Hands and feet were painted, not natural rubber color; bears were soft rubber composition.

Bb8613. Box flat for Berenstain Bears (Test Market) box.

Bb8602 Bb8603 Bb8601 Bb8604

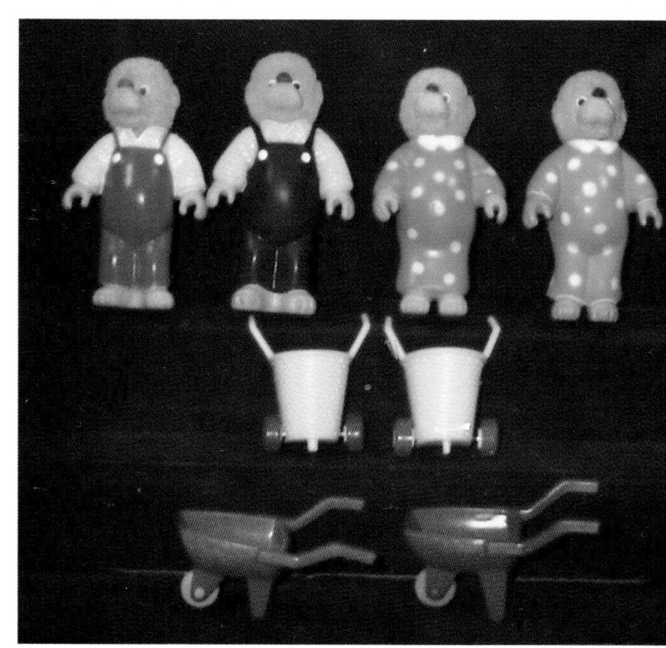

Colorforms Happy Meal, 1986

Boxes:
- ❑ ❑ USA Co8610 **Hm Box - Beach Party,** 1986. $4.00-5.00
- ❑ ❑ USA Co8611 **Hm Box - Camp out,** 1986. $4.00-5.00
- ❑ ❑ USA Co8612 **Hm Box - Picnic Today,** 1986. $4.00-5.00
- ❑ ❑ USA Co8613 **Hm Box - Play Day,** 1986. $4.00-5.00

U-3 Premiums:
- ❑ ❑ USA Co8606 **U-3 Sticker Playset - Grimace At a Beach Party,** 1986. $15.00-20.00
- ❑ ❑ USA Co8607 **U-3 Sticker Playset - Ronald At the Farm,** 1986. $15.00-20.00

Co8610

Co8611

Co8612

Co8613

Co8606 Co8607

Co8601 Co8602 Co8603

Co8604 Co8605

Co8665

Premiums: Sticker Playsets:

❑ ❑ USA Co8601 **Set 1 Beach Party - Grimace**, 1986. $15.00-20.00

❑ ❑ USA Co8602 **Set 2 Picnic Today - Hamburglar**, 1986. $15.00-20.00

❑ ❑ USA Co8603 **Set 3 Play Day - Birdie The Early Bird**, 1986. $15.00-20.00

❑ ❑ USA Co8604 **Set 4 Camp Out - Professor**, 1986. $15.00-20.00

❑ ❑ USA Co8605 **Set 5 Farm - Ronald**, 1986. $15.00-20.00

❑ ❑ USA Co8661 **MC/Cardboard,** 1986. $15.00-25.00

❑ ❑ USA Co8664 **Translite/Sm,** 1986. $15.00-25.00

❑ ❑ USA Co8665 **Translite/Lg,** 1986. $20.00-35.00

Comments: Limited Regional Distribution: USA - December 29, 1986-February 1, 1987. Auction prices realized have increased the price of the Happy Meal premiums. Box flats, as illustrated in USA Co8613, exist for many Happy Meal boxes. Prices of box flats range from $15.00 per one box illustrated on the flat to several hundred dollars for some of the older boxes illustrated on the box flat.

Construx Action Building System Test Market Happy Meal, 1986

Boxes:

❑ ❑ USA Cx8610 **Hm Box - Computer Quick Fix/Repair Station,** 1986. $50.00-75.00

❑ ❑ USA Cx8611 **Hm Box - Mars Landscape,** 1986. $50.00-75.00

Premiums: Construx Kits:

❑ ❑ USA Cx8600 **Set 1 Cylinder,** 1986, Construx Plastic Piece/Decals/Wht with Glow Pieces. $40.00-50.00

❑ ❑ USA Cx8601 **Set 2 Canopy,** 1986, Construx Plastic Piece/Decals/Wht with Glow Pieces. $40.00-50.00

❑ ❑ USA Cx8602 **Set 3 Wing,** 1986, Construx 12 Plastic Pieces/Decals/Wht with Glow Piece. $40.00-50.00

❑ ❑ USA Cx8603 **Set 4 Axel,** 1986, Construx Plastic Piece/Decals/Wht with Glow Pieces. $40.00-50.00

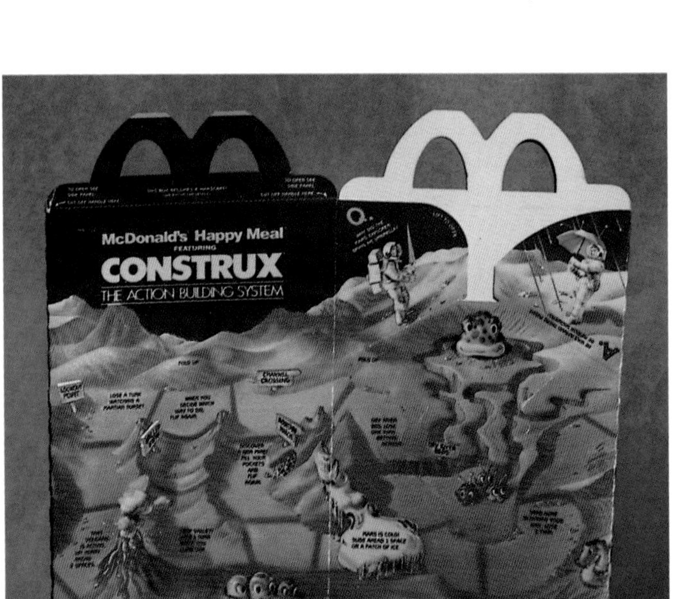

Cx8611

Cx8600

Cx8601

❑ ❑ USA Cx8626 **Display,** 1986. $250.00-350.00
❑ ❑ USA Cx8665 **Translite/Lg,** 1986. $50.00-75.00

Comments: Regional Distribution: USA - 1986. A single space-craft could be built by putting together the four sets. Sets were discontinued after the Challenger disaster. Set 3 contained twelve pieces: three straight white, two curved white, four light blue knots, one glow-n-dark bell, two dark blue panels. Auction prices have dramatically increased the price of these premiums. Not enough transactions to determine market price.

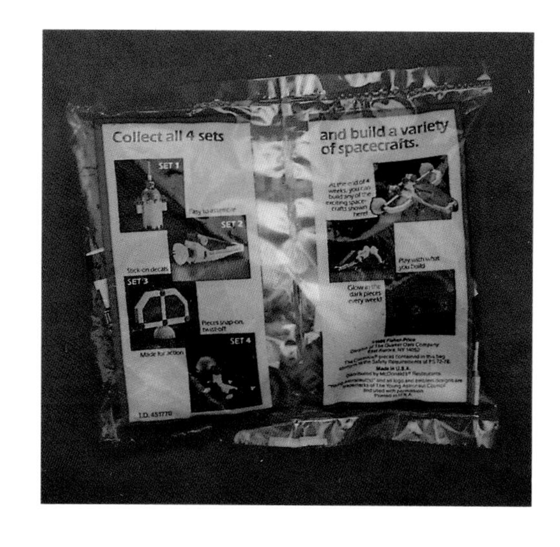

Cx8602

Cx8603

Cx8626

Cx8602 Cx8601 Cx8600 Cx8603

Crayola/Crayon Magic I Test Market Happy Meal, 1986

Boxes:

❑ ❑ USA Cr8610 **Hm Box - Quick Draw Grimace,** 1986. $35.00-40.00

❑ ❑ USA Cr8611 **Hm Box - Quick Draw Ronald,** 1986. $35.00-40.00

Premiums: Stencils:

❑ ❑ USA Cr8601 **Set 1: Circle/Stencil with 12 Cut-Outs and 4 Crayons/red/blue/yellow/green**, 1986, red. $15.00-20.00

❑ ❑ USA Cr8602 **Set 2: Rectangle/Stencil with 22 Cut-Outs and 4 Florescent Crayons,** 1986, red. $15.00-20.00

Cr8610

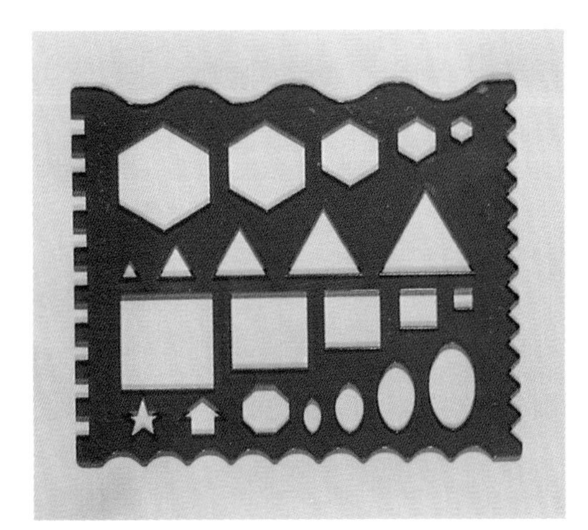

Cr8611

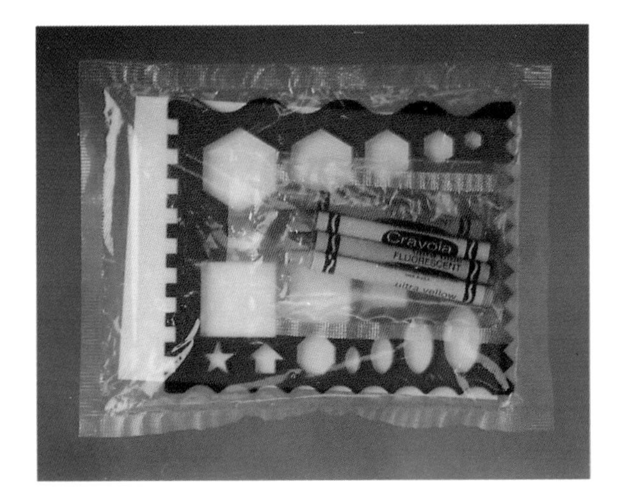

Cr8602

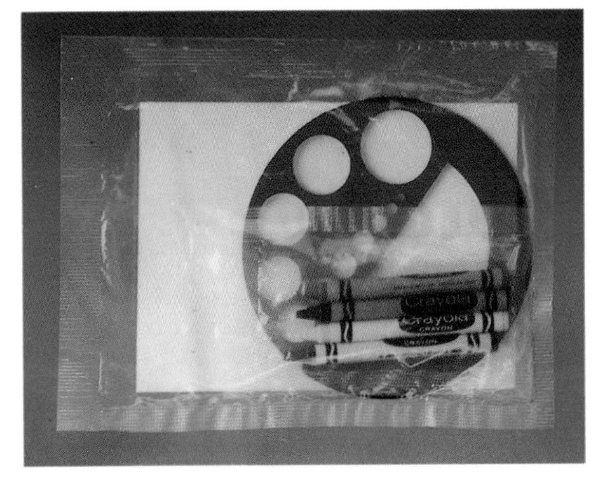

Cr8601

❑ ❑ USA Cr8603 **Set 3: Triangle/Right Triangle/Stencil with 10 Cut-Outs and 1 Thick Grn Crayola Marker**, 1986, blue. $15.00-20.00

❑ ❑ USA Cr8604 **Set 3: Triangle/Isosceles Triangle/Stencil with 10 Cut-Outs and 1 Thick Org Crayola Marker,** 1986, blue. $15.00-20.00

❑ ❑ USA Cr8605 **Set 4: Right Triangle/Stencil with 9 Cut-Outs and thin blue Crayola marker,** 1986, blue. $20.00-35.00

❑ ❑ USA Cr8606 **Set 4: Right Triangle/Stencil with 9 Cut-Outs and thin red Crayola marker,** 1986, blue. $20.00-35.00

❑ ❑ USA Cr8655 **Trayliner/Collect all 4 sets.** $5.00-8.00

❑ ❑ USA Cr8665 **Translite/Lg,** 1986. $40.00-65.00

Comments: Limited Test Market Distribution: USA - 1986. This test Happy Meal used redesigned picture perfect Happy Meal boxes with new titles. Trayliner says, "Collect all 4 sets."

Cr8603

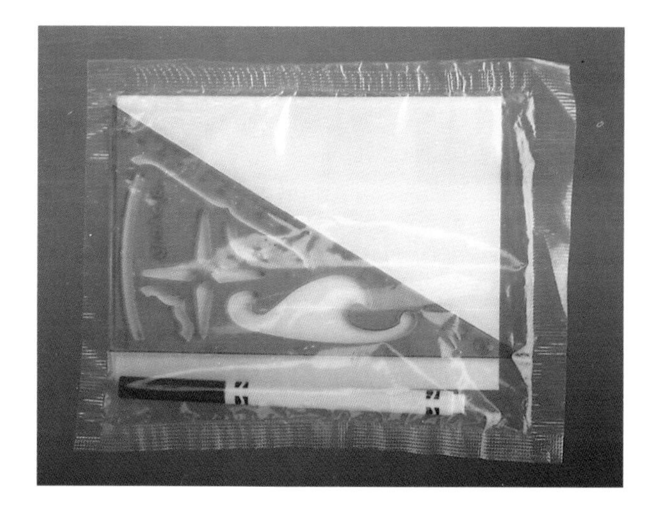

Cr8605

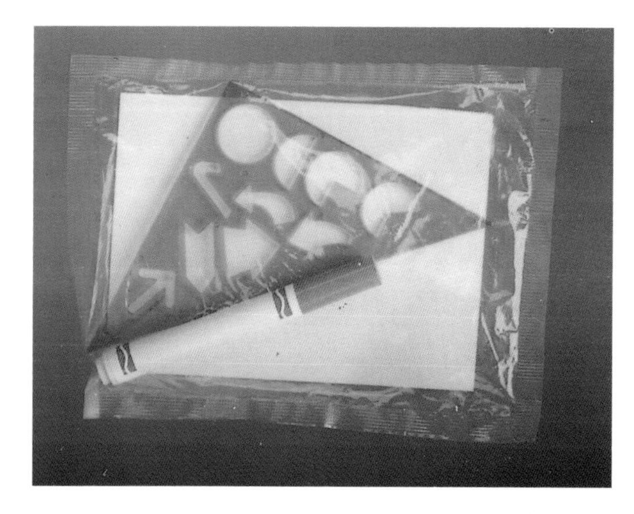

Cr8604

Cr8606

Cr8655

Glo-Tron Spaceship Test Market Happy Meal, 1986

Premiums: Vacuum Form Containers:

- ❑ ❑ USA Sp8601 **Spaceship with 8 Windows,** 1985, Metallic Grey with **Glow Stickers.** $50.00-75.00
- ❑ ❑ USA Sp8602 **Spaceship with Rear Engine,** 1985, Metallic Blue with **Glow Stickers.** $50.00-75.00
- ❑ ❑ USA Sp8603 **Spaceship with Pointed Nose,** 1985, Metallic Green with **Glow Stickers.** $50.00-75.00
- ❑ ❑ USA Sp8604 **Spaceship with 4 Humps,** 1985, Metallic Red with **Glow Stickers.** $50.00-75.00

- ❑ ❑ USA Sp8607 **Sticker Sheet,** 1986. Each $20.00-25.00

- ❑ ❑ USA Sp8665 **Translite/Menu Board,** 1986. $50.00-75.00

Comments: Regional Distribution: USA - 1986. The spaceships have a shiny metallic flake finish. The spaceships were the vacuform Happy Meal containers and the premium. **The stickers are glow-in-dark, not the spaceships.**

Sp8604 Sp8603b Sp8602 Sp8601

Sp8607

Halloween '86 Happy Meal, 1986

Premiums: Halloween Pails with Lids:

- ❑ ❑ USA Ha8601 **McBoo pail dated 1986,** Org Pail/Org Lid. $1.00-2.00
- ❑ ❑ USA Ha8602 **McGoblin pail dated 1986,** Org Pail/Org Lid. $1.00-2.00
- ❑ ❑ USA Ha8603 **McPunk'n pail dated 1986,** Org Pail/Org Lid. $1.00-2.00

- ❑ ❑ USA Ha8661 **MC Insert/Cardboard,** 1986. $7.00-10.00
- ❑ ❑ USA Ha8665 **Translite/Lg,** 1986. $10.00-15.00

Comments: National Distribution: USA - October 13-30, 1986. Lids had six 1/2" holes. Holes were large enough for finger entrapment.

Ha8665

Ha8601 Ha8602 Ha8603

Happy Pail III Happy Meal, 1986

Premiums: Sand Pails:

❑ ❑ USA Hp8690 **Sand Pail: Beach,** 1986, Blu Handle/Top/Yel Shovel. $5.00-10.00
❑ ❑ USA Hp8691 **Sand Pail: Parade,** 1986, Org Handle/Top/Red Rake. $5.00-10.00
❑ ❑ USA Hp8692 **Sand Pail: Picnic,** 1986, Yel Handle/Top/Red Rake. $5.00-10.00
❑ ❑ USA Hp8693 **Sand Pail: Treasure Hunt,** 1986, Red Handle/Top/Yel Shovel. $5.00-10.00
❑ ❑ USA Hp8694 **Sand Pail: Vacation,** 1986, Grn Handle/Top/Red Rake. $5.00-10.00

❑ ❑ USA Hp8626 **Display,** 1986. $35.00-40.00
❑ ❑ USA Hp8641 **Ceiling Dangler,** 1986, with 5 Pails. $20.00-25.00
❑ ❑ USA Hp8660 **Counter Mat,** 1986. $7.00-10.00
❑ ❑ USA Hp8664 **Translite/Sm,** 1986. $10.00-15.00
❑ ❑ USA Hp8665 **Translite/Lg,** 1986. $15.00-20.00

Comments: National Distribution: USA - May 30-July 6, 1986. Pails came with either a yellow shovel or red rake.

Hp8690

Hp8692

Hp8694

Hp8626

Hp8665

High Flying Kite Happy Meal, 1986

Box:

❑ ❑ USA Hi8610 **Hm Box - Kites,** 1986. $100.00-125.00

Premiums: Kites:

❑ ❑ USA Hi8601 **Kite: Hamburglar,** 1986, Blk/Wht Kite/Grn String Handle. $200.00-250.00

Hi8610

Hi8601

Hi8602

❑ ❑ USA Hi8602 **Kite: Ronald,** 1986, Blu/Wht Kite/Grn String Handle. $200.00-250.00

❑ ❑ USA Hi8603 **Kite: Birdie,** 1986, Yel/Wht Kite/Grn String Handle. $250.00-300.00

❑ ❑ USA Hi8664 **Translite/Sm,** 1986. $75.00-100.00

❑ ❑ USA Hi8665 **Translite/Lg,** 1986. $75.00-100.00

Comments: Regional Distribution: USA - 1987 in parts of New England states.

Hi8665

Hi8603

Le8610

Lego Building Sets III Happy Meal, 1986

Boxes:

❑ ❑ USA Le8610 **Hm Box - Capt Crook/Tug Boat/Fry Guy/ Which Came First,** 1986. $4.00-5.00

❑ ❑ USA Le8611 **Hm Box - Grimace/Vacation/Match the Patch,** 1986. $4.00-5.00

❑ ❑ USA Le8612 **Hm Box - Grimace/Golden Gate Bridge/ Animal Power,** 1986. $4.00-5.00

❑ ❑ USA Le8613 **Hm Box - Ronald/Globe/Record Trips Around the World,** 1986. $4.00-5.00

Le8611

Le8612

Le8613

U-3 Premiums:

- ❑ ❑ USA Le8604 **U-3 Bird with Eye,** 1986, Duplo/Red Pkg/5p/ Ages 1 1/2-4. $3.00-4.00
- ❑ ❑ USA Le8605 **U-3 Boat with Sailor,** 1986, Duplo/Blu Pkg/ 5p/Ages 1 1/2-4. $3.00-4.00

Premiums:

- ❑ ❑ USA Le8600 **Set A Race Car,** 1986, Lego/Red Pkg/16p. $3.00-4.00
- ❑ ❑ USA Le8601 **Set B Tanker,** 1986, Lego/Blue Pkg/27p. $3.00-4.00
- ❑ ❑ USA Le8602 **Set C Helicopter,** 1986, Lego/Yellow Pkg/19p. $3.00-4.00
- ❑ ❑ USA Le8603 **Set D Airplane,** 1986, Lego/Green Pkg/ 18p. $3.00-4.00

- ❑ ❑ USA Le8626 **Display/Premiums**, 1986. $75.00-100.00
- ❑ ❑ USA Le8641 **Dangler, 1986, Cardboard.** $25.00-40.00
- ❑ ❑ USA Le8661 **MC Insert/Cardboard,** 1986. $25.00-40.00
- ❑ ❑ USA Le8664 **Translite/Sm,** 1986. $15.00-25.00
- ❑ ❑ USA Le8665 **Translite/Lg,** 1986. $25.00-40.00

Comments: National Distribution: USA - October 31-November 26, 1986. Box flats exist for this Happy Meal. Prices for the flats range upwards, from $10.00-15.00 per box illustrated.

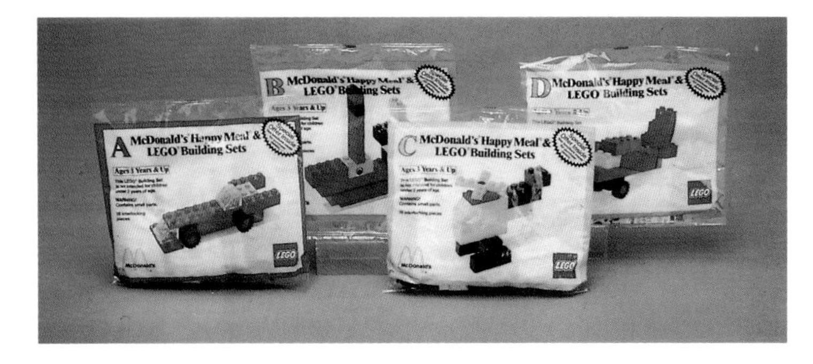

Le8600 Le8601 Le8602 Le8603

Le8604 Le8605

Le8641

Right: Le8665

Mu8600 Mu8601 Mu8603

Muppet Babies I Test Market Happy Meal, 1986

Premiums:

- ❑ ❑ USA Mu8600 **Set 1 Gonzo with Suspenders Crossed in Back/No Shoes/Grn Big Wheels.** $35.00-50.00
- ❑ ❑ USA Mu8601 **Set 2 Fozzie on Yel Horse.** $20.00-25.00
- ❑ ❑ USA Mu8602 **Set 3 Ms. Piggy in Pnk Car/Pnk Ribbon Flat Against Hair.** $35.00-50.00
- ❑ ❑ USA Mu8603 **Set 4 Kermit on Red Skateboard.** $20.00-25.00

Comments: Limited Test Market: USA - August 8-September 7, 1986 in Savannah, Georgia. Gonzo with no shoes/suspenders crossed in the back and Ms. Piggy with ribbon flat against her hair are the two different toys. The others are the same as the national Happy Meal. Test toys were distributed in Canada as well.

Mu8602

Mu8603 Mu8601 Mu8602 Mu8600

Old McDonald's Farm/Barnyard Happy Meal, 1986

Boxes:

- ❑ ❑ USA Ba8615 **Hm Box - Barn,** 1986. $85.00-100.00
- ❑ ❑ USA Ba8616 **Hm Box - House,** 1986. $85.00-100.00

Ba8615

Ba8616

Premiums: Barnyard Animals (MIP: $15.00-20.00; Loose: $4.00-5.00):

❑ ❑ USA Ba8601 **Cow,** 1986, Wht/Brn MIP. $15.00-20.00
❑ ❑ USA Ba8602 **Husband,** 1986, Wht Shirt/Grn Pants with Brn Hat MIP. $15.00-20.00
❑ ❑ USA Ba8603 **Hen,** 1986, Wht MIP. $15.00-20.00
❑ ❑ USA Ba8604 **Pig,** 1986, Beige MIP. $15.00-20.00
❑ ❑ USA Ba8605 **Rooster,** 1986, Wht MIP. $15.00-20.00
❑ ❑ USA Ba8606 **Sheep,** 1986, Wht MIP. $15.00-20.00
❑ ❑ USA Ba8607 **Wife,** 1986, Blu Dress with Yel Hair MIP. $15.00-20.00

❑ ❑ USA Ba8665 **Translite/Lg,** 1986. $65.00-90.00

Ba8601 Ba8602 Ba8603 Ba8604 Ba8605 Ba8606

Comments: Regional Distribution: USA April/May - 1986 in St. Louis, Missouri, and Tennessee. Mint figurines, made by Playmates Co., came in a clear poly bag. All figurines/Mint in Clear Package which say, "Made in Hong Kong." Translite shows five farm animals. However, the hen may have been substituted for the rooster in some New England promotions.

The cow, pig and sheep must say, "Made in Hong Kong" to be original McDonald's figurines. The hen and the rooster came MIP with and without "Made in Hong Kong" logo. Same figurines were sold in retail stores. **Prices quoted are for Mint in the Package (MIP) only. Loose figurines, the same as sold in toy stores are selling loose for $4.00 - 5.00.**

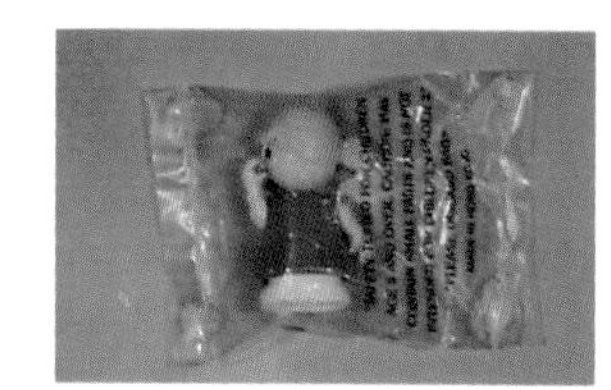

Ba8607

Play-Doh III Happy Meal, 1986

Boxes:

❑ ❑ USA Pl8690 **Hm Box - Circus Animals,** 1986. $4.00-5.00
❑ ❑ USA Pl8691 **Hm Box - Farm Animals,** 1986. $4.00-5.00
❑ ❑ USA Pl8692 **Hm Box - House Pets,** 1986. $4.00-5.00
❑ ❑ USA Pl8693 **Hm Box - Yesterday's Animals,** 1986. $4.00-5.00

Pl8690

Pl8692

Pl8691

Pl8693

PI8676

PI8680

PI8678

PI8682

Premiums: Play-Doh Containers:
❑ ❑ USA PI8675 **Dough: Pink/Hot Pink,** 1984, Plastic Container. $5.00-8.00
❑ ❑ USA PI8676 **Dough: Blue,** 1984, Plastic Container. $5.00-8.00
❑ ❑ USA PI8677 **Dough: Purple,** 1984, Plastic Container. $5.00-8.00
❑ ❑ USA PI8678 **Dough: Red,** 1984, Plastic Container. $5.00-8.00
❑ ❑ USA PI8679 **Dough: Green,** 1984, Plastic Container. $5.00-8.00
❑ ❑ USA PI8680 **Dough: Yellow,** 1984, Plastic Container. $5.00-8.00
❑ ❑ USA PI8681 **Dough: Orange,** 1984, Plastic Container. $5.00-8.00
❑ ❑ USA PI8682 **Dough: White,** 1984, Plastic Container. $5.00-8.00

❑ ❑ USA PI8661 **MC Insert/Cardboard,** 1986. $15.00-20.00
❑ ❑ USA PI8664 **Translite/Sm,** 1986. $10.00-15.00
❑ ❑ USA PI8665 **Translite/Lg,** 1986. $15.00-20.00

Comments: National Distribution: USA - July 7-August 3, 1986. The 2 ounce plastic cans had no McDonald's markings.

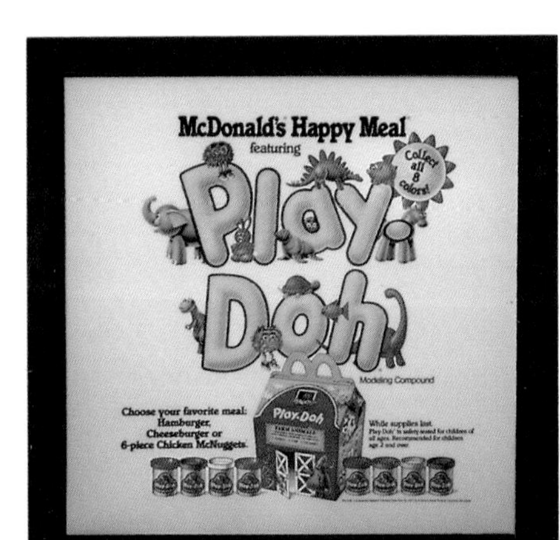

PI8665

Stomper Mini 4 x 4 II Happy Meal, 1986

Boxes:
❑ ❑ USA St8600 **Hm Box - Jalopy Jump,** 1986. $4.00-5.00
❑ ❑ USA St8601 **Hm Box - Quicksand Alley,** 1986. $4.00-5.00
❑ ❑ USA St8602 **Hm Box - Rambunctious Ramp,** 1986. $4.00-5.00
❑ ❑ USA St8603 **Hm Box - Thunderbolt Pass,** 1986. $4.00-5.00

U-3 Premiums:
❑ ❑ USA St8691 **U-3 Chevy Van,** 1986, Yel-Red with Plastic Blk Tires/Blk Rims. $10.00-15.00
❑ ❑ USA St8692 **U-3 Chevy Blazer,** 1986, Yel-Grn with Plastic Blk Tires/Blk Rims. $10.00-15.00
❑ ❑ USA St8693 **U-3 Jeep Renegade,** 1986, Org-Yel with Plastic Blk Tires/Blk Rims. $10.00-15.00
❑ ❑ USA St8694 **U-3 Toyota Tercel SR-5,** 1986, Blu-Yel with Plastic Blk Tires/Blk Rims. $10.00-15.00

St8600 St8603

St8601 St8602

Premiums:

❑ ❑ USA St8675 **Chevy S-10 Pick up,** 1986, Blk/Sil Stripes. $7.00-10.00

❑ ❑ USA St8676 **Chevy S-10 Pick up,** 1986, Yel/Pur Stripes. $7.00-10.00

❑ ❑ USA St8677 **Jeep Renegade 78,** 1986, Maroon/Wht Stripes. $7.00-10.00

❑ ❑ USA St8678 **Jeep Renegade 78,** 1986, Org/Yel Stripes. $7.00-10.00

St8691 St8692 St8693 St8694

❑ ❑ USA St8679 **Chevy Van,** 1986, Red/Yel Stripes. $7.00-10.00

❑ ❑ USA St8680 **Chevy Van,** 1986, Yel/Org Stripes. $7.00-10.00

❑ ❑ USA St8681 **Dodge Rampage Pick up,** 1986, Blu/Maroon Stripes. $7.00-10.00

❑ ❑ USA St8682 **Dodge Rampage Pick up,** 1986, Wht/Blu Stripes. $7.00-10.00

St8677 St8678 St8676 St8675 St8679 St8680

❑ ❑ USA St8683 **Chevy Blazer 4 x 4,** 1986, Yel/Grn Stripes. $7.00-10.00

❑ ❑ USA St8684 **Chevy Blazer 4 x 4,** 1986, Red/Gry Stripes. $7.00-10.00

❑ ❑ USA St8685 **AMC Eagle 74,** 1986, Blk/Gold Stripes. $7.00-10.00

❑ ❑ USA St8686 **AMC Eagle 74,** 1986, Org/Blu Stripes. $7.00-10.00

St8681 St8682 St8683 St8684 St8685 St8686

❑ ❑ USA St8687 **Ford Ranger 23 Pick up,** 1986, Org/Yel Stripes. $7.00-10.00

❑ ❑ USA St8688 **Ford Ranger 23 Pick up,** 1986, Red/Blk Stripes. $7.00-10.00

❑ ❑ USA St8689 **Toyota Tercel SR-5 4 x 4,** 1986, Blu/Yel Stripes. $7.00-10.00

❑ ❑ USA St8690 **Toyota Tercel SR-5 4 x 4,** 1986, Grey/Maroon Stripes. $7.00-10.00

St8687 St8688 St8689 St8690

❑ ❑ USA St8626 **Display with Premiums,** 1986. $150.00-200.00

❑ ❑ USA St8661 **MC Insert/Cardboard,** 1986. $15.00-25.00

❑ ❑ USA St8664 **Translite/Sm,** 1986. $15.00-25.00

❑ ❑ USA St8665 **Translite/Lg,** 1986. $25.00-40.00

❑ ❑ USA St8666 **Translite/Vacuum Formed/Lg,** 1986. $20.00-25.00

St8626

St8664

Tx8610 (front)

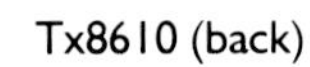
Tx8610 (back)

Comments: National Distribution: USA - August 8-September 7, 1986. St. Louis test market September/October 1985 used four cars. The U-3 premiums came with plastic tires/black hubcaps (rims). The regular premiums came with rubber tires and white hubcaps (rims). **Prices quoted are for Mint in Package (MIP). Loose stomper cars are selling for $2.00-5.00.**

Story of Texas Happy Meal, 1986

Box:
❑ ❑ USA Tx8610 **Hm Box - Alamo/Armadillo.** $——

Premiums: Austin Books:
❑ ❑ USA Tx8601 **Austin/Book - Part 1/The Beginning,** 1986, 46 Pages/KTVV-TV 36 Austin. $——
❑ ❑ USA Tx8602 **Austin/Book - Part 2/Independence,** 1986, 47 Pages/KTVV-TV 36 Austin. $——
❑ ❑ USA Tx8603 **Austin/Book - Part 3/The Frontier,** 1986, 46 Pages/KTVV-TV 36 Austin. $——
❑ ❑ USA Tx8604 **Austin/Book - Part 4/The 20th Century,** 46 Pages/KTVV-TV 36 Austin. $——

Premiums: Houston Books:
❑ ❑ USA Tx8606 **Houston/Book - Part 1/The Beginning,** 1986, 46 Pages/KPRC-TV 2 Houston. $75.00-150.00
❑ ❑ USA Tx8607 **Houston/Book - Part 2/Independence,** 1986, 47 Pages/KPRC-TV 2 Houston. $75.00-150.00
❑ ❑ USA Tx8608 **Houston/Book - Part 3/The Frontier,** 1986, 46 Pages/KPRC-TV 2 Houston. $75.00-150.00
❑ ❑ USA Tx8609 **Houston/Book - Part 4/The 20th Century,** 46 Pages/KPRC-TV 2 Houston. $75.00-150.00

Tx8609 Tx8601

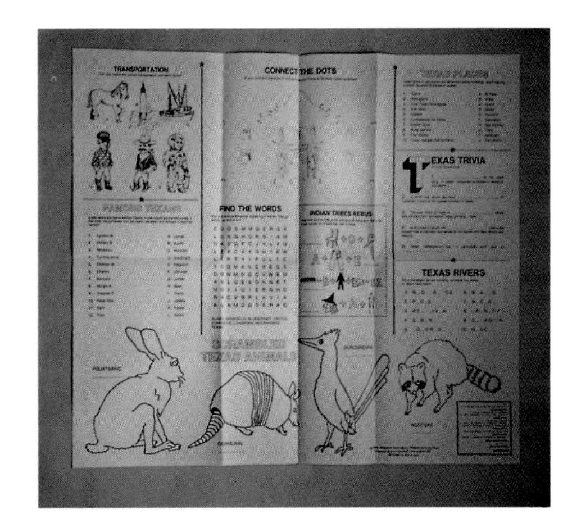

Premiums: Waco Books:

❑ ❑ USA Tx8610 **Waco/Book - Part 1/The Beginning,** 1986, 46 Pages/KXXV 25 TV. $——

❑ ❑ USA Tx8611 **Waco/Book - Part 2/Independence,** 1986, 47 Pages/KXXV 25 TV. $——

❑ ❑ USA Tx8612 **Waco/Book - Part 3/The Frontier,** 1986, 46 Pages/KXXV 25 TV. $——

❑ ❑ USA Tx8613 **Waco/Book - Part 4/The 20th Century,** 46 Pages/KXXV 25 TV. $——

❑ ❑ USA Tx8605 **Map,** 1986, "Discover the Story of Texas" / Size 22" x 24". $250.00-400.00

❑ ❑ USA Tx8655 **Trayliner,** 1986. $——

❑ ❑ USA Tx8665 **Translite/Lg,** 1986. $——

Tx8605

Comments: Regional Distribution - 1986 in Austin, Houston, and Waco, Texas. Books published by Shearer Publishing, Fredericksburg, Texas. Three sets of books were distributed - KTVV-TV-36, KPRC-TV-2, and a Waco, Texas station. Another set of four books was sold by the publisher with part IV(only) having McDonald's information inside the front cover. Not enough transactions to determine price on box and translite. **Auction prices have increased premium prices. Map and books are sold price graded, based on condition and original folding.** Map came folded, unwrapped.

THE STORY OF TEXAS

PART 4: THE 20th CENTURY

"McDonald's of Texas is proud to be part of our state's great heritage. We believe in the power of reading and the influence the ability and desire to read can have on our children's future. McDonald's urges all parents to invest in the future - please give the gift of reading to your kids."

Larry Zimmerman
Regional Vice President
McDonald's Corporation

Tx8605

Tx8665

Ti8615

Tinosaurs Happy Meal, 1986

Box:

❑ ❑ USA Ti8615 **Hm Box - Tinosaurs/8 Premiums,** 1986. $20.00-35.00

Ti8601 Ti8602 Ti8603 Ti8604

Ti8605 Ti8606 Ti8607 Ti8608

Ti8656 (front)

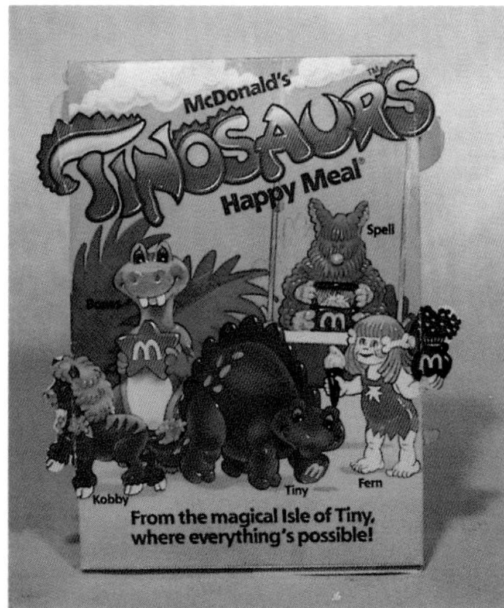

Ti8656 (back)

Premiums: Dinosaur animals (MIP: $12.00-15.00; Loose: $3.00-4.00):

❑ ❑ USA Ti8601 **Bones,** 1986, Grn Dinosaur Holding Orange Star with M. $12.00-15.00

❑ ❑ USA Ti8602 **Dinah,** 1986, Org Dinosaur Holding Pur Heart. $12.00-15.00

❑ ❑ USA Ti8603 **Fern,** 1986, Wht Girl Time Traveler Holding Brn Pot with M. $12.00-15.00

❑ ❑ USA Ti8604 **Jad,** 1986, Pur Dragon with Bead Necklace. $12.00-15.00

❑ ❑ USA Ti8605 **Kobby,** 1986, Pnk Horse with Grn Hair with Wht Feet Kave Kolt. $12.00-15.00

❑ ❑ USA Ti8606 **Link,** 1986, Purp Elf Doing Handstand with Green Outfit. $12.00-15.00

❑ ❑ USA Ti8607 **Spell,** 1986, Blu with Grn Hands and Feet Sitting Gumpies Leader. $12.00-15.00

❑ ❑ USA Ti8608 **Tiny,** 1986, Purp Dinosaur with M on Foot. $12.00-15.00

❑ ❑ USA Ti8656 **Table Tent,** 1986. $8.00-12.00

❑ ❑ USA Ti8660 **Counter Mat,** 1986. $15.00-20.00

❑ ❑ USA Ti8664 **Translite/Sm,** 1986. $25.00-40.00

❑ ❑ USA Ti8665 **Translite/Lg,** 1986. $35.00-50.00

Comments: Regional Distribution: USA - September 12-October 19, 1986 in St. Louis, Missouri. Each figurines has a yellow "M" and came polybagged.

Young Astronauts I '86 Happy Meal, 1986

Boxes:

❑ ❑ USA Yo8610 **Hm Box - Mars Adventure,** 1986. $5.00-8.00

❑ ❑ USA Yo8612 **Hm Box - Repair Station,** 1986. $5.00-8.00

❑ ❑ USA Yo8613 **Hm Box - Space Station,** 1986. $5.00-8.00

❑ ❑ USA Yo8611 **Hm Box - Moonbase/ Blue Letters "Lunar Lookout",** 1986. $8.00-10.00

❑ ❑ USA Yo8614 **Hm Box - Moonbase/ Pink Letters "Lunar Lookout",** 1986. $8.00-10.00

Yo8610

Yo8611

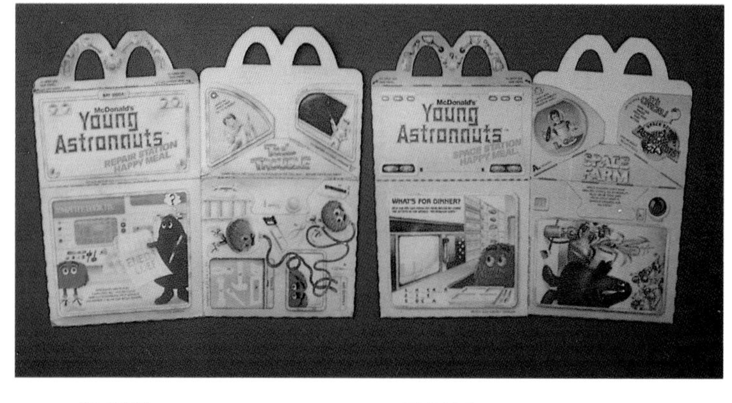

Yo8612 Yo8613

Yo8613

Premiums:

❑ ❑ USA Yo8601 **Space Vehicle: Apollo Command Module,** 1986, grey with sticker sheet. $15.00-20.00

❑ ❑ USA Yo8602 **Space Vehicle: Argo Land Shuttle,** 1986, red with sticker sheet. $15.00-20.00

❑ ❑ USA Yo8603 **Space Vehicle: Cirrus Vtol,** 1986, blue with sticker sheet. $15.00-20.00

❑ ❑ USA Yo8604 **Space Vehicle: Space Shuttle,** 1986, white with sticker sheet. $15.00-20.00

❑ ❑ USA Yo8661 **MC Insert/Cardboard,** 1986. $20.00-35.00

❑ ❑ USA Yo8664 **Translite/Sm,** 1986. $20.00-35.00

❑ ❑ USA Yo8665 **Translite/Lg,** 1986. $25.00-40.00

Comments: Regional Distribution: USA - September 8-October 5, 1986. The "Moonbase" box came in two variations/colors of "Lunar Lookout."

Yo8602 Yo8603 Yo8601 Yo8604

Yo8603 (stickers)

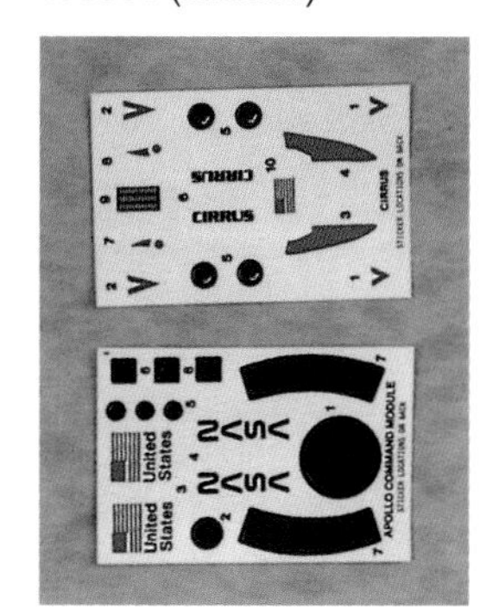

Yo8601 (stickers)

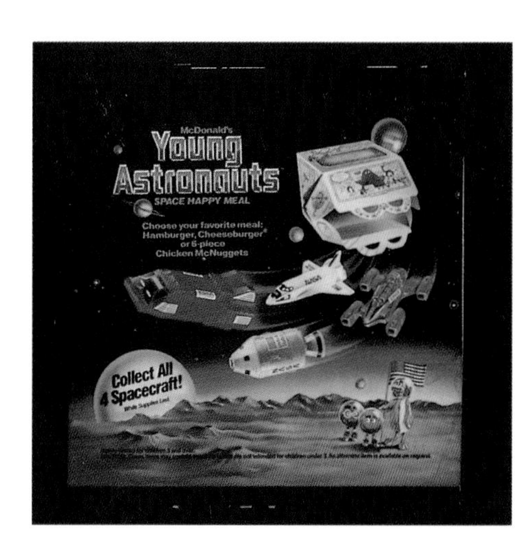

Yo8614

Yo8665

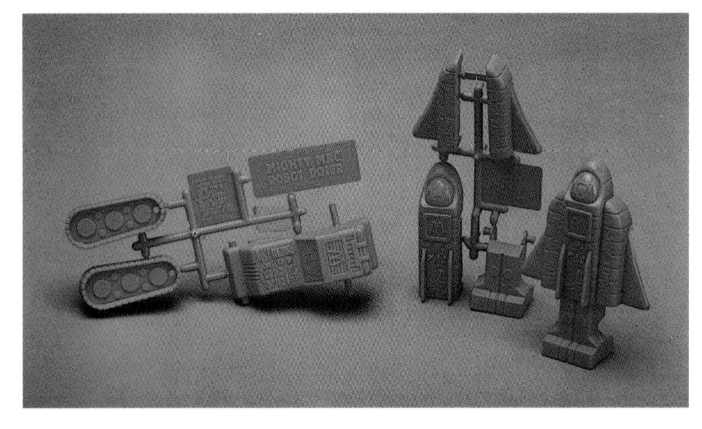

Ge8601 Ge8608

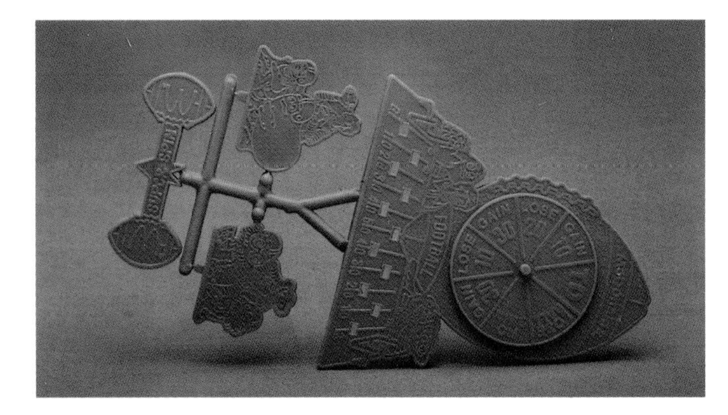

Ge8602

Ge8605

Ge8609 Ge8610 Ge8611 Ge8612

Ge8613 Ge8614

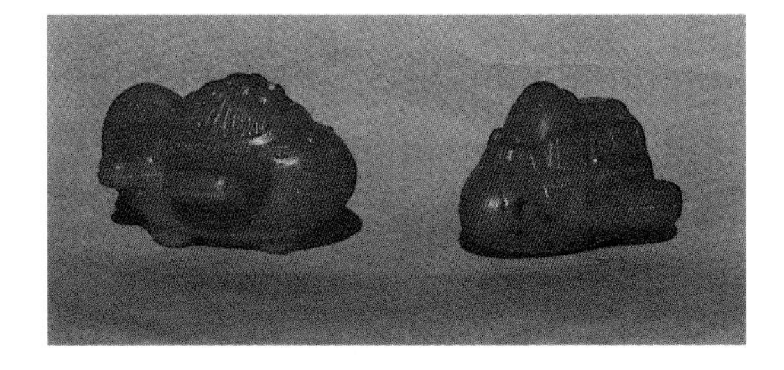

USA Generic Promotions, 1986

❑ ❑ USA Ge8601 **Mighty Mac Robot Dozer,** 1986, Dark Blu/4p. $4.00-7.00

❑ ❑ USA Ge8602 **Football/Super Spin,** 1986, Red or Grn/4p. $4.00-7.00

❑ ❑ USA Ge8603 **Helicopter - Big Mac,** 1986, Blu or Red. $12.00-15.00

❑ ❑ USA Ge8605 **Helicopter - Hello! Copter,** 1986, Blu or Grn or Red. $4.00-6.00

❑ ❑ USA Ge8608 **Grimace Mighty Mac Shuttle,** 1986, Blu/4p. $5.00-7.00

❑ ❑ USA Ge8609 **Finger Puppet - Birdie,** 1986, Yel. $2.50-4.00

❑ ❑ USA Ge8610 **Finger Puppet - Fry Girl,** 1986, Purp. $2.50-4.00

❑ ❑ USA Ge8611 **Finger Puppet - Grimace,** 1986, Pnk. $2.50-4.00

❑ ❑ USA Ge8612 **Finger Puppet - Ronald,** 1986, Grn. $2.50-4.00

❑ ❑ USA Ge8613 **Fry Guy Friendly Flyer,** 1986, Org Floater. $2.00-4.00

❑ ❑ USA Ge8614 **Grimace Smiling Shuttle,** 1986, Org Floater. $2.00-4.00

❑ ❑ USA Ge8615 **Fun Times Magazine: Vol. 8 No. 1 Feb/Mar, 1986.** $2.00-3.00

❑ ❑ USA Ge8616 **Fun Times Magazine: Vol. 8 No. 2 Apr/May, 1986.** $2.00-3.00

Ge8616

☐ ☐ USA Ge8617 **Fun Times Magazine: Vol. 8 No. 3 June/July, 1986.** $2.00-3.00

☐ ☐ USA Ge8618 **Fun Times Magazine: Vol. 8 No. 4 Aug/Sept, 1986.** $2.00-3.00

☐ ☐ USA Ge8619 **Fun Times Magazine: Vol. 8 No. 5 Oct/Nov, 1986.** $2.00-3.00

☐ ☐ USA Ge8620 **Fun Times Magazine: Vol. 8 No. 6 Dec/Jan, 1986.** $2.00-3.00

☐ ☐ USA Ge8621 **Plate: Ronald & friends in space,** 7". $15.00-20.00

☐ ☐ USA Ge8622 **Plate: Ronald & friends with Dinosaurs,** 7". $15.00-20.00

☐ ☐ USA Ge8623 **Plate: Ronald & friends on a train,** 7". $15.00-20.00

Ge8618

Ge8621

Ge8622

Ge8623

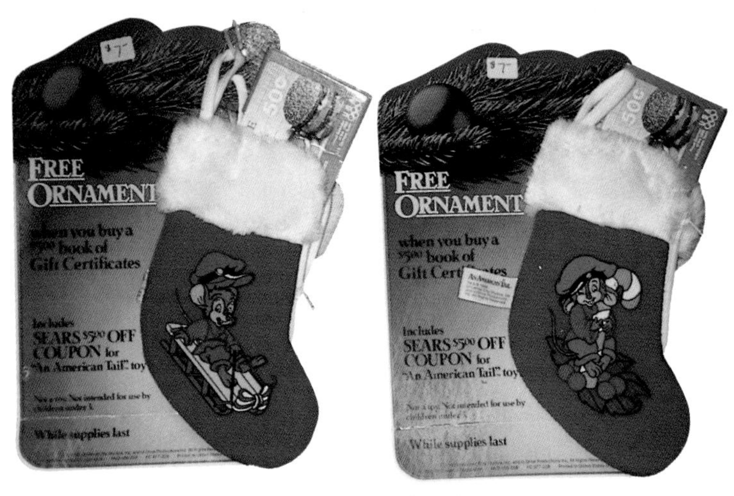

Ge8626 Ge8624

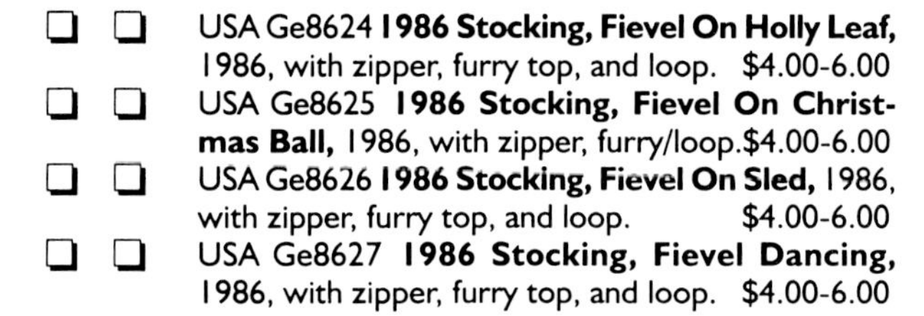

❑ ❑ USA Ge8624 **1986 Stocking, Fievel On Holly Leaf,** 1986, with zipper, furry top, and loop. $4.00-6.00

❑ ❑ USA Ge8625 **1986 Stocking, Fievel On Christmas Ball,** 1986, with zipper, furry/loop. $4.00-6.00

❑ ❑ USA Ge8626 **1986 Stocking, Fievel On Sled,** 1986, with zipper, furry top, and loop. $4.00-6.00

❑ ❑ USA Ge8627 **1986 Stocking, Fievel Dancing,** 1986, with zipper, furry top, and loop. $4.00-6.00

Comments: Regional Distribution: USA - 1986 during Clean-Up weeks. This Is a sampling of generic premiums given away and/or sold at McDonald's restaurants during 1986. The Christmas stockings were given out free with the purchase of $5.00 Gift Certificates in the stores.

Regional Happy Meal promotions are conducted in more selected markets during 1986. The Breakfast Menu includes freshly baked buttermilk biscuits. McDonald's provides a complete food product ingredients listing to the public. Health conscious consumers respond favorably with requests for a more varied menu.

The 9th National O/O Convention was held in Las Vegas, Nevada with the "Back To Our Future" theme.

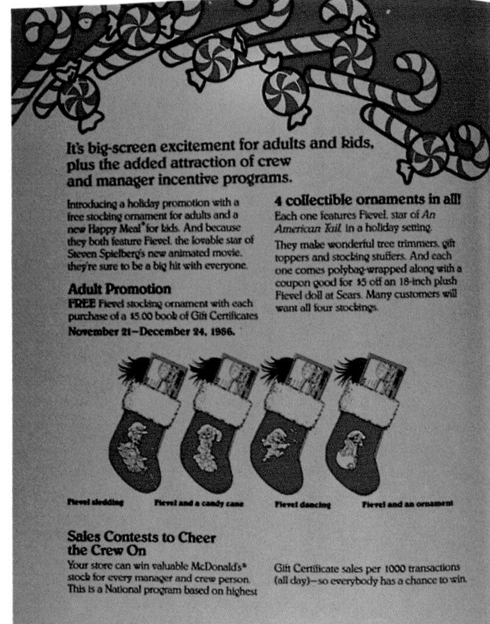

It's big-screen excitement for adults and kids, plus the added attraction of crew and manager incentive programs.

Introducing a holiday promotion with a free stocking ornament for adults and a new Happy Meal® for kids. And because they both feature Fievel, the lovable star of Steven Spielberg's new animated movie, they're sure to be a big hit with everyone.

Adult Promotion

FREE Fievel stocking ornament with each purchase of a $5.00 book of Gift Certificates

November 21–December 24, 1986.

4 collectible ornaments in all!

Each one features Fievel, star of *An American Tail*, in a holiday setting.

They make wonderful tree trimmers, gift toppers and stocking stuffers. And each one comes polybag-wrapped along with a coupon good for $5 off an 18-inch plush Fievel doll at Sears. Many customers will want all four stockings.

Fievel sledding **Fievel and a candy cane** **Fievel dancing** **Fievel and an ornament**

Sales Contests to Cheer the Crew On

Your store can win valuable McDonald's® stock for every manager and crew person. This is a National program based on highest Gift Certificate sales per 1000 transactions (all day)—so everybody has a chance to win.

1987

Berenstain Bears II Happy Meal, 1987
Bigfoot/Without Arches Happy Meal, 1987
Bigfoot/With Arches Happy Meal, 1987
Boats 'N Floats Happy Meal, 1987
Castlemakers/Sand Castle Happy Meal, 1987
Changeables '87 Happy Meal, 1987
Crayola II Happy Meal, 1987
Design-O-Saurs Happy Meal, 1987
Disney Favorites Happy Meal, 1987
Fraggle Rock/Doozers I Test Market Happy Meal, 1987
Good Friends Happy Meal, 1987
Halloween '87 Happy Meal, 1987
Kissyfur Happy Meal, 1987
Little Engineer Happy Meal, 1987
Lunch Box/Characters Promotion, 1987
McDonaldland Band Happy Meal, 1987
McDonaldland TV Lunch Box/Lunch Bunch Happy Meal, 1987
Metrozoo Happy Meal, 1986
Muppet Babies II Happy Meal, 1987
Potato Head Kids I Happy Meal, 1987
Real Ghostbusters I Happy Meal, 1987
Runaway Robots Happy Meal, 1987
Super Summer I Test Market Happy Meal, 1987
Zoo Face I Test Market Happy Meal, 1987
USA Generic Promotions, 1987

- **Tossed salads are added to the national menu**
- **McKids clothing introduced**
- **Ronald's Playplace introduced**

Berenstain Bears II Happy Meal, 1987

Boxes:

❑ ❑ USA Bb8730 **Hm Box - Barn Dance,** 1987. $4.00-5.00
❑ ❑ USA Bb8731 **Hm Box - Bear Country General Store,** 1987. $4.00-5.00
❑ ❑ USA Bb8732 **Hm Box - Bear Country School,** 1987. $4.00-5.00
❑ ❑ USA Bb8733 **Hm Box - Tree House/Clean as a Whistle,** 1987. $4.00-5.00

Bb8731

Bb8732

Bb8730

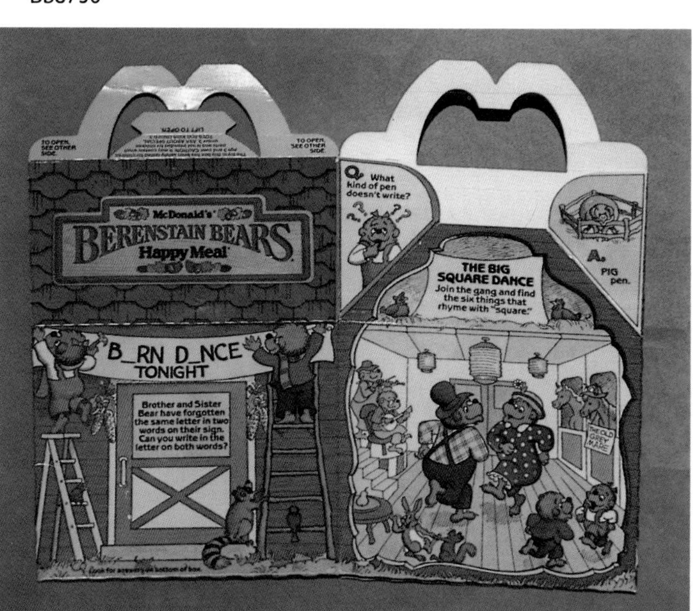

Bb8733

Bb8721 Bb8722 Bb8723 Bb8720 Bb8724 Bb8725

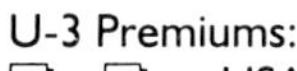

U-3 Premiums:

❑ ❑ USA Bb8724 **U-3 Mama,** 1987, with Paper Punch Outs/No Flocking. $8.00-10.00

❑ ❑ USA Bb8725 **U-3 Papa,** 1987, with Paper Punch Outs/No Flocking. $8.00-10.00

Premiums:

❑ ❑ USA Bb8720 **Set 1 Sister,** 1987, with Red Wagon/Flocked. $3.00-4.00

❑ ❑ USA Bb8721 **Set 2 Papa,** 1987, with Brn Wheelbarrow/ Flocked. $3.00-4.00

❑ ❑ USA Bb8722 **Set 3 Brother,** 1987, with Grn/Yel Scooter/ Flocked. $3.00-4.00

❑ ❑ USA Bb8723 **Set 4 Mama,** 1987, with Dress and Yel Shop Cart/Flocked. $3.00-4.00

❑ ❑ USA Bb8726 **Display/Prem,** 1987. $125.00-175.00

❑ ❑ USA Bb8750 **Button/Crew,** 1987, Berenstain Bears Happy Meals. $5.00-7.00

❑ ❑ USA Bb8764 **Translite/Sm,** 1987. $12.00-15.00

❑ ❑ USA Bb8765 **Translite/Lg,** 1987. $15.00-20.00

❑ ❑ USA Bb8795 **Pin,** 1987, Square/Grn McDonald's Presents Berenstain Bears. $3.00-5.00

Comments: National Distribution: USA - October 30-November 29, 1987. Premium markings - "S&J Berenstain China" or "McDonalds."

Bb8750

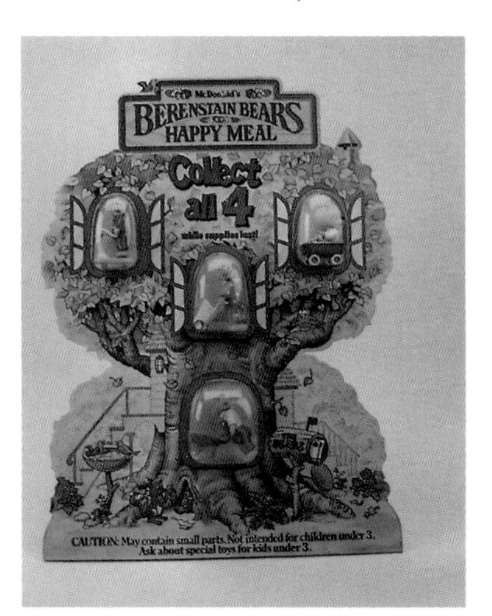

Bb8726

Bb8765

Bi8725

Bigfoot/Without Arches Happy Meal, 1987

Box:

❑ ❑ USA Bi8725 **Hm Box - Bigfoot Game with Crossword Puzzle,** 1987. $8.00-12.00

Premiums: Bigfoot cars WITHOUT Arches:

❑ ❑ USA Bi8701 **Ford Bronco,** 1987, Grn 1 1/2" Whls. $6.00-8.00

❑ ❑ USA Bi8702 **Ford Pickup,** 1987, Purp 1 1/2" Whls. $6.00-8.00

❑ ❑ USA Bi8703 **Ms Ford Pickup,** 1987, Turq 1 1/2" Whls. $6.00-8.00

❑ ❑ USA Bi8704 **Shuttle Ford,** 1987, Red with Wht 1 1/2" Whls. $6.00 -8.00

❑ ❑ USA Bi8705 **Ford Bronco,** 1987, Org with 1" Whls. $6.00 -8.00

❑ ❑ USA Bi8706 **Ford Pickup,** 1987, Light Blue 1" Whls. $6.00-8.00

Bi8704 Bi8703 Bi8702 Bi8701

Bi8708 Bi8707 Bi8706 Bi8705

❑ ❑ USA Bi8707 **Ms Ford Pickup,** 1987, Pnk 1" Whls. $6.00-8.00

❑ ❑ USA Bi8708 **Shuttle Ford,** 1987, Blk-Sil 1" Whls. $6.00-8.00

Comments: Regional Distribution: USA - 1987 in St. Louis, Missouri. Each Bigfoot car came polybagged with no McDonald's logo markings. Note: in other regions, See Bigfoot/With Arches Happy Meal, 1987.

Bigfoot/With Arches Happy Meal, 1987

Premiums: Bigfoot cars WITH Arches:

❑ ❑ USA Bi8709 **Ford Bronco,** 1987, Grn 1 1/2" Whls. $8.00-10.00

❑ ❑ USA Bi8710 **Ford Pickup,** 1987, Purp 1 1/2" Whls. $8.00-10.00

❑ ❑ USA Bi8711 **Ms Ford Pickup,** 1987, Turq 1 1/2" Whls. $8.00-10.00

❑ ❑ USA Bi8712 **Shuttle Ford,** 1987, Red with Wht 1 1/2" Whls. $8.00-10.00

❑ ❑ USA Bi8713 **Ford Bronco,** 1987, Org with 1" Whls. $8.00-10.00

❑ ❑ USA Bi8714 **Ford Pickup,** 1987, Light Blu 1" Whls. $8.00 -10.00

❑ ❑ USA Bi8715 **Ms Ford Pickup,** 1987, Pnk with 1" Whls. $8.00-10.00

❑ ❑ USA Bi8716 **Shuttle Ford,** 1987, Blk-Sil 1" Whls. $8.00-10.00

❑ ❑ USA Bi8763 **Window Decal,** 1987. $10.00-15.00

❑ ❑ USA Bi8764 **Translite/Sm,** 1987. $15.00-25.00

❑ ❑ USA Bi8765 **Translite/Lg,** 1987. $20.00-35.00

Comments: Regional Distribution: USA - 1987 in parts of Florida and Buffalo, New York regions. Each Bigfoot car had McDonald's logo markings. For cars without logo/arches, see Bigfoot/Without Arches Happy Meal, 1987. **Price range quoted is for Mint in Package (MIP). Loose Bigfoot cars are selling for $2.00 - 3.00.**

Boats 'N Floats Happy Meal, 1987

Premiums: Vacuum Form Containers:

❑ ❑ USA Bo8700 **Hm Container - Grimace Ski Boat,** 1986, Pur. $6.00-8.00

❑ ❑ USA Bo8701 **Hm Container - Fry Kids on Raft,** 1986, Grn. $6.00-8.00

❑ ❑ USA Bo8702 **Hm Container - McNugget Buddies Life Boat,** 1986, Org. $6.00-8.00

❑ ❑ USA Bo8703 **Hm Container - Birdie on Raft,** 1986, Yel. $6.00-8.00

❑ ❑ USA Bo8704 **Sticker Sheet - Ski Boat,** 1986.$3.00-5.00

❑ ❑ USA Bo8705 **Sticker Sheet - Fry Kids on Raft,** 1986. $3.00-5.00

Bi8711 Bi8715 Bi8710 Bi8714

Bi8712 Bi8716 Bi8709 Bi8713

Bi8765

Bo8701 Bo8700 Bo8702 Bo8703

Bo8765

❑ ❑ USA Bo8706 **Sticker Sheet - Life Boat,** 1986. $3.00-5.00
❑ ❑ USA Bo8707 **Sticker Sheet - Birdie on Raft,** 1986. $3.00-5.00

❑ ❑ USA Bo8764 **Translite/Sm,** 1986. $12.00-15.00
❑ ❑ USA Bo8765 **Translite/Lg,** 1986. $15.00-25.00

Comments: National Distribution: USA - August 7-September 3, 1987. Vacuum formed boats serve as the food container and premium.

Castlemakers/Sand Castle Happy Meal, 1987

Premiums: Vacuum Form Containers:
❑ ❑ USA Ca8700 **Sand Mold - Cylindrical,** 1987, Yel/8". $20.00-25.00
❑ ❑ USA Ca8701 **Sand Mold - Domed,** 1987, Blu/8". $20.00-25.00
❑ ❑ USA Ca8702 **Sand Mold - Rectangle,** 1987, Red/9". $20.00-25.00
❑ ❑ USA Ca8703 **Sand Mold - Square,** 1987, Dk Blu/5 1/2". $20.00-25.00

❑ ❑ USA Ca8764 **Translite/Sm,** 1987. $25.00-40.00
❑ ❑ USA Ca8765 **Translite/Lg,** 1987. $35.00-50.00

Comments: Regional Distribution: USA - 1987 in Michigan, Illinois, and Houston, Texas. McDonald's logo "M" was molded into each top section.

Ca8700 Ca8703 Ca8701 Ca8702

Ch8710

Changeables '87 Happy Meal, 1987

Boxes:
❑ ❑ USA Ch8710 **Hm Box - 5 Changeables without Milk Shake Picture,** 1987. $8.00-10.00
❑ ❑ USA Ch8711 **Hm Box - 6 Changeables with Milk Shake,** 1987. $4.00-5.00

Premiums: Changeables WITHOUT Painted Hands:

- ❑ ❑ USA Ch8701 **Big Mac Sandwich,** 1987, Robotic Form/ Blu-Yel. $4.00-5.00
- ❑ ❑ USA Ch8702 **Chicken McNuggets,** 1987, Robotic Form/ Blu-Red. $4.00-5.00
- ❑ ❑ USA Ch8703 **Egg McMuffin,** 1987, Robotic Form/ Pnk-Red-Blu. $4.00-5.00
- ❑ ❑ USA Ch8704 **Large French Fries,** 1987, Robotic Form/ Blu-Yel. $4.00-5.00
- ❑ ❑ USA Ch8705 **Quarter Pounder,** 1987, Robotic Form/ Yel-Blu. $4.00-5.00
- ❑ ❑ USA Ch8706 **Milk Shake,** 1987, Robotic Form Opens from Top/Red-Blu. $5.00-8.00

- ❑ ❑ USA Ch8755 **Trayliner/5 Premiums.** $2.00-3.00
- ❑ ❑ USA Ch8764 **Translite/5 Premiums/Lg,** 1987. $15.00-20.00

Ch8711 (front)

Ch8701 Ch8702 Ch9703 Ch8704 Ch8705 Ch8706

Ch8711 (back)

Ch8755

Ch8765

- ❑ ❑ USA Ch8765 **Translite/5 Premiums/Sm,** 1987. $10.00-15.00
- ❑ ❑ USA Ch8766 **Translite/6 Premiums/Lg,** 1987. $15.00-25.00
- ❑ ❑ USA Ch8767 **Translite/6 Premiums/Sm,** 1987. $10.00-15.00

Comments: Regional Distribution: USA - 1987. The 1987 Changeables do not have painted hands. See 1989 New Food Changeables for painted characteristics.

Crayola II Happy Meal, 1987

Boxes:

- ❑ ❑ USA Cr8735 **Hm Box - Whack on the Track/Secret Destinations/Org Band,** 1987. $4.00-5.00
- ❑ ❑ USA Cr8736 **Hm Box - Road Maze/Early Bird Crossword/Yel Band,** 1987. $4.00-5.00
- ❑ ❑ USA Cr8737 **Hm Box - Crazy Face/Space Creatures/Purp Band,** 1987. $4.00-5.00
- ❑ ❑ USA Cr8738 **Hm Box - Absent Apples/Red Band,** 1987. $4.00-5.00

Ch8767

Cr8736

Cr8735

Cr8737

U-3 Premium:

❑ ❑ USA Cr8729 **U-3 Ronald on Fire Engine,** 1986, Red Cardboard with **4 Crayons.** $8.00-10.00

Premiums:

❑ ❑ USA Cr8725 **Set 1 Hamburglar in Steam Engine,** 1986, Org Stencil/7 Figs with **4 Reg Crayons.** $3.00-5.00

❑ ❑ USA Cr8726 **Set 2 Earlybird in Car,** 1986, Yel Stencil with 9 Figs with **Thick Grn or Org Marker.** $3.00-5.00

❑ ❑ USA Cr8727 **Set 3 Grimace in Rocket,** 1986, Pur Stencil with 10 Figs with **4 Fluorescent Crayons.** $3.00-5.00

❑ ❑ USA Cr8728 **Set 4 Ronald on Tractor,** 1986, Red Stencil with 7 Figs with **Thin Blu or Red Magic Marker.** $3.00-5.00

❑ ❑ USA Cr8727 **Hanging Banner,** 1987, Crayola. $20.00-25.00

❑ ❑ USA Cr8750 **Button,** 1986, Collect Crayola/Rectangle. $5.00-8.00

Cr8738

Cr8729

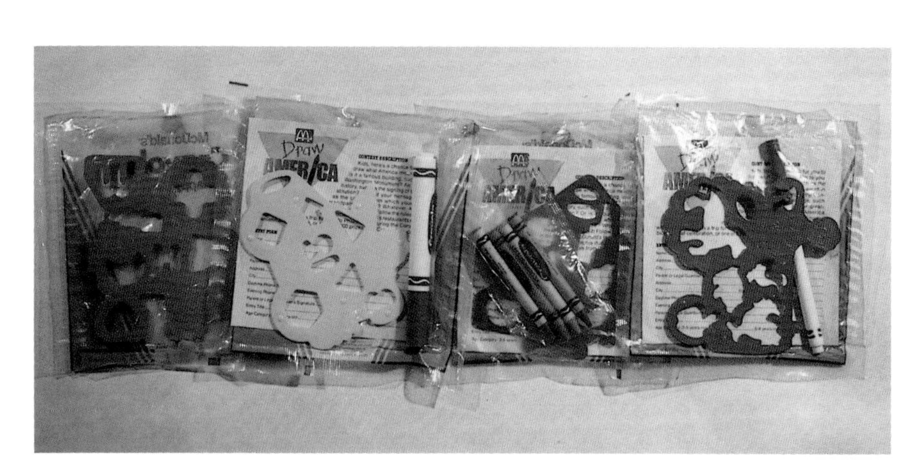

Cr8725 Cr8726 Cr8727 Cr8728

Cr8727

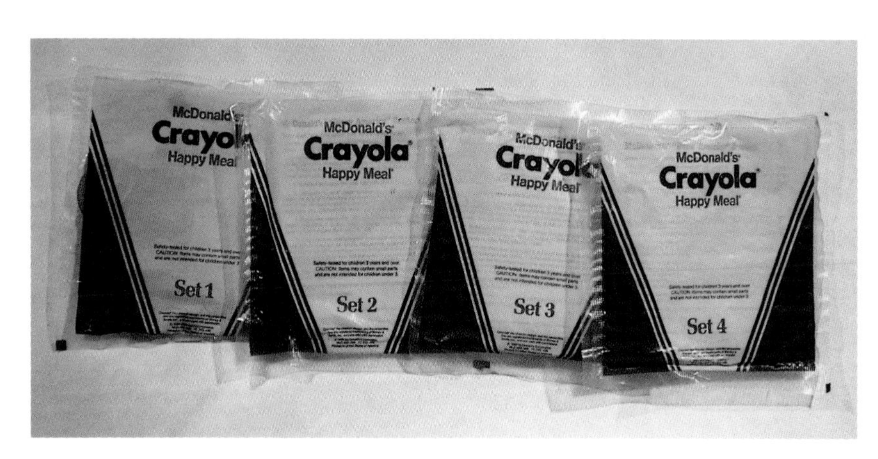

Cr8750

Cr8760

❑ ❑ USA Cr8760 **Counter Mat,** 1986. $7.00-10.00
❑ ❑ USA Cr8761 **MC Insert/Cardboard,** 1986. $15.00-25.00
❑ ❑ USA Cr8763 **Menu Board Lug-On with 1 Premium,** 1986. $20.00-35.00
❑ ❑ USA Cr8764 **Translite/Sm,** 1986. $10.00-20.00
❑ ❑ USA Cr8765 **Translite/Lg,** 1986. $20.00-25.00

Comments: National Distribution: USA - March 20-April 26, 1987. Evansville, Indiana conducted test market in the summer of 1985.

Cr8765

Design-O-Saurs Happy Meal, 1987

Premiums:

❑ ❑ USA De8710 **Set 1 Ronald on Tyronaldsaurus Rex Tyrannosaurus**, 5p Red, 1987. $5.00-8.00
❑ ❑ USA De8711 **Set 2 Grimace on Grimacesaur Pterodactyl**, 5p Purp, 1987. $5.00-8.00
❑ ❑ USA De8712 **Set 3 Fry Guy on Brontofry Guy Brontosaurus**, 5p Grn, 1987. $5.00-8.00
❑ ❑ USA De8713 **Set 4 Hamburglar on Tricerahamburglar Triceratops**, 5p Org, 1987. $5.00-8.00

Comments: Limited Regional Distribution: USA - July 2-August 6, 1987. McDonald's premium order form identified the promotion as a sustaining "Happy Meal"/Mix and Match. Premium markings - "Design-O-Saurs."

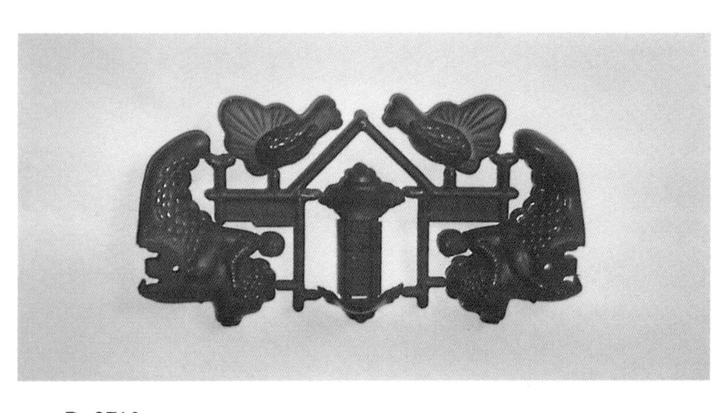

De8710

De8712 De8711 De8713

Design-O-Saurs blue book.

Disney Favorites Happy Meal, 1987

Boxes:
- ❑ ❑ USA Di8710 **Hm Box - Cinderella with Godmother,** 1987. $4.00-5.00
- ❑ ❑ USA Di8711 **Hm Box - Cinderella & Prince Dancing,** 1987. $4.00-5.00

Premiums: Books:
- ❑ ❑ USA Di8700 **Book: Cinderella,** 1987, Book/Paint with Water Coupons. $3.00-5.00
- ❑ ❑ USA Di8701 **Book: Lady and the Tramp,** 1987, Book/Sticker. $3.00-5.00
- ❑ ❑ USA Di8702 **Book: Dumbo,** 1987, Book/Press-Out Book. $3.00-5.00

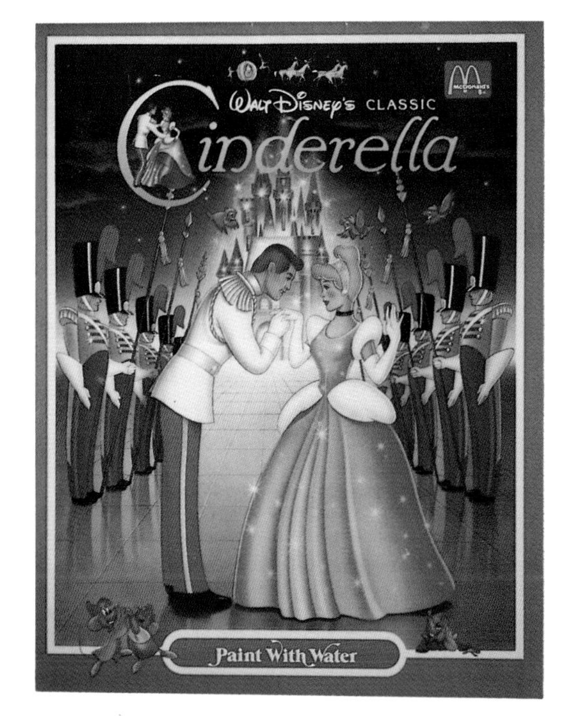

Di8700

Di8710

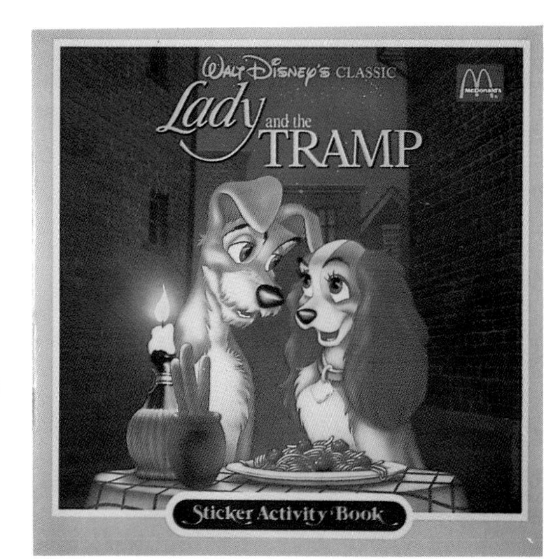

Di8701

Di8711

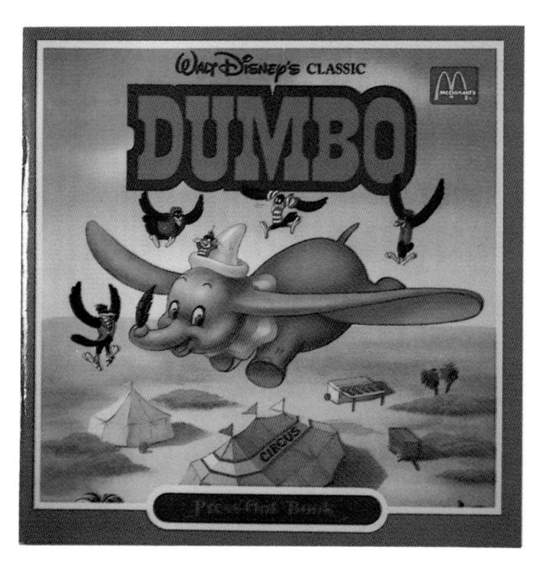

Di8702

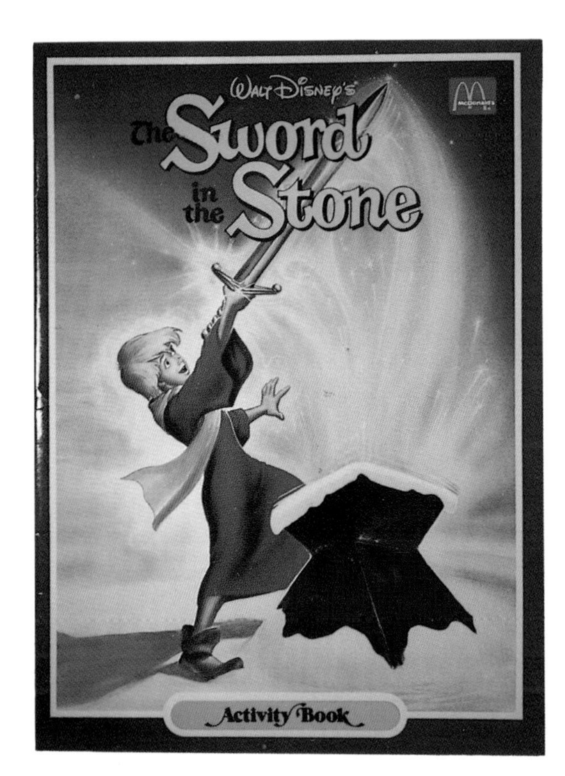

Di8703

☐ ☐ USA Di8703 **Book: The Sword in the Stone,** 1987, Book/Activity. $3.00-5.00

☐ ☐ USA Di8763 **Menu Board Prem Lug-On,** 1987. $15.00-20.00
☐ ☐ USA Di8764 **Translite/Sm,** 1987. $10.00-15.00
☐ ☐ USA Di8765 **Translite/Lg,** 1987. $15.00-25.00

Comments: National Distribution: USA - November 30-December 24, 1987.

DI8765

Disney Favorites blue book.

Fraggle Rock/Doozer I Test Market Happy Meal, 1987

Premiums:

☐ ☐ USA Fr8700 **Set I Gobo in Carrot**, Wheels Offset/Wobbles/Blu Pkg, 1987. $35.00-50.00
☐ ☐ USA Fr8701 **Set 2 Bulldozer in Bulldozer**, Friends/Red Pkg, 1987. $50.00-75.00
☐ ☐ USA Fr8702 **Set 3 Red Fraggle in Radish**, Wheels Offset/Wobbles/Grn Pkg, 1987. $35.00-50.00

☐ ☐ USA Fr8703 **Set 4 Cotterpin Dozer in Fork Lift**, Friends/Pur Pkg, 1987. $50.00-75.00

Comments: Limited Regional Distribution: USA - April 24-May 28, 1987 in West Virginia. Premium markings - "Henson Associates Inc China." **Loose Sets 1 and 3 are selling for $15.00 - 25.00 each. Loose Sets 2 and 4 are selling for 25.00 - 40.00 each.**

Fr8700

Fr8701

Fr8700 Fr8702

Fr8703

Go8700

Good Friends Happy Meal, 1987

Boxes:
- ❑ ❑ USA Go8700 **Hm Box - A Clean Sweep/Hamburglar with Pen Pals,** 1987. $3.00-5.00
- ❑ ❑ USA Go8701 **Hm Box - Snapshot Shuffle/W Merry Go Ronald,** 1987. $3.00-5.00
- ❑ ❑ USA Go8764 **Translite/Sm,** 1987. $4.00-5.00
- ❑ ❑ USA Go8765 **Translite/Lg,** 1987. $10.00-15.00

Comments: Limited Regional Distribution: USA -1987. Backup Happy Meal - generic premiums given in 1987.

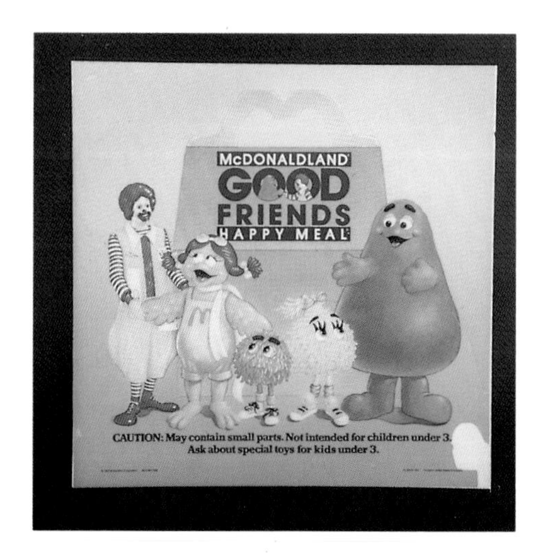

Go8765

Go8701

Halloween '87 Happy Meal, 1987

Premiums: Halloween Pails with Lids
- ❑ ❑ USA Ha8701 **McBoo,** 1986, Org Pail/Org Lid with Grid. $1.00-2.00
- ❑ ❑ USA Ha8702 **McGoblin,** 1986, Org Pail/Org Lid with Grid. $1.00-2.00
- ❑ ❑ USA Ha8703 **McPunk'n,** 1986, Org Pail/Org Lid with Grid. $1.00-2.00

- ❑ ❑ USA Ha8764 **Translite/Sm,** 1987. $4.00-5.00
- ❑ ❑ USA Ha8765 **Translite/Lg,** 1987. $10.00-15.00

Comments: National Distribution: USA - October 16-31, 1987. USA Ha8615-17 were given out with redesigned lids. The 1987 lids had grids on the six 1/2" holes to prevent finger entrapment.

Ha8701 Ha8702 Ha8703

Ki8715

Ki8708 Ki8707 Ki8702 Ki8701

Ki8704 Ki8706 Ki8703 Ki8705

Kissyfur Happy Meal, 1987

Box:
❑ ❑ USA Ki8715 **Hm Box - Gus Juggling Apples,** 1987. $25.00-40.00

Premiums: Flocked:
❑ ❑ USA Ki8701 **Beehonie Rabbit** - Flocked. $25.00-40.00
❑ ❑ USA Ki8702 **Duane Pig** - Pink/Flocked. $25.00-40.00
❑ ❑ USA Ki8707 **Lennie Wart Hog** - Brown/Flocked. $25.00-40.00
❑ ❑ USA Ki8708 **Toot Beaver** - Grey/Flocked. $25.00-40.00

Premiums: Non-flocked:
❑ ❑ USA Ki8703 **Floyd Male Alligator** - Green/Not Flocked/Smooth. $10.00-15.00
❑ ❑ USA Ki8704 **Gus Father Bear** - Big Bear/Brown/Smooth with Arms Extended up. $10.00-15.00
❑ ❑ USA Ki8705 **Jolene Female Alligator** - Not Flocked/Smooth. $10.00-15.00
❑ ❑ USA Ki8706 **Kissyfur Baby Bear** - Little Bear/Brown/Smooth/Arms Folded. $10.00-15.00

❑ ❑ USA Ki8755 **Trayliner.** $4.00-5.00
❑ ❑ USA Ki8760 **Counter Mat,** 1987. $15.00-20.00
❑ ❑ USA Ki8764 **Translite/Sm,** 1987. $20.00-25.00
❑ ❑ USA Ki8765 **Translite/Lg,** 1987. $25.00-40.00

Comments: Regional Distribution: USA - April - May 1987. Premium markings - "1985 Phil Mendez."

Ki8755

Little Engineer Happy Meal, 1987

Boxes:
❑ ❑ USA Li8710 **Hm Box - Round House Train Garage/Repair Pairs,** 1986. $4.00-5.00
❑ ❑ USA Li8711 **Hm Box - Station/Waiting Room,** 1986. $4.00-5.00
❑ ❑ USA Li8712 **Hm Box - Trestle/A Place for Everything,** 1986. $4.00-5.00
❑ ❑ USA Li8713 **Hm Box - Tunnel/Tunnel Project,** 1986. $4.00-5.00

Li8710 Li8711

Li8712 Li8713

Li8707 Li8708 Li8705 Li8706

Li8701

Li8702 Li8703 Li8704

U-3 Premiums:

❑ ❑ USA Li8705 **U-3 Fry Guy Happy Car,** 1985, Team McDonald's Grn Floater. $5.00-7.00

❑ ❑ USA Li8706 **U-3 Fry Guy Happy Car,** 1985, Team McDonald's Yel Floater. $5.00-7.00

❑ ❑ USA Li8707 **U-3 Grimace Happy Taxi Company,** 1985, Grn Floater. $4.00-5.00

❑ ❑ USA Li8708 **U-3 Grimace Happy Taxi Company,** 1985, Yel Floater. $4.00-5.00

Premiums: Train Engines (MIP: $15.00-40.00 with stickers; Loose: $3.00-5.00):

❑ ❑ USA Li8700 **Birdie's Sunshine Special Train Engine**, 1986, Yel 3p. $15.00-20.00

❑ ❑ USA Li8701 **Fry Girl Express Train Engine**, 1986, Blu 3p with Stickers. $25.00-40.00

❑ ❑ USA Li8702 **Fry Guy Flyer Train Engine**, 1986, Day Glo Orange 3p with Sticker. $15.00-20.00

❑ ❑ USA Li8703 **Grimace Purple Streak Train Engine**, 1986, Purple 3p with Stickers. $15.00-20.00

❑ ❑ USA Li8704 **Ronald's Railway Train Engine**, 1986, Red 3p with Stickers. $15.00-20.00

❑ ❑ USA Li8763 **Translite/Drive-Thru Insert,** 1986. $10.00-15.00

❑ ❑ USA Li8764 **Translite/Sm,** 1986. $15.00-20.00

❑ ❑ USA Li8765 **Translite/Lg,** 1986. $20.00-25.00

Comments: Regional Distribution: USA - February 9-March 15, 1987. Auction prices realized have increased premium prices. **Loose train engines with good stickers are selling for $7.00 - 10.00. Loose train engines with no stickers are selling for $5.00 - 7.00.**

Li8765

Lunch Box/Characters Promotion, 1987

Premiums: Lunch Boxes:

❑ ❑ USA Lu8700 **Lunch Box - Grimace Batting with Ron/Friends,** 1987, Lunch Box/Purp/Wht. $8.00-10.00

❑ ❑ USA Lu8701 **Lunch Box - Ronald Playing Football,** 1987, Lunch Box/Blu/Wht. $8.00-10.00

❑ ❑ USA Lu8702 **Lunch Box - Ronald and Friends Rainbow,** 1987, Lunch Box/Yel/Wht. $8.00-10.00

❑ ❑ USA Lu8703 **Lunch Box - Ronald Flying Bubble Spaceship,** 1987, Lunch Box/Red/Wht. $8.00-10.00

Comments: Regional Distribution: USA - 1987 in southern states.

Lu8700 Lu8701 Lu8702 Lu8703

McDonaldland Band Happy Meal, 1987

Boxes:

❑ ❑ USA Mc8710 **Hm Box - Band Concert/One Man Band,** 1986. $4.00-5.00

❑ ❑ USA Mc8711 **Hm Box - Can You Find/Instrument Rhyme Time,** 1986. $4.00-5.00

❑ ❑ USA Mc8712 **Hm Box - Jam Session/Scavenger Hunt,** 1986. $4.00-5.00

❑ ❑ USA Mc8713 **Hm Box - Professor and Hamburglar/Band Leader,** 1986. $4.00-5.00

Premiums: Musical Instruments:

❑ ❑ USA Mc8700 **Trumpet,** 1986, Fry Kid/Green. $2.50-4.00

❑ ❑ USA Mc8701 **Kazoo,** 1986, Birdie/Pink. $1.00-1.50

❑ ❑ USA Mc8702 **Siren,** 1986, Hamb/Org. $1.00-1.50

❑ ❑ USA Mc8703 **Train Engine Whistle,** 1986, Ronald/Purp. $1.00-1.50

❑ ❑ USA Mc8704 **Pan Pipes,** 1986, Ronald/Yel. $1.50-2.00

❑ ❑ USA Mc8705 **Harmonica,** 1986, Ronald/Red. $2.50-4.00

❑ ❑ USA Mc8706 **Boat Whistle,** 1986, Fry Kid/Blu. $1.00-1.50

❑ ❑ USA Mc8707 **Saxophone,** 1986, Grimace/Purp. $1.00-1.50

Mc8710

Mc8711

Mc8712

Mc8713

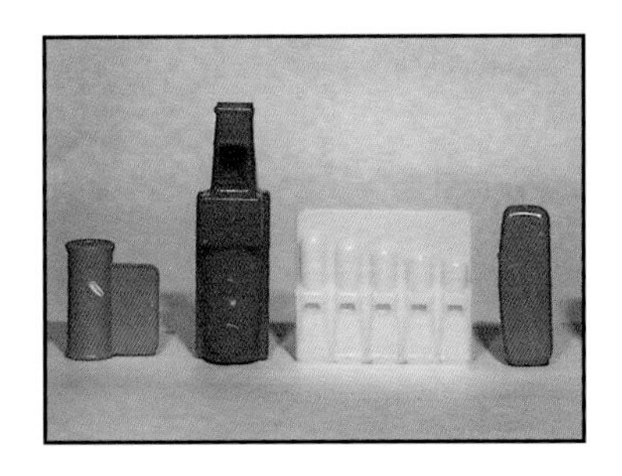

Mc8702 Mc8703 Mc8704 Mc8705

Mc8764

❑ ❑ USA Mc8761 **MC Insert/Cardboard,** 1986. $15.00-25.00
❑ ❑ USA Mc8764 **Translite/Sm,** 1986. $15.00-20.00
❑ ❑ USA Mc8765 **Translite/Lg,** 1986. $20.00-25.00

Comments: Regional Distribution: USA - April 27-June 4, 1987. The pan pipes and purple Ronald train whistle were also used as U-3 premiums. None were polybagged. During the 1990s, the premiums were sold in retail stores on blister packs and in Halloween treat bags by retail stores, thus decreasing prices. Box flats exist for each of the Happy Meal Boxes, note pictures of USA Mc8710 and USA Mc8711.

McDonaldland TV Lunch Box/Lunch Bunch Happy Meal, 1987

Premiums: Lunch Boxes:
❑ ❑ USA Lb8710 **Lunch Box - Blue with TV on Back**, 1987, Stickers. $7.00-10.00
❑ ❑ USA Lb8711 **Lunch Box - Green with TV on Back**, 1987, Stickers. $7.00-10.00
❑ ❑ USA Lb8712 **Lunch Box - Red with TV on Back**, 1987, Stickers. $7.00-10.00
❑ ❑ USA Lb8713 **Lunch Box - Yellow with TV on Back**, 1987, Stickers. $7.00-10.00

❑ ❑ USA Lb8716 **Sticker Sheet - Blue Box,** 1986. $4.00-5.00
❑ ❑ USA Lb8717 **Sticker Sheet - Green Box,** 1986. $4.00-5.00
❑ ❑ USA Lb8718 **Sticker Sheet - Red Box,** 1986. $4.00-5.00
❑ ❑ USA Lb8719 **Sticker Sheet - Yellow Box,** 1986. $4.00-5.00

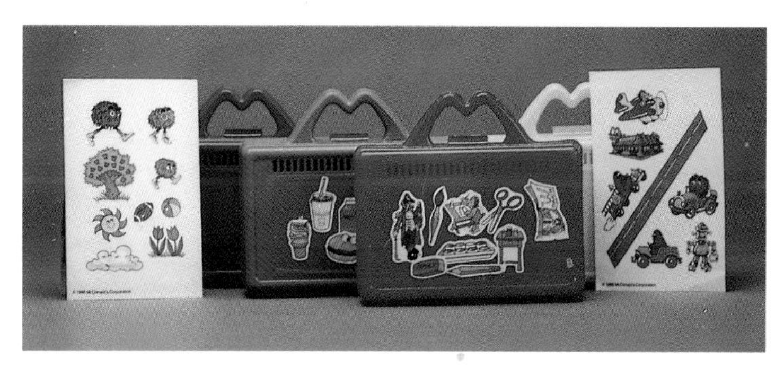

Lb8716 Lb8711 Lb8712 Lb8719

❑ ❑ USA Lb8765 **Translite/Lg**, 1987. $35.00-50.00

Comments: Limited Regional Distribution: USA - 1987 in New England states. Sticker sheets carried a 1986 date with "Safety Tips for Back to School" on the back.

Lb8765

Metrozoo Happy Meal, 1987

Box:
❑ ❑ USA Me8715 **Hm Box - Elephant/Tiger "One of the World's Great Zoos"**, 1986. $——

Me8715

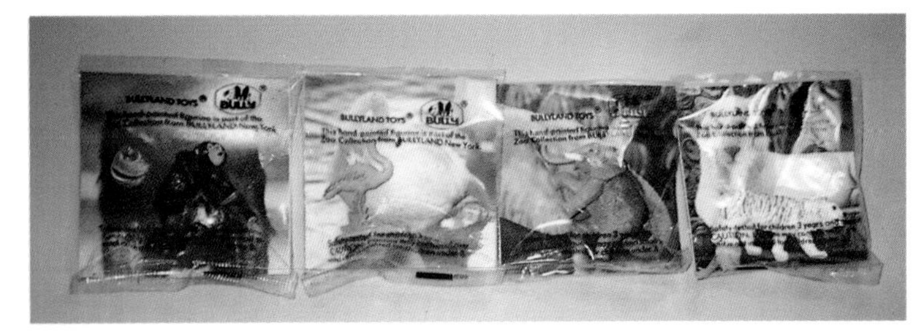

Me8701 Me8703 Me8702 Me8704

Premiums:

- ❑ ❑ USA Me8701 **Chimp,** Blk with Tan Face with Left Leg Behind, 1986. $200.00-250.00
- ❑ ❑ USA Me8702 **Elephant,** Grey with Wht Tusks, 1986. $200.00-250.00
- ❑ ❑ USA Me8703 **Flamingo,** Pink with Yel Bill, 1986. $200.00-250.00
- ❑ ❑ USA Me8704 **Tiger,** Wht with Rust Stripes, 1986. $200.00-250.00

- ❑ ❑ USA Me8765 **Translite/Lg,** 1986. $——

Comments: Limited Regional Distribution: USA - February/March 1987 in the South Florida area. Premium markings - "Bully ...West Germany." MIP includes figurine and local map with Miami Zoo location. Auction prices indicate price range of $200.00-250.00 per premium. Availability may increase or decrease price in the long run. Loose, out of package, the Metro Zoo figures sell for $20.00-25.00 each.

Me8701

Me8704

Me8702

Me8703

Me8765

Muppet Babies II Happy Meal, 1987

Boxes:

- ❑ ❑ USA Mu8730 **Hm Box - Fozzie Bear,** 1987. $4.00-5.00
- ❑ ❑ USA Mu8731 **Hm Box - Gonzo,** 1987. $4.00-5.00
- ❑ ❑ USA Mu8732 **Hm Box - Kermit,** 1987. $4.00-5.00
- ❑ ❑ USA Mu8733 **Hm Box - Miss Piggy's,** 1987. $4.00-5.00

Mu8730

Mu8731

Mu8732

Mu8733

Mu8701 Mu8702 Mu8703 Mu8704

U-3 Premiums:

❑ ❑ USA Mu8705 **U-3 Kermit on Skates.** $3.00-5.00
❑ ❑ USA Mu8706 **U-3 Ms Piggy on Skates.** $3.00-5.00

Premiums:

❑ ❑ USA Mu8701 **Set 1 Gonzo on Green Big Whls**, 2p, 1987. $2.00-3.00
❑ ❑ USA Mu8702 **Set 2 Fozzie on Yellow Horse**, 2p, 1987. $2.00-3.00
❑ ❑ USA Mu8703 **Set 3 Ms Piggy in Pink Car**, Pnk Ribbon/2p, 1987. $2.00-3.00
❑ ❑ USA Mu8704 **Set 4 Kermit on Red Skateboard**, 2p, 1987. $2.00-3.00

❑ ❑ USA Mu8726 **Display/Premiums,** 1987. $125.00-175.00
❑ ❑ USA Mu8741 **Dangler/Each Character,** 1987. $5.00-8.00
❑ ❑ USA Mu8745 **Register Topper,** 1987. $10.00-15.00
❑ ❑ USA Mu8750 **Button-Gonzo,** 1987, Rectangle with Heart. $4.00-5.00
❑ ❑ USA Mu8751 **Button-Muppet Babies,** 1987, with Heart. $4.00-5.00
❑ ❑ USA Mu8764 **Translite/Sm,** 1987. $10.00-15.00
❑ ❑ USA Mu8765 **Translite/Lg,** 1987. $15.00-25.00

Comments: National Distribution: USA - June 5-July 9, 1987. Premium markings - "Ha! 1986 China."

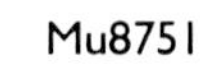

Mu8751

Mu8726

Mu8765

Potato Head Kids I Happy Meal, 1987

Box:

❑ ❑ USA Po8715 **Hm Box - 12 Potato Head Kids,** 1986. $25.00-40.00

Premiums: Potato Head Kids:

❑ ❑ USA Po8700 **Big Chip - Red Baseball Hat/Blue Shoes,** 1986. $20.00-25.00

❑ ❑ USA Po8701 **Dimples - Org Hat/Pnk Shoes,** 1986. $20.00-25.00

❑ ❑ USA Po8702 **Lolly - Yel Rumpled Hat/Daisy with Curls/Grn Shoes,** 1986. $20.00-25.00

❑ ❑ USA Po8703 **Lumpy - Sideways Grn Baseball Cap/Yel Shoes,** 1986. $20.00-25.00

❑ ❑ USA Po8704 **Potato Dumpling - Ruffle Blu Baby Hat with Bow/Pink Shoes,** 1986. $20.00-25.00

❑ ❑ USA Po8705 **Potato Puff - Pink Baby Bonnet Hat/Light Purp Shoes,** 1986. $20.00-25.00

❑ ❑ USA Po8706 **Slick - Purp Derby Hat/Umbrella/Wht Shoes,** 1986. $20.00-25.00

❑ ❑ USA Po8707 **Slugger - Blu Batter's Helmet Hat/Yel Shoes,** 1986. $20.00-25.00

❑ ❑ USA Po8708 **Smarty Pants - Yel Hat/Bow/Glasses/Purp Shoes,** 1986. $20.00-25.00

Po8715

Po8700

Po8701

Po8702

Po8703

Po8704

Po8705

Po8706

Po8707

Po8708

Po8709

Po8710

Po8711

❑ ❑ USA Po8709 **Spike - Blue King Hat/Brn Shoes/3p,** 1986. $20.00-25.00
❑ ❑ USA Po8710 **Spud - Wht Cowboy Hat/Boots/Org Shoes/3p,** 1986. $20.00-25.00
❑ ❑ USA Po8711 **Tulip - Purp Hat/Blue Shoes,** 1986. $20.00-25.00

❑ ❑ USA Po8741 **Dangler,** 1986. $75.00-100.00
❑ ❑ USA Po8745 **Register Topper/1 Premium Blister Pack,** 1986. $25.00-35.00
❑ ❑ USA Po8765 **Translite/Lg,** 1986. $75.00-100.00

Comments: Regional Distribution: USA - February 20-March 19, 1987 in regional areas of Texas, Oklahoma, and New Mexico. Packaged in polybag which said, "Made in China. Safety Tested for Children of All Ages. Recommended for Ages 2 & Up." Hasbro sold the same premiums at retail outlets. The colors of hats and shoes do not always match the colors on Happy Meal box.

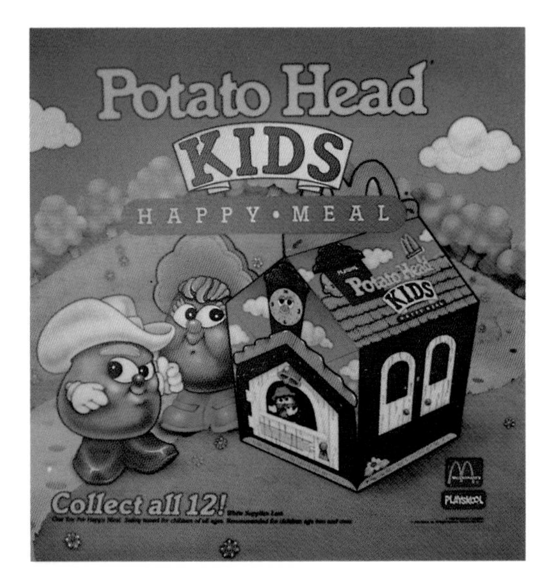

Po8745

Re8710

Real Ghostbusters I Happy Meal, 1987

Boxes:
❑ ❑ USA Re8710 **Hm Box - Headquarters,** 1987. $4.00-5.00
❑ ❑ USA Re8711 **Hm Box - Museum,** 1987. $4.00-5.00
❑ ❑ USA Re8712 **Hm Box - Public Library,** 1987. $4.00-5.00
❑ ❑ USA Re8713 **Hm Box - Schoolhouse,** 1987. $4.00-5.00

Re8711

U-3 Premium:

- ❑ ❑ USA Re8706 **U-3 Ruler/Note Pad,** 1987, X-O-Graphic/ Marshmallow Pad. $10.00-15.00

Premiums:

- ❑ ❑ USA Re8701 **Pencil Case,** 1987, Real Ghbs Containment Chamber. $4.00-5.00
- ❑ ❑ USA Re8702 **Ruler,** 1987, 6" X-O-Graphic Ghbs. $5.00-7.00
- ❑ ❑ USA Re8703 **Note Pad/Eraser,** 1987, Marshmallow Pad/ Ghbs Eraser. $7.00-10.00
- ❑ ❑ USA Re8704 **Pencil/Pencil Topper,** 1987, with Grn Slimer Topper. $7.00-10.00
- ❑ ❑ USA Re8705 **Pencil Sharpener,** 1987, Wht Ghbs. $5.00-7.00

Re8702

Re8712

Re8713

Re8701 Re8706 Re8704 Re8705 Re8703

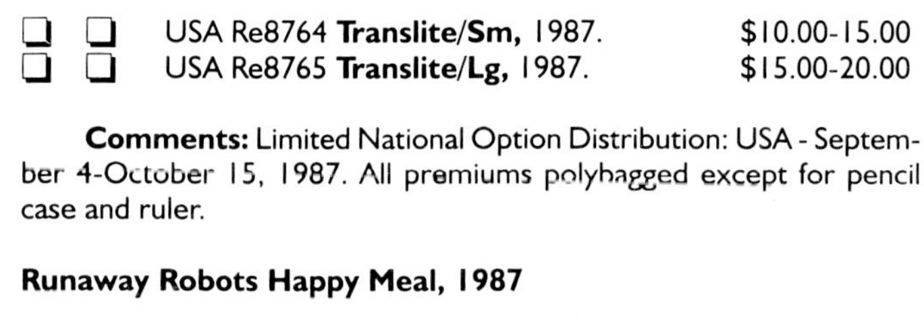

❑ ❑ USA Re8764 **Translite/Sm,** 1987. $10.00-15.00
❑ ❑ USA Re8765 **Translite/Lg,** 1987. $15.00-20.00

Comments: Limited National Option Distribution: USA - September 4-October 15, 1987. All premiums polybagged except for pencil case and ruler.

Runaway Robots Happy Meal, 1987

Box:
❑ ❑ USA Ru8710 **Hm Box - Six Runaway Robots,** 1987. $15.00-25.00

Premiums: Robot type cars (MIP: $10.00-12.00; Loose: $3.00-4.00):
❑ ❑ USA Ru8700 **Beak,** 1987, Blue Robot with Whls. $10.00-12.00
❑ ❑ USA Ru8701 **Bolt,** 1987, Purple Robot with Whls. $10.00-12.00
❑ ❑ USA Ru8702 **Coil,** 1987, Green Robot with Whls. $10.00-12.00
❑ ❑ USA Ru8703 **Flame,** 1987, Red Robot with Whls. $10.00-12.00
❑ ❑ USA Ru8704 **Jab,** 1987, Yel Robot with Whls. $10.00-12.00
❑ ❑ USA Ru8705 **Skull,** 1987, Blk Robot with Whls. $10.00-12.00

❑ ❑ USA Ru8755 **Trayliner,** 1987. $3.00-5.00

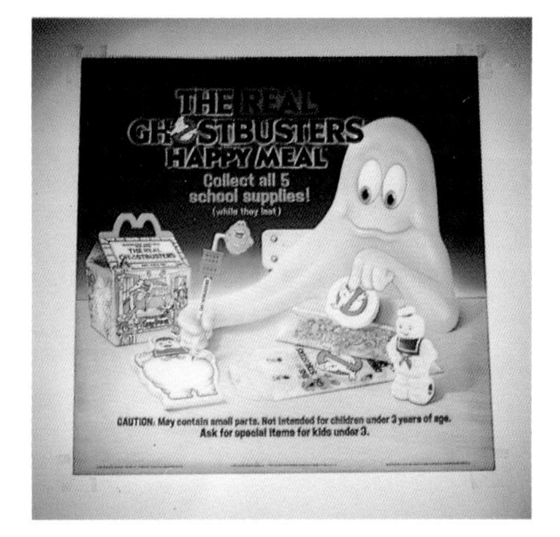

Re8765

Ru8710

Ru8700 Ru8701 Ru8702 Ru8703 Ru8704 Ru8705

Ru8755

❑ ❑ USA Ru8760 **Counter Mat,** 1987. $15.00-20.00
❑ ❑ USA Ru8764 **Translite/Sm,** 1987. $15.00-25.00
❑ ❑ USA Ru8765 **Translite/Lg,** 1987. $25.00-40.00

Comments: Regional Distribution: USA - February 6-March 22, 1987 in St. Louis, Missouri/ Nebraska/Michigan/Maine/Massachusetts/ Tennessee/Alabama. Premium markings - "Colburn China." Loose, runaway robots sell for $4.00-5.00 each.

Ru8765

Ru8760

Super Summer I Test Market Happy Meal, 1987

Bag:
❑ ❑ USA Su8730 **Hm Bag - Super Summer with No Characters on Bag,** 1987. $10.00-15.00

Premiums: Beach Toys:
❑ ❑ USA Su8701 **Sailboat,** 1987, Grimace/inflatable/dated 1987, not 1988. $20.00-25.00
❑ ❑ USA Su8726 **Beach Ball,** 1987, Wht/Dated 1987. $20.00-25.00
❑ ❑ USA Su8731 **Watering Can,** Nd, with McDonald's Logo/ Blue. $35.00-50.00

Comments: Test Distribution: USA - May 22-June 25, 1987 in Fresno, California. USA Su8701 MIP package says "Made in Taiwan" and not "Contents Made in China." Bottom of Happy Meal bag shows three premiums.

Su8730

Su8731

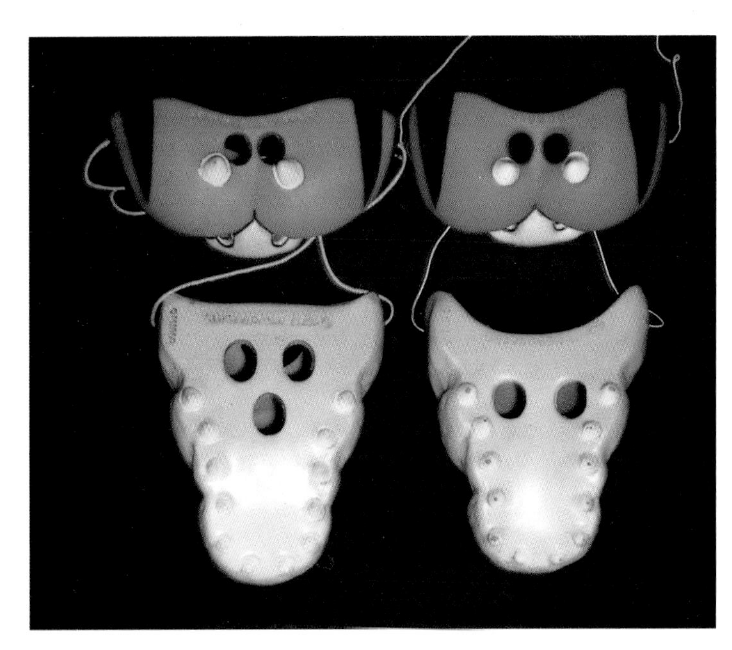

Top right: Zo8703; bottom right: Zo8701

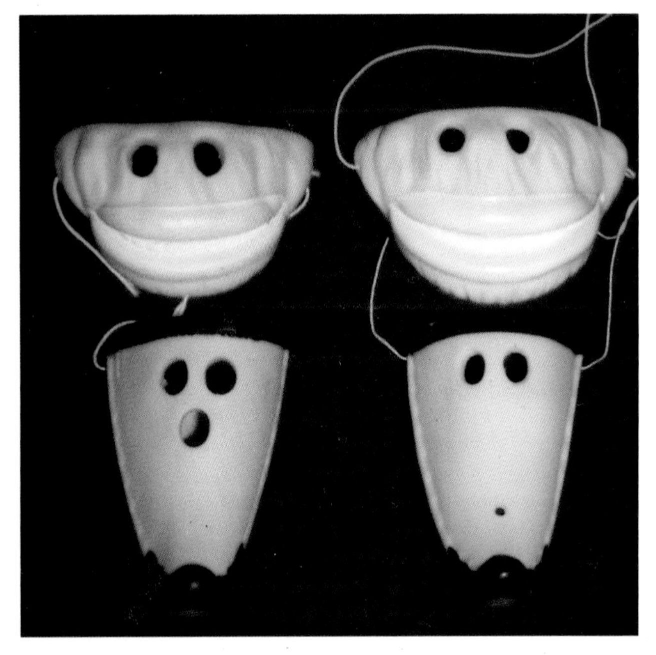

Top right: Zo8702; bottom right: Zo8704

Zoo Face I Test Market Happy Meal, 1987

U-3 Premium: Paper Face Mask:
❑ ❑ USA Zo8705 **U-3 Tiger,** 1987, Paper Mask. $35.00-40.00

Premiums: Face Masks:
❑ ❑ USA Zo8701 **Alligator,** 1987, with Paas Make-Up Kit Dated 1987/2 Small Air Holes. $25.00-35.00
❑ ❑ USA Zo8702 **Monkey,** 1987, with Paas Make-Up Kit Dated 1987/Thin Elastic. $25.00-35.00
❑ ❑ USA Zo8703 **Tiger,** 1987, with Paas Make-Up Kit Dated 1987/Thin Elastic. $25.00-35.00
❑ ❑ USA Zo8704 **Toucan,** 1987, with Paas Make-Up Kit Dated 1987/2 Small Air Holes. $25.00-35.00

Comments: Very limited Regional Test: USA - October 2-31, 1987 in Evansville, Indiana. The test market package has no McDonald's number listed and is dated 1987, not 1988.

USA Generic Promotions, 1987

❑ ❑ USA Ge8701 **Cookie Cutter/Fun Mold-Fry Kid,** 1987, Grn or Red. $1.00-1.50
❑ ❑ USA Ge8703 **Cookie Cutter/Fun Mold-Ronald,** 1987, Grn or Red/with Balloons $1.00-1.50

❑ ❑ USA Ge8705 **Super Sticker Squares,** 1987, 9 Scenes/100 Stickers. $2.00-3.00

❑ ❑ USA Ge8706 **Fry Guy on Brontofry Guy Brontosaurus,** 5p Yel, 1987. $4.00-5.00
❑ ❑ USA Ge8707 **Hamburglar on Tricerahamburglar Triceratops**, 5p Grn or Purp, 1987. $5.00-8.00

❑ ❑ USA Ge8708 **Cookie Box: Grimace - Set of 5 character cookie boxes: Grimace on the yellow brick walk/Can you find/purple box.** $.50-1.00

Ge8701 Ge8703

Ge8705

Ge8707

Ge8706

❑ ❑ USA Ge8709 **Cookie Box: Fry Kids - Set of 5 character cookie boxes: Fry Kids on the yellow brick walk/Can you find 6/blue box.** $.50-1.00

❑ ❑ USA Ge8710 **Cookie Box: Hamburglar - Set of 5 character cookie boxes: Hamburglar on the yellow brick walk/Help Hamburglar find/red box.** $.50-1.00

❑ ❑ USA Ge8711 **Cookie Box: Ronald - Set of 5 character cookie boxes: Ronald on the yellow brick walk/Help Ronald find/green box.** $.50-1.00

❑ ❑ USA Ge8723 **Cookie Box: Birdie - Set of 5 character cookie boxes: Birdie on the yellow brick road/orange box.** $.50-1.00

❑ ❑ USA Ge8712 **Record: Mac Tonight 45 RPM**: "McDonald's cordially invites you to attend the premiere showing of their newest television personality...MacTonight," Friday, August 21, 1987, Marriott Crocker Room. 45 RPM record with jacket (also came with ticket for admittance). Regionally distributed. $20.00-25.00

❑ ❑ USA Ge8713 **Fun Times Magazine: Vol. 9 No. 1 Feb/ Mar, 1987.** $2.00-3.00

❑ ❑ USA Ge8714 **Fun Times Magazine: Vol. 9 No. 2 Apr/ May, 1987.** $2.00-3.00

❑ ❑ USA Ge8715 **Fun Times Magazine: Vol. 9 No. 3 June/ July, 1987.** $2.00-3.00

❑ ❑ USA Ge8716 **Fun Times Magazine: Vol. 9 No. 4 Aug/ Sept, 1987.** $2.00-3.00

Ge8714

Ge8711 Ge8709 Ge8723

Ge8716

Ge8711 Ge8708 Ge8710 Ge8723

Ge8719

Ge8720

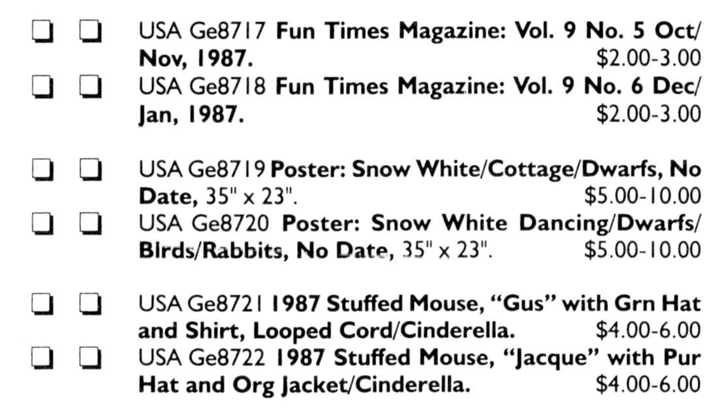

❑ ❑ USA Ge8717 **Fun Times Magazine: Vol. 9 No. 5 Oct/Nov, 1987.** $2.00-3.00

❑ ❑ USA Ge8718 **Fun Times Magazine: Vol. 9 No. 6 Dec/Jan, 1987.** $2.00-3.00

❑ ❑ USA Ge8719 **Poster: Snow White/Cottage/Dwarfs, No Date,** 35" x 23". $5.00-10.00

❑ ❑ USA Ge8720 **Poster: Snow White Dancing/Dwarfs/BIrds/Rabbits, No Date,** 35" x 23". $5.00-10.00

❑ ❑ USA Ge8721 **1987 Stuffed Mouse, "Gus" with Grn Hat and Shirt, Looped Cord/Cinderella.** $4.00-6.00

❑ ❑ USA Ge8722 **1987 Stuffed Mouse, "Jacque" with Pur Hat and Org Jacket/Cinderella.** $4.00-6.00

Comments: Regional Distribution: USA - 1987. These toys are a sampling of premiums given away during 1987. Tossed salads received a fanfare welcome introduction in 1987. McKids clothing was introduced at Sears. Ronald's Play Place, a new kind of playland, was introduced; features included long tubes for crawling, ball pits, and slides. The Snow White posters were sold in the stores for 99 cents during the Happy Meal promotion period. The stuffed Christmas ornaments were given out free with the purchase of Gift Certificates in the stores.

Hi Ho, Hi Ho, It's Off to Work We Go...Selling Posters to Help the Kids!
CREW INCENTIVES AND GIFTS
A. Crew Visor. One size fits all with adjustable strap. White visor with imprinted message. Minimum order: 6 pcs. $2.49/Ea. #51375
B. Crew Apron. White with suggestive selling message on front will let all your customers know of this worthy cause. One size fits all. Minimum order: 6 pcs. $2.99/Ea. #51376
C. Crew Button. This big 3" button allows any person on your crew team to display their support, and make your customer aware of this offer. Packed 50 per case. Minimum order: 1 case. $12.95/Case #51377
D. Lovable 7" Dolls. These cute and cuddly 7" plush Seven Dwarf Dolls would make a great gift or crew incentive to your crew. Available in sets of seven only. $39.95/Set #51378
F. T-Shirt. Featuring the Seven Dwarfs and special copy. Shirt is 50/50 blend for easy care. Adult sizes only. Minimum order: 12 pcs. #51380 (S) #51381 (M) #51382 (L) #51383 (XL) $4.29/Ea.
Snow White
Together We Can Help!
SNOW WHITE POSTER
Together We Can Help!
Cinderella
FREE ORNAMENT
from Walt Disney's Classic Movie
Cinderella
Get one when you buy a $5.00 book of Gift Certificates
GIFT CERTIFICATE
McDonald's

Ba8810

Ba8811

Ba8812

1988

Bambi Happy Meal, 1988
Big Top Happy Meal, 1988
Black History Happy Meal, 1988
CosMc Crayola Happy Meal, 1988
Duck Tales I Happy Meal, 1988
Duck Tales II Happy Meal, 1988
Flintstone Kids Happy Meal, 1988
Fraggle Rock II Happy Meal, 1988
Garfield I Test Market Happy Meal, 1988
Halloween '88 Happy Meal, 1988
Hot Wheels Happy Meal, 1988
Luggage Tag Promotion, 1988
Mac Tonight Travel Toys Happy Meal, 1990/1989/1988
Matchbox Super GT Happy Meal, 1988
McNugget Buddies Happy Meal, 1988
Moveables/McDonaldland Happy Meal, 1988
New Archies Happy Meal, 1988
Oliver & Company Happy Meal, 1988
Olympic Pin/Sports II Clip-On Buttons/Happy Meal, 1988
On the Go II Lunch Box/Bags Happy Meal, 1988
Peter Rabbit's Happy Meal, 1988
Sailors Happy Meal, 1988
Sea World of Ohio Happy Meal, 1988
Sea World of Texas I Happy Meal, 1988
Sport Ball Test Market Happy Meal, 1990/1988
Storybook Muppet Babies Happy Meal, 1988
Super Summer II Happy Meal, 1988
Turbo Macs I Test Market Happy Meal, 1988
Zoo Face II/Halloween '88 Happy Meal, 1988
USA Muppet Babies Holiday Promotion, 1988
USA Generic Promotions, 1988

- **"Good Time, Great Taste of McDonald's" ad slogan**
- **"Good Time, Great Taste, That's Why This is My Place"**
- **"CosMc" character joins Ronald's McDonald's Cast of Characters**
- **McKids Specialty Stores introduced**
- **10th National O/O Convention held**
- **"Sharing the Dream" - O/O advertising theme**

Bambi Happy Meal, 1988

Boxes:
❑ ❑ USA Ba8810 **Hm Box - Fall/Squirrel with Thumper/Bambi,** 1988. $5.00-7.00
❑ ❑ USA Ba8811 **Hm Box - Spring/Bambi with Mother,** 1988. $5.00-7.00
❑ ❑ USA Ba8812 **Hm Box - Summer/Bambi and Faline with Owl/Frog,** 1988. $5.00-7.00
❑ ❑ USA Ba8813 **Hm Box - Winter/Bambi and Thumper,** 1988. $5.00-7.00

U-3 Premiums:
❑ ❑ USA Ba8805 **U-3 Bambi,** 1988, **with Butterfly on Tail/No Moving Legs.** $7.00-10.00
❑ ❑ USA Ba8806 **U-3 Bambi,** 1988, **without Butterfly on Tail/No Moving Legs.** $15.00-20.00
❑ ❑ USA Ba8807 **U-3 Thumper,** 1988, Rabbit without Moving Arms/Legs. $7.00-10.00

Premiums:

- ❑ ❑ USA Ba8801 **Set 1 Bambi,** 1988, Deer. $4.00-5.00
- ❑ ❑ USA Ba8802 **Set 2 Flower,** 1988, Skunk. $4.00-5.00
- ❑ ❑ USA Ba8803 **Set 3 Friend Owl,** 1988, Owl. $4.00-5.00
- ❑ ❑ USA Ba8804 **Set 4 Thumper,** 1988, Rabbit. $4.00-5.00

- ❑ ❑ USA Ba8826 **Display/Premiums,** 1988. $95.00-125.00
- ❑ ❑ USA Ba8841 **Dangler/Each Character,** 1988. $8.00-10.00
- ❑ ❑ USA Ba8864 **Translite/Sm,** 1988. $15.00-25.00
- ❑ ❑ USA Ba8865 **Translite/Lg,** 1988. $20.00-25.00

Comments: National Distribution: USA - July 8 - August 4, 1988. Premium markings: "Disney China." The U-3 Bambi without butterfly on tail was released in Las Vegas, Nevada and Buffalo, New York among other areas.

Ba8805 Ba8801 Ba8803
Ba8806 Ba8807 Ba8802 Ba8804

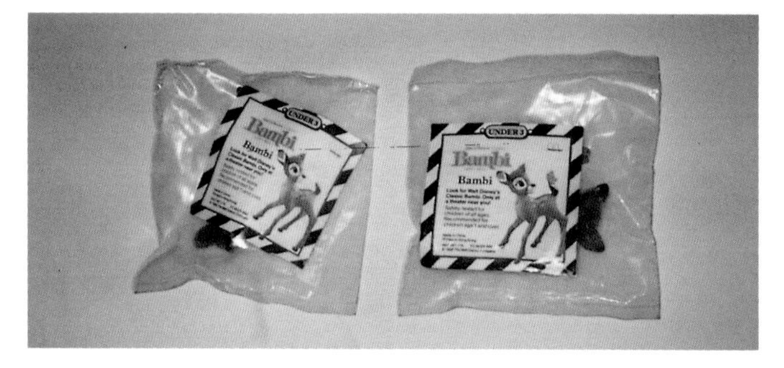

Ba8805

Ba8806

Big Top Happy Meal, 1988

Box:

- ❑ ❑ USA Bi8801 **Hm Box - Big Top,** 1988. $2.00-3.00

Comments: Regional Distribution: USA - 1988. Given at McDonald's birthday parties with generic premiums.

Ba8813

Ba8826

Ba8865

Bi8801

BI8805 (front)

Black History Happy Meal, 1988

Box:

❑ ❑ USA BI8805 **Hm Box - Black History.** $——

Premiums: Coloring Books:

❑ ❑ USA BI8800 **Coloring Book - Little Martin Jr. Coloring Book Vol. 1**, 1988. $——

❑ ❑ USA BI8801 **Coloring Book - Little Martin Jr. Coloring Book Vol. 2**, 1988. $——

❑ ❑ USA BI8865 **Translite/Lg,** 1988. $——

Comments: Very limited Regional Distribution: USA - 1988 in Detroit, Michigan in six stores. The coloring books are not marked with McDonald's logo. Auction prices realized indicate the Happy Meal box and coloring books are very rare. A complete set exchanged hands at $2,600.00 in the last four years. Within the last year, a Black History Happy Meal box changed hands at $1000.00 and a Black History coloring book changed hands for $500.00. Not enough transactions to determine a fair price for buyer and seller.

BI8800

BI8805 (back)

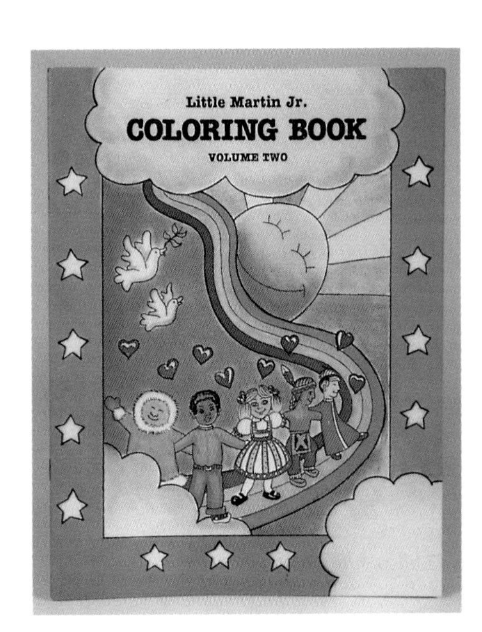

BI8801

BI8865

CosMc Crayola Happy Meal, 1988

Boxes:

❑ ❑ USA Co8810 **Hm Box - Launch Pad/Grimace with Space Scramble,** 1987. $4.00-5.00

❑ ❑ USA Co8811 **Hm Box - Lunar Base with Planet Roundup,** 1987. $4.00-5.00

❑ ❑ USA Co8812 **Hm Box - Planets/Craters with Towing Troubles,** 1987. $4.00-5.00

❑ ❑ USA Co8813 **Hm Box - Martians with Fuzzy Space Friends,** 1987. $4.00-5.00

U-3 Premium:

❑ ❑ USA Co8806 **U-3 Spaceship Color**, 1987, Activity Sheet/2 Metallic/Florescent Crayons. $5.00-8.00

Premiums: Crayola Sets:

❑ ❑ USA Co8801 **Set 1 Crayons,** 1987, 4/Red/Blu/Copper/Sil with Coloring Page. $5.00-7.00

❑ ❑ USA Co8802 **Set 2 Marker,** 1987, Red Marker with Planet Coloring Page. $5.00-7.00

❑ ❑ USA Co8803 **Set 3 Chalk,** 1987, 4 Pastel Chalks with Chalk Board. $5.00-7.00

Co8801

Co8810 Co8811

Co8802

Co8812 Co8813

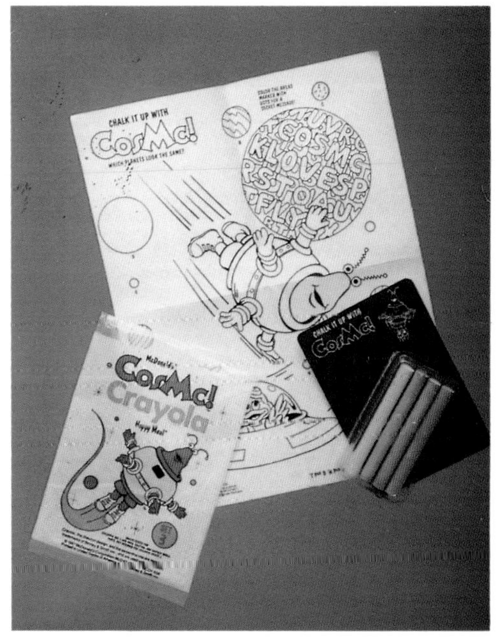

Co8803

Co8804

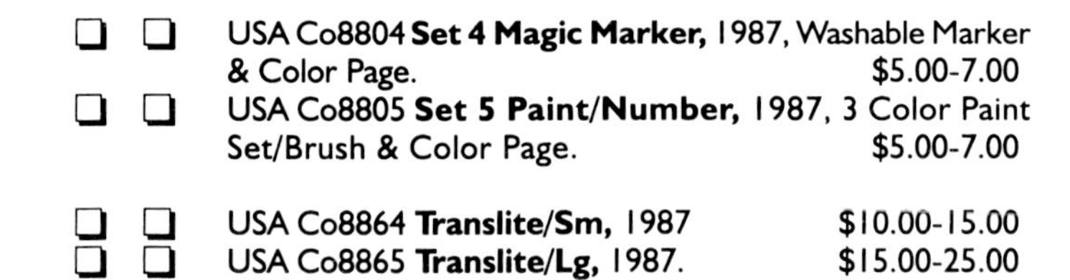

❑ ❑ USA Co8804 **Set 4 Magic Marker,** 1987, Washable Marker & Color Page. $5.00-7.00

❑ ❑ USA Co8805 **Set 5 Paint/Number,** 1987, 3 Color Paint Set/Brush & Color Page. $5.00-7.00

❑ ❑ USA Co8864 **Translite/Sm,** 1987 $10.00-15.00

❑ ❑ USA Co8865 **Translite/Lg,** 1987. $15.00-25.00

Comments: National Distribution: USA - April 15-May 12, 1988.

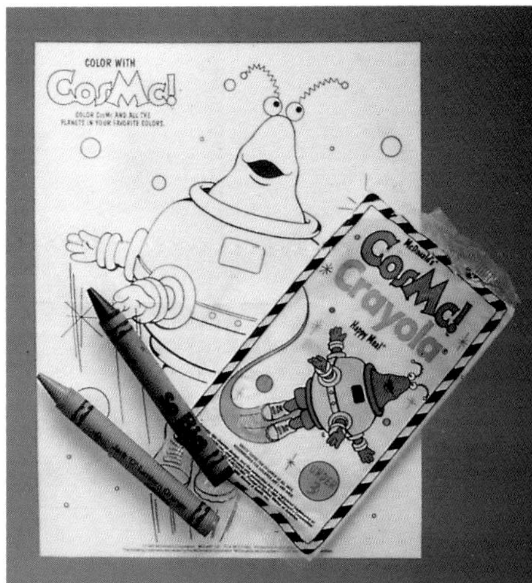

Co8806

Co8805

Co8865

Duck Tales I Happy Meal, 1988

Boxes:

❑ ❑ USA Dt8810 **Hm Box - City of Gold,** 1987. $4.00-5.00

❑ ❑ USA Dt8811 **Hm Box - Cookie of Fortune,** 1987. $4.00-5.00

❑ ❑ USA Dt8812 **Hm Box - Hula Hoopla,** 1987. $4.00-5.00

❑ ❑ USA Dt8813 **Hm Box - Westward Dough,** 1987. $4.00-5.00

Dt8811

Dt8812 Dt8813

U-3 Premium:
❑ ❑ USA Dt8805 **U-3 Magic Motion Map,** 1987, Paper. $3.00-4.00

Premiums:
❑ ❑ USA Dt8801 **Quacker/Whistle,** 1987, Orange. $3.00-4.00
❑ ❑ USA Dt8802 **Magnifying Glass,** 1987, Grn with Decal. $3.00-4.00
❑ ❑ USA Dt8803 **Spy Glass,** 1987, Yellow with Decal Vertical or Horizontal. $2.00-3.00
❑ ❑ USA Dt8804 **Watch/Wrist/Encoder,** 1987, with Secret Compartment/Blu with Decal. $3.00-4.00

❑ ❑ USA Dt8864 **Translite/Sm,** 1987. $10.00-15.00
❑ ❑ USA Dt8865 **Translite/Lg,** 1987. $20.00-25.00

Comments: National Distribution: USA - February 5-March 10, 1988. U-3 was the only premium polybagged.

Dt8865

Dt8805 Dt8801 Dt8803 Dt8802 Dt8804

Duck Tales II Happy Meal, 1988

Box:
❑ ❑ USA Du8840 **Hm Box - Duck Tales Press-Out,** 1988. $5.00-7.00

U-3 Premium (MIP $20.00-25.00; Loose: $7.00-10.00):
❑ ❑ USA Du8834 **U-3 Huey on Skates,** 1988, Yel-Org-Grn/1p. $20.00-25.00

Premiums:
❑ ❑ USA Du8830 **Set 1 Uncle Scrooge in Red Car**, 2p, 1988. $4.00-5.00
❑ ❑ USA Du8831 **Set 2 Webby on Tricycle**, 2p Blu-Pnk-Wht, 1988. $5.00-7.00
❑ ❑ USA Du8832 **Set 3 Launchpad in Org Plane**, Org-Brn-Blu, 1988. $3.00-5.00
❑ ❑ USA Du8833 **Set 4 Huey Duey Louie on Ski Boat WITH Wheels**, Yel-Grn, 1988. $3.00-5.00
❑ ❑ USA Du8835 **Set 4 Huey Duey Louie on Ski Boat WITHOUT Wheels**, Yel-Grn, 1988. $10.00-20.00

❑ ❑ USA Du8864 **Translite/Sm,** 1988. $10.00-15.00
❑ ❑ USA Du8865 **Translite/Lg,** 1988. $15.00-25.00

Comments: Regional Distribution: USA - 1988 in Texas/Michigan/Maryland and New Jersey. Premium markings "Disney China C3" or "McDonald's China 1988." The ski boat without wheels had very limited distribution.

Du8840

Du8831 Du8833 Du8832 Du8830 Du8834

Du8865

Flintstone Kids Happy Meal, 1988

Box:

❑ ❑ USA Fl8810 **Hm Box - Drive in Country,** 1987. $15.00-20.00

U-3 Premium:

❑ ❑ USA Fl8805 **U-3 Dino,** 1987, Purple Dinosaur. $10.00-15.00

Premiums:

❑ ❑ USA Fl8801 **Barney in Blu Mastodon Car**, 1987. $12.00-15.00

❑ ❑ USA Fl8802 **Betty in Org Pterydoctil Car**, 1987. $12.00-15.00

❑ ❑ USA Fl8803 **Fred in Grn Alligator Car**, 1987. $12.00-15.00

❑ ❑ USA Fl8804 **Wilma in Purple Dragon Car**, 1987. $12.00-15.00

Fl8805 Fl8801 Fl8802 Fl8803 Fl8804

❑ ❑ USA Fl8864 **Translite/Sm,** 1987. $20.00-25.00

❑ ❑ USA Fl8865 **Translite/Lg,** 1987. $20.00-35.00

Comments: Regional Distribution: USA - 1988 in New England and parts of Florida. Premium markings - "1988 H_b Prod. Inc. China."

Fl8810

Fl8864

Fraggle Rock II Happy Meal, 1988

Boxes:

❑ ❑ USA Fr8830 **Hm Box - Radish Tops/Meet Mokey Fraggle,** 1987. $2.00-3.00

❑ ❑ USA Fr8831 Hm Box - **Party Picks/Vote/Meet Boober Fraggle,** 1987. $2.00-3.00

❑ ❑ USA Fr8832 Hm Box - **Radishes in Cave/Meet Gobo Fraggle,** 1987. $2.00-3.00

❑ ❑ USA Fr8833 Hm Box - **Swimming Hole Blues/Meet Red Fraggle,** 1987. $2.00-3.00

Fr8830

Fr8831 Fr8833

Fr8832

Fr8833

Fr8821 Fr8820 Fr8823 Fr8822

U-3 Premiums:

❑ ❑ USA Fr8824 **U-3 Gobo Holding Large Carrot.** $5.00-6.00

❑ ❑ USA Fr8825 **U-3 Red Holding Large Radish.** $5.00-6.00

Premiums:

❑ ❑ USA Fr8820 **Set 1 Gobo,** 1987, in Carrot Car/Org. $3.00-4.00

❑ ❑ USA Fr8821 **Set 2 Red,** 1987, in Radish Car/Red. $3.00-4.00

❑ ❑ USA Fr8822 **Set 3 Mokey,** 1987, in Eggplant Car/Purp. $3.00-4.00

❑ ❑ USA Fr8823 **Set 4 Wembly/Boober,** 1987, in Pickle Car/Grn. $3.00-4.00

❑ ❑ USA Fr8826 **Display/with Premiums,** 1988. $50.00-75.00

❑ ❑ USA Fr8841 **Dangler,** 1988. Each $8.00-10.00

❑ ❑ USA Fr8845 **Register Topper,** 1988. Each $10.00-15.00

Fr8825 Fr8824

Fr8826

Fr8845

Fr8850 Fr8851

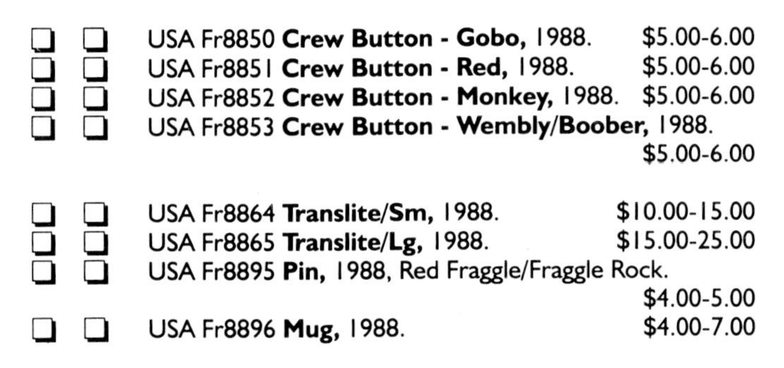

❑	❑	USA Fr8850 **Crew Button - Gobo,** 1988.	$5.00-6.00
❑	❑	USA Fr8851 **Crew Button - Red,** 1988.	$5.00-6.00
❑	❑	USA Fr8852 **Crew Button - Monkey,** 1988.	$5.00-6.00
❑	❑	USA Fr8853 **Crew Button - Wembly/Boober,** 1988.	$5.00-6.00
❑	❑	USA Fr8864 **Translite/Sm,** 1988.	$10.00-15.00
❑	❑	USA Fr8865 **Translite/Lg,** 1988.	$15.00-25.00
❑	❑	USA Fr8895 **Pin,** 1988, Red Fraggle/Fraggle Rock.	$4.00-5.00
❑	❑	USA Fr8896 **Mug,** 1988.	$4.00-7.00

Comments: National Distribution: USA March 11-April 7, 1988. Premium markings - "Henson Associates Inc China."

Fr8852 Fr8853

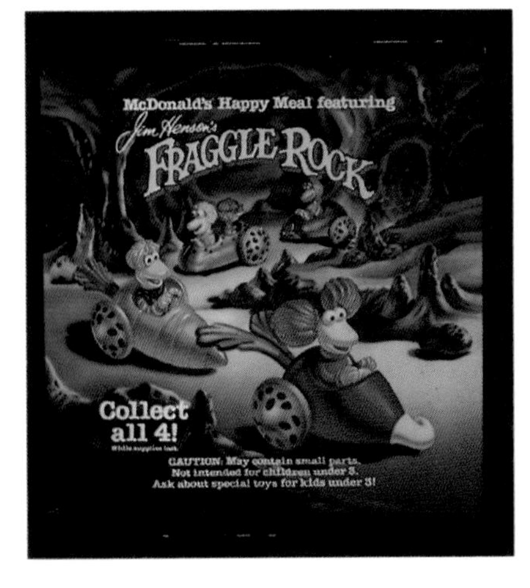

Fr8865

Fr8896

Ga8830

Garfield I Test Market Happy Meal, 1988

Bag:

❑ ❑ USA Ga8830 **Garfield in vehicle/Safari.** $25.00-40.00

Premiums (MIP: $50.00-60.00; Loose: $20.00-35.00):

❑ ❑ USA Ga8801 **Garfield on Skateboard,** 1988, on Pnk Skateboard/W Pnk Helmet/Wht Shirt. $50.00-60.00
❑ ❑ USA Ga8802 **Garfield on Big Wheels,** 1988, Grn/Pur Whls/Big Whls/Yel Cap-Backwards. $50.00-60.00
❑ ❑ USA Ga8803 **Garfield on Scooter,** 1988, on Red Scooter/Grn Whls/Gry Vest/Big Smile. $50.00-60.00
❑ ❑ USA Ga8804 **Garfield in Car,** 1988, in Dark Blu Car/Yel Trim/Tires/Purp Hat/Scarf. $50.00-60.00

Comments: Limited Regional Distribution: USA - July 1988 in Erie, Pennsylvania and Charleston, South Carolina. Premium markings - "United Feat. Synd. China H6." **Loose test market Garfield figurines with complete accessories are selling for $20.00 - 35.00 each.**

Ga8801 Ga8802 Ga8803 Ga8804

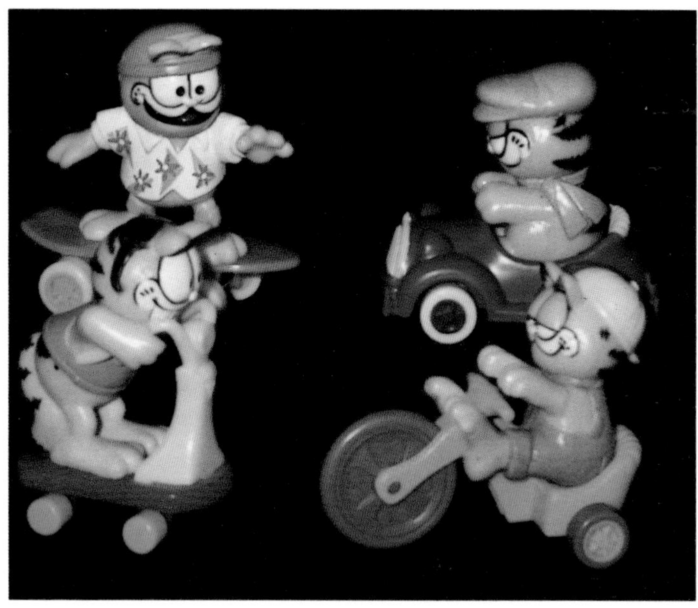

Halloween '88 Happy Meal, 1988

Comments: National Distribution: USA - 1988. See Zoo Face II Happy Meal, 1988.

Hot Wheels Happy Meal, 1988

Box:

❑ ❑ USA Hw8820 **Hm Box - Speed Ramp Punch out 12 Cars Listed,** 1987. $15.00-20.00

Hw8820

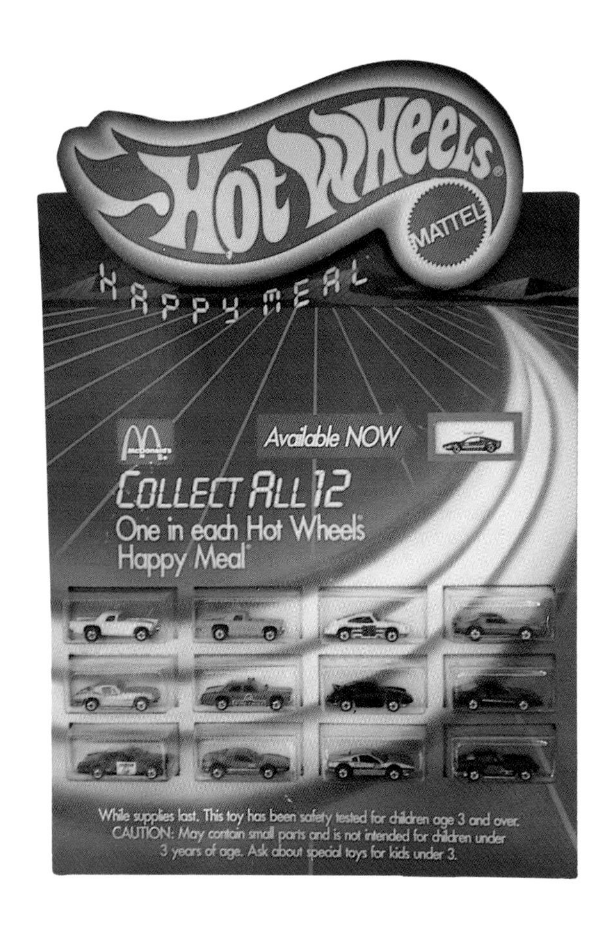

Premiums: Cars

❑ ❑ USA Hw8800 **57 T-Bird - Turq,** 1988. $10.00-15.00
❑ ❑ USA Hw8801 **57 T-Bird - Wht,** 1988. $10.00-15.00

❑ ❑ USA Hw8802 **80s Firebird - Blu,** 1988. $10.00-15.00
❑ ❑ USA Hw8803 **80s Firebird - Blk/No. 3972.** $10.00-15.00

❑ ❑ USA Hw8804 **Fire Chief - Red,** 1988. $10.00-15.00

❑ ❑ USA Hw8805 **P-911 Turbo - Black/No. 3968.** $10.00-15.00

Lu8801 Lu8802 Lu8803 Lu8804

❑ ❑ USA Hw8806 **P-911 Turbo - Wht,** 1988. $10.00-15.00

❑ ❑ USA Hw8807 **Sheriff Patrol - Blk,** 1988. $10.00-15.00

❑ ❑ USA Hw8808 **Split Window '63 - Black,** 1988. $10.00 -15.00
❑ ❑ USA Hw8809 **Split Window '63 - Silver,** 1988. $10.00-15.00

❑ ❑ USA Hw8810 **Street Beast - Red,** 1988. $10.00-15.00
❑ ❑ USA Hw8811 **Street Beast - Silver,** 1988. $10.00-15.00

❑ ❑ USA Hw8812 **CJ7 Jeep - Wht/Org/No. 3953.** $10.00-15.00
❑ ❑ USA Hw8813 **CJ7 Jeep - Yel/Org/No. 3954.** $10.00-15.00

❑ ❑ USA Hw8814 **Corvette Stingray - Wht/Red Stripe/No. 3973.*** $10.00-15.00
❑ ❑ USA Hw8815 **Corvette Stingray - Yel/Org Stripe/No. 3974.*** $10.00-15.00

❑ ❑ USA Hw8816 **Thunder Streek - Burgan/Purp/No. 3998.*** $10.00-15.00
❑ ❑ USA Hw8817 **Thunder Streek - Blu/Yel/No. 3999.*** $10.00-15.00

❑ ❑ USA Hw8818 **Fire Eater - Red Truck/Blu/No. 4000.*** $10.00-15.00
❑ ❑ USA Hw8819 **Fire Eater - Yel Truck/Blu/No. 4001.*** $10.00-15.00

❑ ❑ USA Hw8826 **Display with 12 Cars,** 1988. $250.00-300.00
❑ ❑ USA Hw8864 **Translite/Sm,** 1988. $20.00-25.00
❑ ❑ USA Hw8865 **Translite/Lg,** 1988. $25.00-35.00

Comments: Regional Distribution: USA - 1988 in Texas/Connecticut and Virginia. Different cars may have been given in different regions. An asterisk (*) denotes cars given out primarily on the East Coast. **Loose Hot Wheels cars are selling for $5.00 - 7.00.**

Luggage Tag Promotion, 1988

Premiums: Luggage Tags:

❑ ❑ USA Lu8801 **Luggage Tag: Birdie,** 1988, Pnk Tag/Strap. $4.00-5.00
❑ ❑ USA Lu8802 **Luggage Tag: Grimace,** 1988, Purp Tag/Strap. $4.00-5.00
❑ ❑ USA Lu8803 **Luggage Tag: Hamburglar,** 1988, Blk Tag/Strap. $4.00-5.00
❑ ❑ USA Lu8804 **Luggage Tag: Ronald,** 1988, Red Tag/Strap. $4.00-5.00

Comments: Regional Distribution: USA - 1988 during Clean-Up weeks.

Mac Tonight Travel Toys Happy Meal, 1990/1989/1988

Box and Bag:

❑ ❑ USA Ma8815 **Hm Box - On the Road/World Tour/ Cream Color Box,** 1988. $4.00-5.00

❑ ❑ USA Ma8830 **Hm Bag - Mac/Vehicles,** 1989. $1.00-2.00

U-3 Premium:

❑ ❑ USA Ma8808 **U-3 Skateboard,** 1988, Mac on Skateboard. $10.00-15.00

Premiums:

❑ ❑ USA Ma8800 **Set 1 Jeep/Off Roader,** 1988, 4 Wheeler with Mac Grn. $4.00-6.00

❑ ❑ USA Ma8801 **Set 2 Sports Car,** 1988, Porsche with Mac Red. $4.00-6.00

❑ ❑ USA Ma8802 **Set 3 Surf Ski - with NO wheels,** 1988, Skiboat with Mac (1988). $10.00-20.00

❑ ❑ USA Ma8803 **Set 3 Surf Ski - with WHITE wheels,** 1988, Skiboat with Mac (1990). $4.00-6.00

❑ ❑ USA Ma8804 **Set 4 Scooter,** 1988, Scooter with Mac Blk. $4.00-6.00

❑ ❑ USA Ma8805 **Set 5 Motorcycle,** 1988, Red with Mac Tonight. $4.00-6.00

❑ ❑ USA Ma8806 **Set 6 Airplane - Mac with BLUE sunglasses,** 1988, Blu Plane with Mac. $4.00-6.00

❑ ❑ USA Ma8807 **Set 6 Airplane - Mac with BLACK sunglasses,** 1988, Blu Plane with Mac. $4.00-6.00

❑ ❑ USA Ma8826 **Display/Premiums,** 1988. $150.00-200.00

❑ ❑ USA Ma8862 **Translite/Sm,** 1988, Make It Mac Tonight. $12.00-20.00

❑ ❑ USA Ma8863 **Translite/Lg,** 1988, Make It Mac Tonight. $20.00-35.00

❑ ❑ USA Ma8864 **Translite/Sm,** 1988, Happy Meal Toys. $12.00-20.00

❑ ❑ USA Ma8865 **Translite/Lg,** 1988, Happy Meal Toys. $20.00-35.00

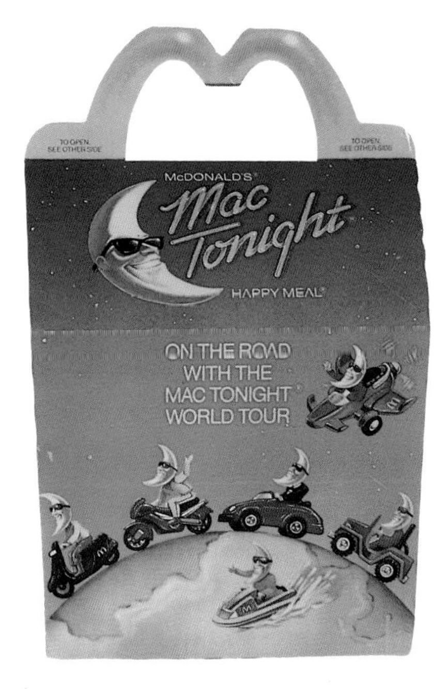

Ma8815

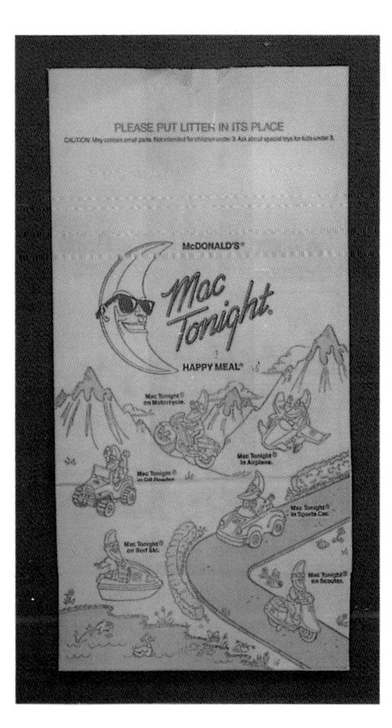

Ma8830

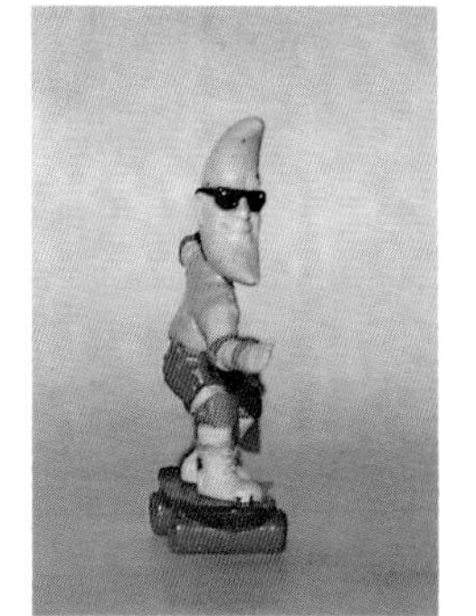

Ma8808

Ma8865

Ma8802

Ma8800 Ma8801 Ma8803 Ma8804

Ma8866

Ma8805 Ma8807 Ma8806

Ma8867

❑ ❑ USA Ma8866 **Translite/X-O-Graphic,** 1988, Make it Mac Tonight/**White Frame.** $25.00-40.00

❑ ❑ USA Ma8867 **Translite/X-O-Graphic,** 1988, Mac Tonight/**Red Frame.** $25.00-40.00

Comments: Regional Distribution: USA - 1988 and September 7-October 4, 1990 in California. Happy Meal was test marketed in 1988 and September 1989. St. Louis, Missouri test market used Happy Meal bag in early 1990. Chicago test market used Happy Meal bag in September 1990. Other Mac Tonight figurines were sold in retail stores. These figurines marked with McDonald's logo, were not part of the Happy Meal promotion. USA Ma8816 Happy Meal bag was used in 1989 with 1988 premiums. USA Ma8803 Surf Ski with Wheels was issued in 1990 in parts of New Jersey.

Mb8825

Matchbox Super GT Happy Meal, 1988

Box:

❑ ❑ USA Mb8825 **Hm Box - Super GT/8 Cars,** 1988.$15.00-20.00

Premiums: Cars

❑ ❑ USA Mb8801 **BR 27/28 Almond with Black Stripes.** $8.00-10.00

❑ ❑ USA Mb8802 **BR 23/24 Beige Starfire with Red Stripes.** $8.00-10.00

❑ ❑ USA Mb8803 **BR 37/38 Blue #8 Red Flame Graphics.** $8.00-10.00

❑ ❑ USA Mb8804 **BR 9/10 Metallic Blue/Yel-Grn Graphics.** $8.00-10.00

❑ ❑ USA Mb8805 **BR 27/28 Grn #8 Wht-Yel Graphics.** $8.00-10.00

❑ ❑ USA Mb8806 **BR 21/22 Orange #6 with Blu-Wht Stripes.** $8.00-10.00

❑ ❑ USA Mb8807 **BR 7/8 Orange Turbo with Wht Flame Graphics.** $8.00-10.00

❑ ❑ USA Mb8808 **BR 35/36 Silver/Grey #3.** $8.00-10.00

❑ ❑ USA Mb8809 **BR 5/6 Silver Grey with Grn/Red Stripes.** $8.00-10.00

❑ ❑ USA Mb8810 **BR 21/22 Silver Grey Super GT with Blk/Blu Stripe.** $8.00-10.00

❑ ❑ USA Mb8811 **BR 9/10 White #1 with Red Flame Graphics.** $8.00-10.00

❑ ❑ USA Mb8812 **BR 19/20 White #18 with Red Stripe.** $8.00-10.00

❑ ❑ USA Mb8813 **BR 33/34 White #45 Racer with Red Stripes.** $8.00-10.00

❑ ❑ USA Mb8814 **BR 7/8 White with Yel-Blu Graphics.** $8.00-10.00

❑ ❑ USA Mb8815 **BR 31/32 Yellow #19 with Red-Blu Stripes.** $8.00-10.00

❑ ❑ USA Mb8816 **BR 29/30 Mustard Yellow with Dark Blu-Org Stripes.** $8.00-10.00

❑ ❑ USA Mb8826 **Counter Card with 16 Premiums,** 1988. $200.00-250.00

❑ ❑ USA Mb8864 **Translite/Sm,** 1988. $15.00-25.00

❑ ❑ USA Mb8865 **Translite/Lg,** 1988. $20.00-30.00

Comments: Regional Distribution: USA - 1988 in Oklahoma region.

McNugget Buddies Happy Meal, 1988

Boxes:

❑ ❑ USA Nu8820 **Hm Box - Apartments,** 1988. $3.00-5.00

❑ ❑ USA Nu8821 **Hm Box - Beauty Shop,** 1988. $3.00-5.00

❑ ❑ USA Nu8822 **Hm Box - Gardens,** 1988. $3.00-5.00

❑ ❑ USA Nu8823 **Hm Box - Post Office,** 1988. $3.00-5.00

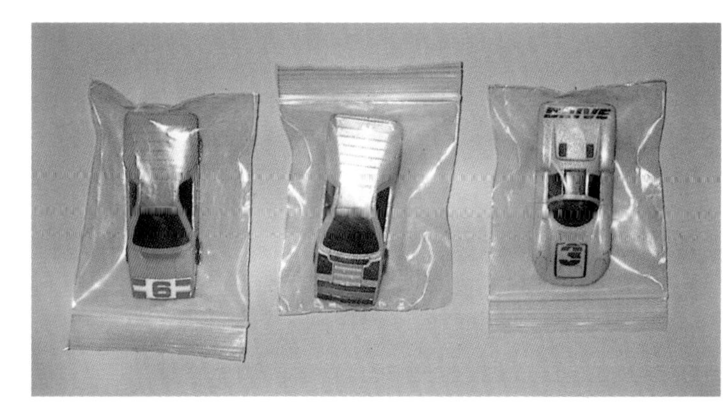

Mb8806 Mb8810 Mb8808

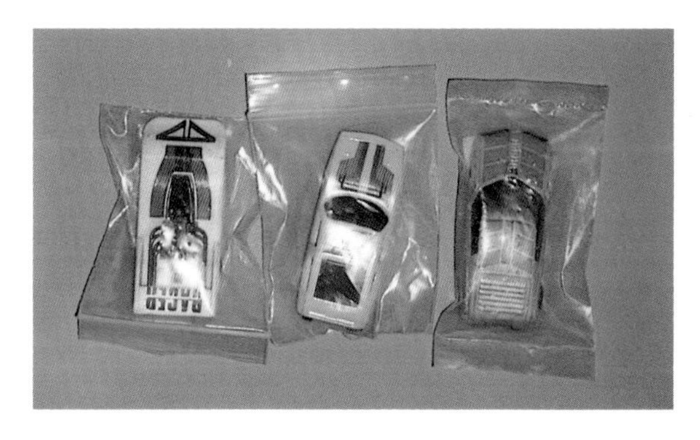

Mb8813 Mb8815 Mb8809

Nu8820 Nu8821

Nu8822 Nu8823

Nu8800 Nu8812 Nu8811 Nu8802 Nu8801 Nu8803 Nu8805 Nu8804 Nu8806 Nu8807 Nu8808 Nu8809

U-3 Premiums:

- ❑ ❑ USA Nu8811 **U-3 Slugger,** 1988, Brn with Baseball Glove/2p. $5.00-7.00
- ❑ ❑ USA Nu8812 **U-3 Daisy,** 1988, Brn Bear with Daisy Flower on Hat/2p. $12.00-15.00

Premiums:

- ❑ ❑ USA Nu8800 **Cowpoke McNugget - Cowboy Hat with Scarf/3p,** 1988. $5.00-6.00
- ❑ ❑ USA Nu8801 **First Class - Hat with Letter Belt/3p,** 1988. $5.00-6.00
- ❑ ❑ USA Nu8802 **Sarge McNugget - Hat with Cuffs/Radio Belt/3p,** 1988. $5.00-6.00
- ❑ ❑ USA Nu8803 **Drummer McNugget - Drum Major Hat with Drum Belt/3p,** 1988. $5.00-6.00
- ❑ ❑ USA Nu8804 **Corny McNugget - Straw Hat with Red Popcorn Belt/3p,** 1988. $5.00-6.00
- ❑ ❑ USA Nu8805 **Corny McNugget - Straw Hat with Beige Popcorn Belt/3p,** 1988. $5.00-6.00
- ❑ ❑ USA Nu8806 **Sparky McNugget - Fire Hat with Hatchet/Extinguish Belt/3p,** 1988. $5.00-6.00
- ❑ ❑ USA Nu8807 **Boomerang McNugget - Ausie Hat with Boomerang/3p,** 1988. $5.00-6.00
- ❑ ❑ USA Nu8808 **Volley McNugget - Head Band E Tennis Belt/3p,** 1988. $5.00-6.00
- ❑ ❑ USA Nu8809 **Snorkel McNugget - Mask with Knife/Lite Belt/3p,** 1988. $5.00-6.00
- ❑ ❑ USA Nu8810 **Rocker McNugget - Orange/reddish Hair & Guitar Belt/3p,** 1988. $5.00-6.00

- ❑ ❑ USA Nu8826 **Display/Premiums,** 1988. $100.00-125.00
- ❑ ❑ USA Nu8855 **Trayliner,** 1988. $2.00-3.00
- ❑ ❑ USA Nu8863 **Menu Board/Lug-On,** 1988. $10.00-15.00
- ❑ ❑ USA Nu8864 **Translite/Sm,** 1988. $10.00-15.00
- ❑ ❑ USA Nu8865 **Translite/Lg,** 1988. $15.00-25.00
- ❑ ❑ USA Nu8895 **Pin - McNugget Buddies.** $4.00-5.00

Nu8811 Nu8812

Nu8855

Nu8865

Nu8826

Nu8895

Comments: National Distribution: USA - December 30, 1988-January 26, 1989. Loose McNugget Buddies are selling for $3.00 each or more.

Moveables/McDonaldland Happy Meal, 1988

Box:

❑ ❑ USA Mo8810 **Hm Box - Ship/Move into Homes,** 1988. $8.00-12.00

Premiums: Figurines (MIP: $10.00-12.00; Loose: $4.00-5.00):

❑ ❑ USA Mo8800 **Birdie,** 1988, 4" Hard Rubber. $10.00-12.00

❑ ❑ USA Mo8801 **Captain,** 1988, 4" Hard Rubber. $10.00-12.00

❑ ❑ USA Mo8802 **Fry Girl,** 1988, 2" Hard Rubber. $10.00-12.00

❑ ❑ USA Mo8803 **Hamburglar,** 1988, 4" Hard Rubber. $10.00-12.00

❑ ❑ USA Mo8804 **Professor,** 1988, 4" Hard Rubber. $10.00-12.00

❑ ❑ USA Mo8805 **Ronald,** 1988, 4" Hard Rubber. $10.00-12.00

❑ ❑ USA Mo8864 **Translite/Sm,** 1988. $15.00-20.00

❑ ❑ USA Mo8865 **Translite/Lg,** 1988. $20.00-25.00

Comments: Regional Distribution: USA - 1988 in St. Louis, Missouri. **Loose McDonaldland figurines are selling for $4.00 - 5.00 Each.** Ronald moveable is the most difficult to obtain.

Mo8800 Mo8801 Mo8802 Mo8803 Mo8804 Mo8805

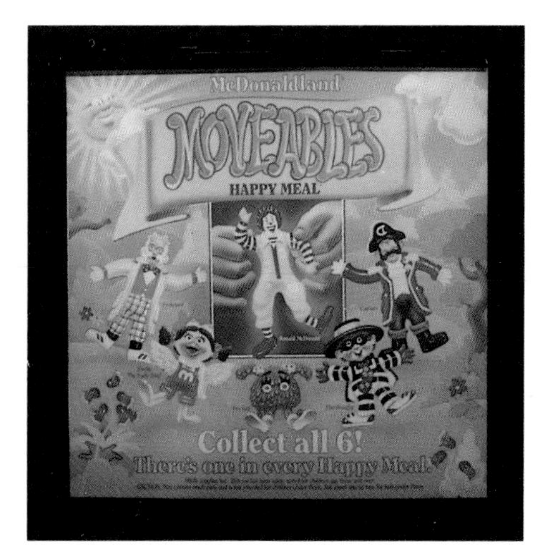

Mo8865

Mo8810

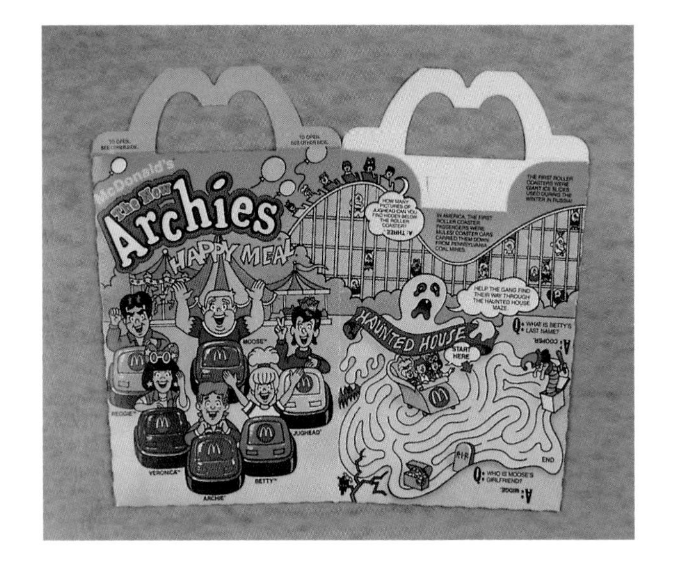

Ne8810

New Archies Happy Meal, 1988

Box:
❑ ❑ USA Ne8810 **Hm Box - Archies/Fun Park,** 1988. $20.00-25.00

Premiums: Figures in cars (MIP: $10.00-20.00; Loose: $4.00-5.00):
❑ ❑ USA Ne8800 **Archie,** 1988, in Red Bumper Car. $15.00-20.00
❑ ❑ USA Ne8801 **Betty,** 1988, in Blu Bumper Car. $10.00-12.00
❑ ❑ USA Ne8802 **Jughead,** 1988, in Yel Bumper Car. $10.00-12.00
❑ ❑ USA Ne8803 **Moose,** 1988, in Pnk Bumper Car. $10.00-12.00
❑ ❑ USA Ne8804 **Reggie,** 1988, in Grn Bumper Car. $10.00-12.00
❑ ❑ USA Ne8805 **Veronica,** 1988, in Purp Bumper Car. $10.00-12.00

❑ ❑ USA Ne8864 **Translite/Sm,** 1988. $15.00-30.00
❑ ❑ USA Ne8865 **Translite/Lg,** 1988. $25.00-40.00

Comments: Regional Distribution: USA - Spring-1988 in St. Louis, Missouri. **Loose Archies figurines in cars are selling for $3.00 - 5.00 each.**

Ne8800 Ne8801 Ne8804 Ne8803 Ne8802 Ne8805

Oliver & Company Happy Meal, 1988

Oc8810

Boxes:
❑ ❑ USA Oc8810 **Hm Box - Funny Bones,** 1988. $3.00-5.00
❑ ❑ USA Oc8811 **Hm Box - Noisy Neighborhood,** 1988. $3.00-5.00
❑ ❑ USA Oc8812 **Hm Box - Shadow Scramble,** 1988. $3.00-5.00
❑ ❑ USA Oc8813 **Hm Box - Tricky Trike,** 1988. $3.00-5.00

Oc8811 Oc8812 Oc8813

Premiums: Finger Puppets:

❑ ❑ USA Oc8800 **Set 1 Oliver the Kitten.** $2.00-3.00
❑ ❑ USA Oc8801 **Set 2 Francis the Bulldog.** $2.00-3.00
❑ ❑ USA Oc8802 **Set 3 Georgette the French Poodle.** $2.00-3.00
❑ ❑ USA Oc8803 **Set 4 Dodger the Dog with Goggles.** $2.00-3.00

❑ ❑ USA Oc8844 **Crew Poster,** 1988, Company's Coming! $4.00-5.00
❑ ❑ USA Oc8864 **Translite/Sm,** 1988. $10.00-15.00
❑ ❑ USA Oc8865 **Translite/Lg,** 1988. $15.00-25.00
❑ ❑ USA Oc8895 **Pin: Oliver & Company,** 1988, Round. $3.00-4.00

Comments: National Distribution - November 25-December 22, 1988. Premium markings - "Disney China P7."

Oc8800 Oc8801 Oc8802 Oc8803

Oc8812

Oc8844

Oc8865

Oc8813

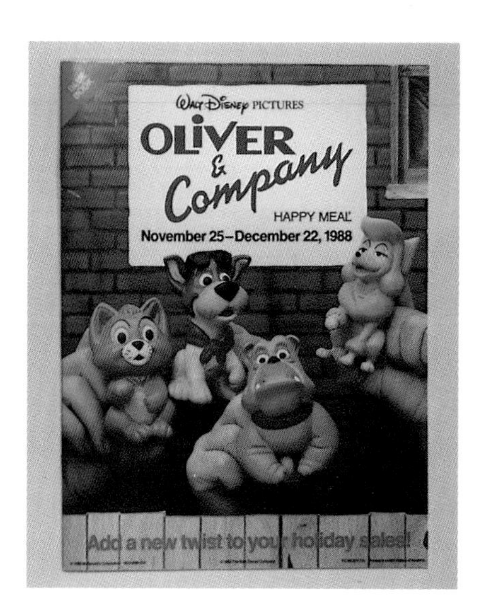

Oc8895

Olympic Pin/Sports II Clip-On Buttons/Happy Meal, 1988

Boxes:

❑ ❑ USA O18860 **Hm Box - Hilarious Hurdles,** 1988. $3.00-5.00

❑ ❑ USA O18861 **Hm Box - Order on the Court,** 1988. $3.00-5.00

Premiums: Clip-on Buttons:

❑ ❑ USA O18850 **Button/Olympic Pin - Hamburglar Jumping Hurdles ,** 1988, clip on. $5.00-7.00

❑ ❑ USA O18851 **Button/Olympic Pin - Birdie Performing Gymnastics,** 1988, clip on. $5.00-7.00

❑ ❑ USA O18852 **Button/Olympic Pin - Grimace Kicking Soccer Ball,** 1988, clip on. $5.00-7.00

❑ ❑ USA O18853 **Button/Olympic Pin - Fry Girl Diving, 1988,** clip on. $5.00-7.00

❑ ❑ USA O18854 **Button/Olympic Pin - Ronald Bicycling,** 1988, clip on. $5.00-7.00

❑ ❑ USA O18855 **Button/Olympic Pin - CosMc Playing Basketball,** 1988, clip on. $5.00-7.00

❑ ❑ USA O18864 **Translite/Sm,** 1988. $15.00 -20.00

❑ ❑ USA O18865 **Translite/Lg,** 1988. $20.00-25.00

Comments: Limited National Distribution - September 9-29, 1988 in Alabama and Georgia. After additional research using the Blue Book, Happy Meal name is: "Olympic Pin Happy Meal." Also called: "Team McDonaldland Happy Meal" on the Continuity Box '88.

O18860

O18861

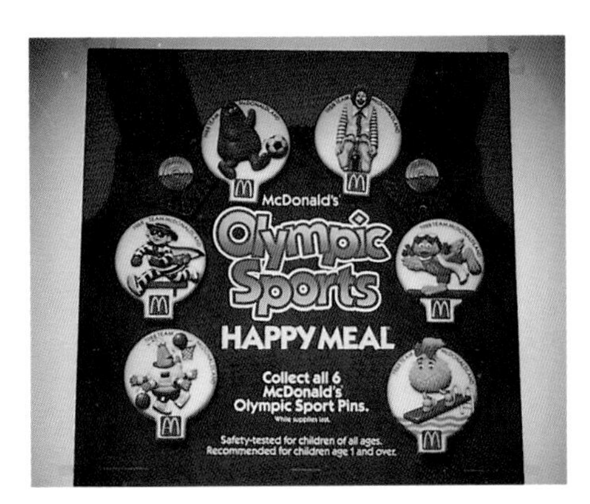

O18865

O18851 O18855 O18853

O18852 O18850 O18854

Olympic Pin blue book

On the Go II Lunch Box/Bags Happy Meal, 1988

U-3 Premium:
❑ ❑ USA On8805 **U-3 Lunch Box,** 1988, Ron in School/Blue with Sticker Picture. $7.00-10.00

Premiums: Lunch Boxes & Lunch Bags:
❑ ❑ USA On8801 **Red Lunch Box - Grimace with Ronald Raised Scene with Bulletin Board.** $3.00-5.00
❑ ❑ USA On8803 **Grn Lunch Box - Grimace with Ronald Raised Scene with Bulletin Board.** $3.00-5.00

❑ ❑ USA On8802 **Yellow Lunch Bag - Ronald**, 1988, Soft Bag. $2.00-3.00
❑ ❑ USA On8804 **Blue Lunch Bag - Grimace**, 1988, Soft Bag. $2.00-3.00

❑ ❑ USA On8806 **Sticker Sheet/Each,** 1988, for Grn Lunch Box. $3.00-5.00
❑ ❑ USA On8807 **Sticker Sheet/Each,** 1988, for Red Lunch Box. $3.00-5.00

❑ ❑ USA On8864 **Translite/Sm,** 1988. $10.00-15.00
❑ ❑ USA On8865 **Translite/Lg,** 1988. $20.00-25.00

Comments: National Distribution: USA - August 12 - September 8, 1988.

On8801 On8803

On8865

On the Go II blue book.

Pe8817

Peter Rabbit's Happy Meal, 1988

Box:
❑ ❑ USA Pe8817 **Hm Box, 1988, Mr. McGregor's Garden.** $100.00-125.00

Premiums: Small Books (MIP: $15.00-20.00; Loose: $7.00-10.00):
❑ ❑ USA Pe8810 **Book: Tale of Benjamin Bunny,** 1988, Book/Paper/Beatrix Potter. $15.00-20.00
❑ ❑ USA Pe8811 **Book: Tale of Flopsy Bunnies,** 1988, Book/Paper/Beatrix Potter. $15.00-20.00
❑ ❑ USA Pe8812 **Book: Tale of Peter Rabbit,** 1988, Book/Paper/Beatrix Potter. $15.00-20.00
❑ ❑ USA Pe8813 **Book: Tale of Squirrel Nutkin,** 1988, Book/Paper/Beatrix Potter. $15.00-20.00

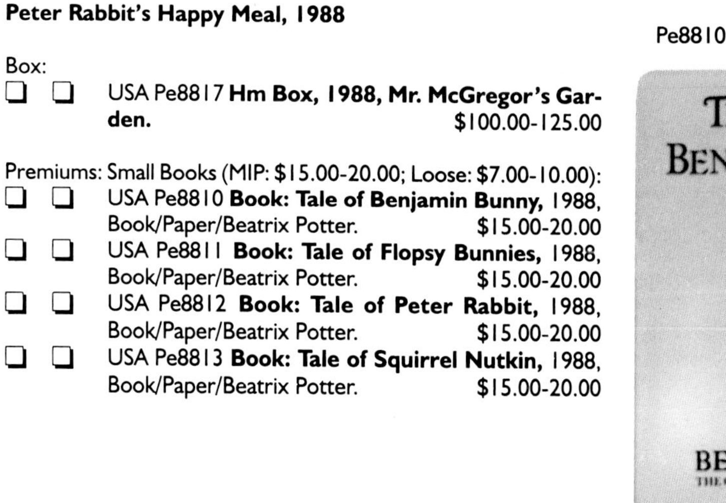

Pe8810

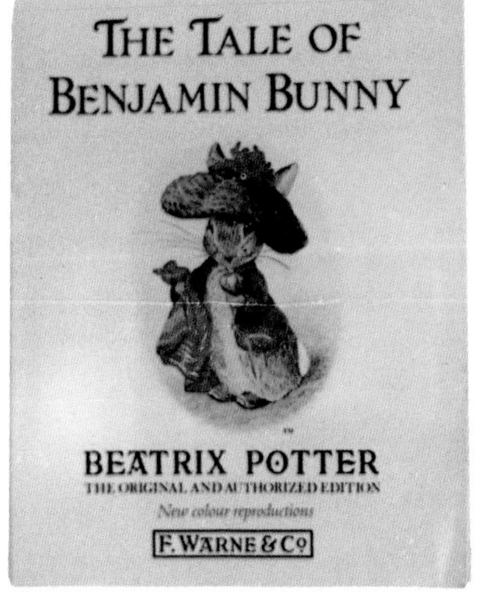

Pe8812

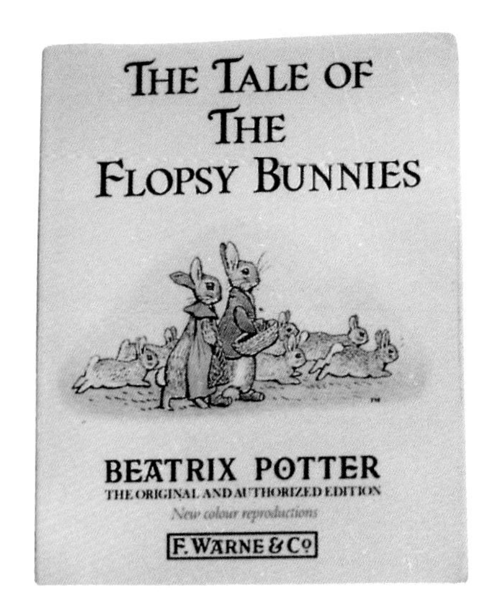

Pe8811

Pe8813

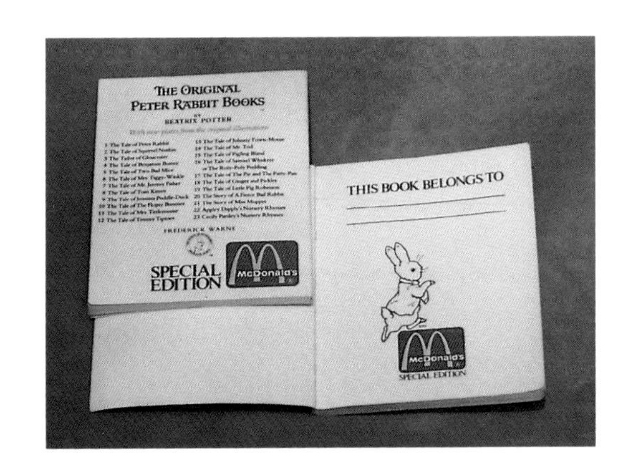

Pe8865

- ❑ ❑ USA Pe8864 **Translite/Sm,** 1988. $50.00-65.00
- ❑ ❑ USA Pe8865 **Translite/Lg,** 1988. $75.00-100.00

Comments: Regional Distribution: USA - Fall 1988 in parts of New York and Pennsylvania. Peter Rabbit book prices vary according to condition of book. Soiled, torn and books with pages removed typically sell for $5.00-8.00.

Sailors Happy Meal, 1988

Boxes:

- ❑ ❑ USA Sa8810 **Hm Box - Fry Guy Afloat,** 1987. $3.00-5.00
- ❑ ❑ USA Sa8811 **Hm Box - Houseboat Is Whose,** 1987. $3.00-5.00

Sa8811

Sa8810

❑ ❑ USA Sa8812 **Hm Box - Island Eyes,** 1987. $3.00-5.00
❑ ❑ USA Sa8813 **Hm Box - Ronald Fishing/Crossword,** 1987. $3.00-5.00

U-3 Premiums:
❑ ❑ USA Sa8804 **U-3 Grimace Speedboat,** 1987, Blu/Floater. $5.00-8.00
❑ ❑ USA Sa8805 **U-3 Fry Guy on Tube,** 1987, Blu/Floater. $5.00-8.00

Premiums: Water Vehicles:
❑ ❑ USA Sa8800 **Grimace Submarine,** 1987, Pur Top/Bottom/Prop/3p. $5.00-7.00
❑ ❑ USA Sa8801 **Fry Kids Ferry,** 1987, Grn Top/Bottom/Ferry Car/3p. $7.00-10.00
❑ ❑ USA Sa8802 **Hamburglar Pirate Ship,** 1987, Blu Top/Bottom/Sail/3p. $5.00-7.00
❑ ❑ USA Sa8803 **Ronald McDonald Airboat,** 1987, Red Bottom/Ronald/Prop/3p. $5.00-7.00

❑ ❑ USA Sa8864 **Translite/Sm,** 1987. $10.00-15.00
❑ ❑ USA Sa8865 **Translite/Lg,** 1987. $15.00-25.00

Comments: National Distribution: USA - January 1 - 28, 1988.

Sa8812

Sa8801 Sa8800 Sa8802 Sa8803

Sa8813

Sa8805 Sa8804

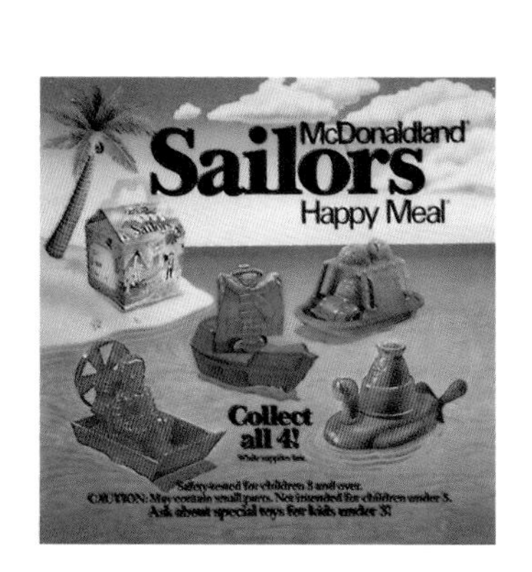

Sa8865

Se8805

Se8800 Se8801 Se8802

Sw8810

Sw8803 Sw8800 Sw8802 Sw8801

Sea World of Ohio Happy Meal, 1988

Box:

❑ ❑ USA Se8805 **Hm Box - 3 Kids on Whale,** 1988. $50.00-75.00

Premiums (MIP: $20.00-45.00; Loose: $7.00-10.00):

❑ ❑ USA Se8800 **Dolly Dolphin,** 1988, Grey-Wht Dolphin with Wreath on Forehead on blue base/3". $25.00-35.00

❑ ❑ USA Se8801 **Penny Penguin,** 1988, Blk-Wht Penguin with red/yellow scarf on neck/yellow web feet/3". $30.00-45.00

❑ ❑ USA Se8802 **Shamu the Whale,** 1988, Blk-Wht Orca Whale on blue base/3". $20.00-25.00

❑ ❑ USA Se8864 **Translite/Sm,** 1988. $20.00-30.00

❑ ❑ USA Se8865 **Translite/Lg,** 1988. $25.00-40.00

Comments: Regional Distribution: USA - Spring 1988 in Cleveland, Ohio. Premium markings - "Made in China, 1987, Sea World, Inc." on the bottom of figurine -- not on the side! Figurines marked on the side were sold in Sea World gift shops.

Sea World of Texas I Happy Meal, 1988

Box:

❑ ❑ USA Sw8810 **Hm Box - Ronald in Yel Sub/without Coupon on Box,** 1988. $10.00-15.00

Premiums: Stuffed Sea Animals:

❑ ❑ USA Sw8800 **Stuffed Dolphin,** 1988, Grey-Wht with Blk Eyes 6". $20.00-25.00

❑ ❑ USA Sw8801 **Stuffed Penguin,** 1988, Blk-Wht with Org Beak/Feet 6". $20.00-25.00

❑ ❑ USA Sw8802 **Stuffed Walrus,** 1988, Brown-Wht Face/Tusks 6". $20.00-25.00

❑ ❑ USA Sw8803 **Stuffed Whale,** 1988, Black-Wht 6". $20.00-25.00

❑ ❑ USA Sw8826 **Counter Display with Premiums**, 1988. $125.00-150.00
❑ ❑ USA Sw8864 **Translite/Sm,** 1988. $25.00-40.00
❑ ❑ USA Sw8865 **Translite/Lg,** 1988. $25.00-50.00

Comments: Regional Distribution: USA - Summer 1988 in San Antonio, Texas. Premium markings - detailed cloth label says, "Sea World J3 Inc Korea." **Loose plush Dolphin, Penguin, Walrus and Whale range $8.00 - 10.00.** Auction prices have exceeded listed prices.

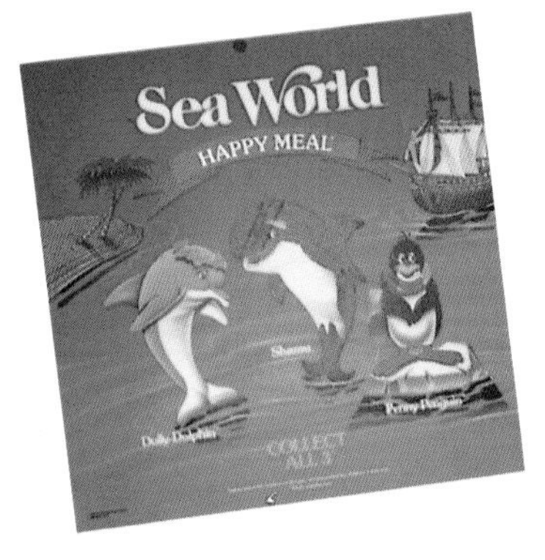

Se8864

Sw8864

Sport Ball Test Market Happy Meal, 1990/1988

Boxes:
❑ ❑ USA Sp8810 **Hm Box - Clear the Court,** 1988. $10.00-15.00
❑ ❑ USA Sp8811 **Hm Box - Match Point,** 1988. $10.00-15.00

U-3 Premium:
❑ ❑ USA Sp8805 **U-3 Baseball,** 1988, 3" Hard Plastic/M Molded Both Sides. $35.00-40.00

Premiums: Small Sport Premiums
❑ ❑ USA Sp8801 **Football,** 1988, Yel/Red with Red M/Ronald McDonald's Signature. $20.00-25.00
❑ ❑ USA Sp8802 **Tennis Ball,** 1988, Sponge/M Cut into Ball. $20.00-25.00
❑ ❑ USA Sp8803 **Baseball,** 1988, Hard Plastic/Wht/M Molded into Ball. $20.00-25.00
❑ ❑ USA Sp8804 **Basketball,** 1988, Org Ball with Org Hoop/ Wht Net. $35.00-40.00

❑ ❑ USA Sp8826 **Display/Premiums/Motion,** 1988, $150.00-200.00
❑ ❑ USA Sp8864 **Translite/Sm,** 1988, $20.00-30.00
❑ ❑ USA Sp8865 **Translite/Lg,** 1988, $30.00-40.00

Comments: Regional Distribution: USA - 1988/1990 in Springfield, Missouri.

Sp8810

Sp8811

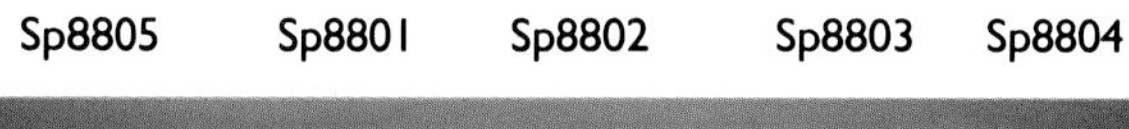

Sp8805 Sp8801 Sp8802 Sp8803 Sp8804

Storybook Muppet Babies Happy Meal, 1988

Boxes:

❑ ❑ USA St8860 **Hm Box - Library with Scary Pit,** 1988. $2.00-4.00

❑ ❑ USA St8861 **Hm Box - Nursery with Treasure Map,** 1988. $2.00-4.00

❑ ❑ USA St8862 **Hm Box - Picnic with Gold Pan,** 1988. $2.00-4.00

Premiums: Books:

❑ ❑ USA St8850 **Book: Just Kermit and Me,** 1988, Muppet Babies. $2.00-3.00

❑ ❑ USA St8851 **Book: The Legend of Gimme Gulch,** 1988, Muppet Babies. $2.00-3.00

❑ ❑ USA St8852 **Book: Baby Piggy the Living Doll,** 1988, Muppet Babies. $2.00-3.00

❑ ❑ USA St8864 **Translite/Sm,** 1988. $10.00-15.00

❑ ❑ USA St8865 **Translite/Lg,** 1988. $20.00-25.00

❑ ❑ USA St8867 **Menu Board Lug-On,** 1988. $10.00-12.00

Comments: National Distribution: USA - October 28 - November 17, 1988.

St8860

St8861

St8852 St8851 St8850

St8862

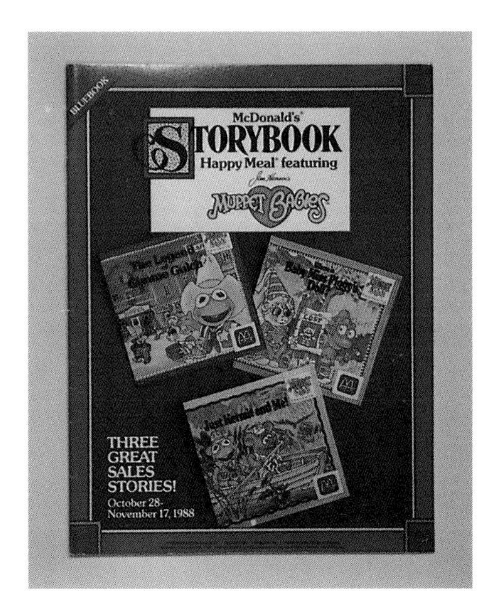

Storybook blue book.

Super Summer II Happy Meal, 1988

Bag:

❑ ❑ USA Su8835 **Hm Bag - Picnic Puzzler with Grimace Grilling/Wht/Grn,** 1987. $1.00-2.00

Premiums: Sand Toys:

❑ ❑ USA Su8825 **Pail/Sand,** 1987, Wht Sand Pail with Yel Rake. $1.50-2.50

❑ ❑ USA Su8826 **Beach Ball,** 1987, Wht/Fry Kids/Made in China. $1.00-1.50

❑ ❑ USA Su8827 **Boat/Sail,** 1988, Inflatable Sailboat/Grimace. $1.00-1.50

❑ ❑ USA Su8828 **Sand/Fish Mold,** 1988, Blue/Fish Shaped. $3.00-5.00

❑ ❑ USA Su8829 **Castle Mold,** 1987, Wht Castle Pail with Red Shovel/Sand Sifter Lid. $1.50-2.50

❑ ❑ USA Su8831 **Watering Can,** 1988, without McDonald's Logo/Blue. $40.00-75.00

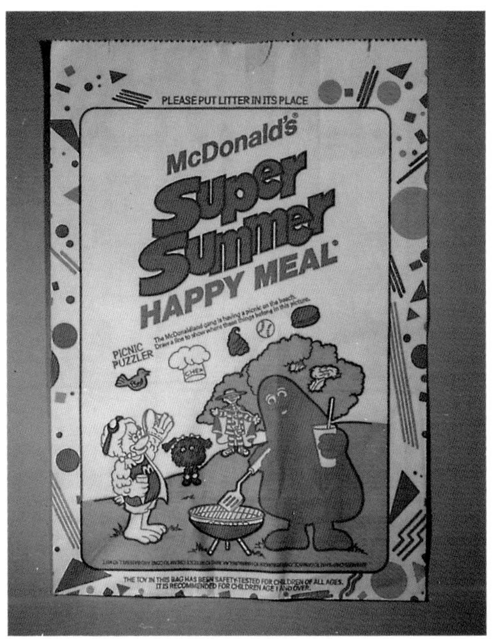

Su8835

Su8825 Su8829

Su8827

Su8828 Su8826

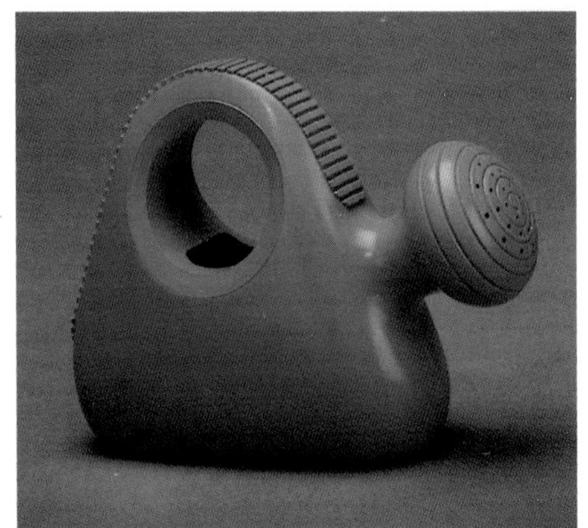

Su8831

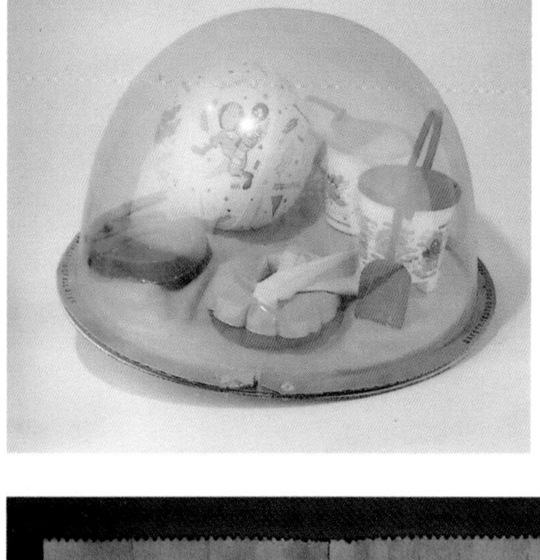

❑ ❑ USA Su8844 **Display/Floor/Premiums with Fish Mold,** 1988. $65.00-80.00
❑ ❑ USA Su8864 **Translite/with Fish Mold/Sm,** 1988.$10.00-15.00
❑ ❑ USA Su8865 **Translite/with Fish Mold/Lg,** 1988.$10.00-15.00
❑ ❑ USA Su8866 **Translite/with Watering Can/Sm,** 1988. $10.00-15.00
❑ ❑ USA Su8867 **Translite/with Watering Can/Lg,** 1988. $10.00-15.00

Comments: National Distribution: USA - May 20-June 23, 1988. California test market: May-June 1987. Pail came with either a red rake or yellow shovel. In the national promotion the sand mold was substituted for the watering can. USA Su8826 MIP says, "Contents Made in China/Contents Printed in Taiwan."

Su8865

Su8867

Turbo Macs I Test Market Happy Meal, 1988

Box:

❑ ❑ USA Tu8830 **Hm Box - Ronald/Red Race Car,** 1988. $5.00-7.00

U-3 Premium:

❑ ❑ USA Tu8824 **U-3 Ronald in Rubber Red Car**, 1988, Yel Wheels/Lg Arches. $5.00-8.00

Premiums:

❑ ❑ USA Tu8820 **Birdie in Pink Car**, Brn Hair/Small Yel Arches, 1988. $4.00-7.00

❑ ❑ USA Tu8821 **Grimace in White Car**, Lg Yel M on Front of Car, 1988. $4.00-7.00

❑ ❑ USA Tu8822 **Hamburglar in Yellow Car**, Lg Red M on Front of Car, 1988. $4.00-7.00

❑ ❑ USA Tu8823 **Ronald in Red Car with TEAR DROP Under Eyes**, Lg Yel M on Front of Car, 1988. $5.00-8.00

Tu8824 Tu8820 Tu8821 Tu8822 Tu8823

❑ ❑ USA Tu8826 **Display/Premiums/Motion,** 1988. $125.00-150.00

❑ ❑ USA Tu8864 **Translite/Sm,** 1988. $15.00-20.00

❑ ❑ USA Tu8865 **Translite/Lg,** 1988. $20.00-25.00

Comments: Regional Distribution: USA - 1988/1990 in California throughout the southern belt to Florida. USA Tu8820-23 came polybagged with insert card. U-3 USA Tu8824 came polybagged without insert card and polybagged in a clear bag.

Tu8830

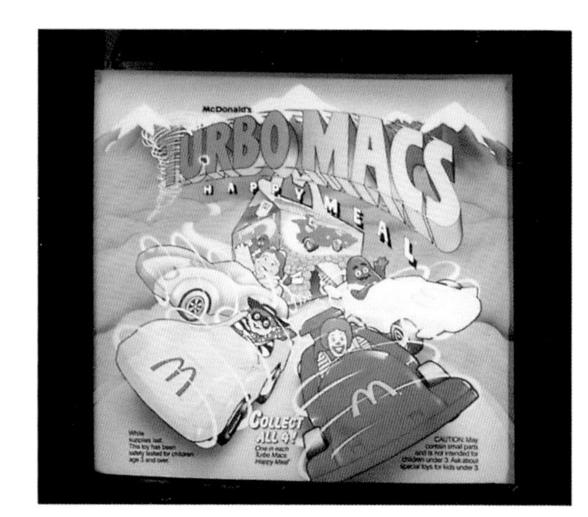

Tu8865

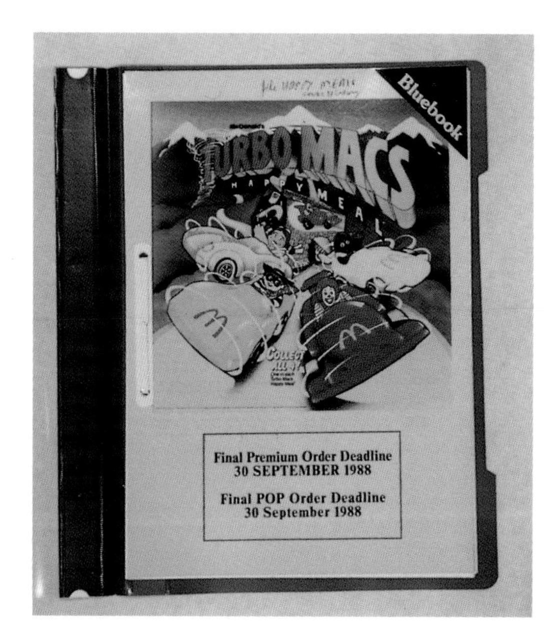

Turbo Mac blue book.

Zoo Face II/Halloween '88 Happy Meal, 1988

❑ ❑ USA Zo8810 **Hm Box - Ape House,** 1988. $3.00-4.00
❑ ❑ USA Zo8811 **Hm Box - Bird House,** 1988. $3.00-4.00
❑ ❑ USA Zo8812 **Hm Box - Lion House,** 1988. $3.00-4.00
❑ ❑ USA Zo8813 **Hm Box - Reptile House,** 1988. $3.00-4.00

U-3 Premiums:
❑ ❑ USA Zo8805 **U-3 Monkey 3D Face Mask**, 1988, Org. $10.00-15.00
❑ ❑ USA Zo8806 **U-3 Tiger 3D Face Mask**, 1988, Yel. $10.00-15.00

Premiums: Masks with Make-Up Kits:
❑ ❑ USA Zo8801 **Set 1 Toucan,** 1988, with Paas Make-Up Kit. $3.00-5.00
❑ ❑ USA Zo8802 **Set 2 Monkey,** 1988, with Paas Make-Up Kit. $3.00-5.00
❑ ❑ USA Zo8803 **Set 3 Tiger,** 1988, with Paas Make-Up Kit. $3.00-5.00
❑ ❑ USA Zo8804 **Set 4 Alligator,** 1988, with Paas Make-Up Kit. $3.00-5.00

❑ ❑ USA Zo8864 **Translite/Sm,** 1988. $10.00-15.00
❑ ❑ USA Zo8865 **Translite/Lg,** 1988. $15.00-25.00

Zo8810

Zo8811

Zo8812

Zo8813

Zo8865

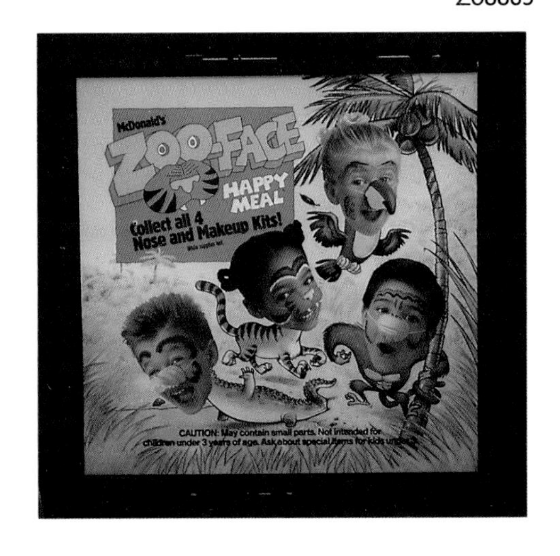

Comments: National Distribution: USA - September 30-October 27, 1988. As a result of the test market, holes in the Toucan were enlarged and the holes in the Alligator increased to three. The Toucan and Alligator were reshaped and heavier elastic string provided.

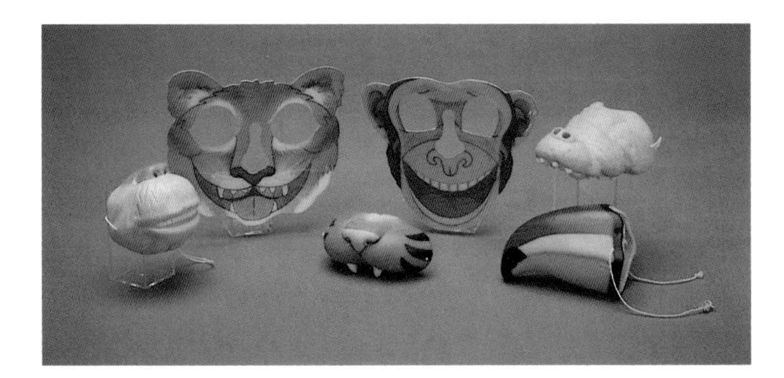

Zoo Faces blue book.

USA Muppet Babies Holiday Promotion, 1988

❑ ❑ USA Mu8801 **Doll: Fossie,** 1987, Brn/Gold Stuffed Doll. $3.00-5.00
❑ ❑ USA Mu8802 **Doll: Kermit,** 1987, Grn/Wht/Red Stuffed Doll. $3.00-5.00
❑ ❑ USA Mu8803 **Doll: Miss Piggy,** 1987, Red/Wht/Pnk Stuffed Doll. $3.00-5.00
❑ ❑ USA Mu8864 **Translite/Sm,** 1988. $5.00-8.00

Comments: Regional Distribution: USA - 1988. Dolls were sold for: $1.99 during Holiday promotion. They came MIP, wrapped in cellophane with a paper tag.

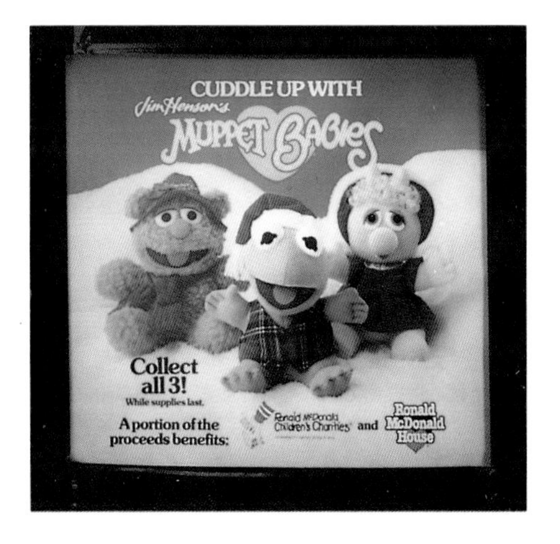

Mu8864

Mu8803 Mu8801 Mu8802

USA Generic Promotions, 1988

❑ ❑ USA Ge8810 **Hm Box,** 1988, **Continuity Program.** $——

❑ ❑ USA Ge8811 **Hm Box,** 1988, **Ron/Birdie/Fry Kids at the Beach.** $7.00-10.00

Ge8810

Ge8810

Ge8811

☐ ☐ USA Ge8801 **Dress-Up McNuggets,** 1988, Paper/Stickers/1p. $1.00-1.25

☐ ☐ USA Ge8802 **Zipper Pull - Birdie,** 1988, Yel/Pnk with Pnk Latch. $4.00-5.00
☐ ☐ USA Ge8803 **Zipper Pull - Grimace,** 1988, Purp with Purp Latch. $4.00-5.00
☐ ☐ USA Ge8804 **Zipper Pull - Hamburglar,** 1988, Blk/Wht with Blk Latch. $4.00-5.00
☐ ☐ USA Ge8805 **Zipper Pull - Ronald,** 1988, Yel/Red/Wht with Red Latch. $4.00-5.00

☐ ☐ USA Ge8806 **Pen - Birdie,** 1988, Pnk Birdie/Pnk Pen. $5.00-7.00
☐ ☐ USA Ge8807 **Pen - Fry Girl,** 1988, Pnk/Girl. $5.00-8.00
☐ ☐ USA Ge8808 **Pen - Fry Guy,** 1988, Grn/Guy. $5.00-8.00

☐ ☐ USA Ge8810 **Photo Card: Ronald with Birdie the Earlybird, Fry Kids, and Grimace,** 1988. $3.00-5.00
☐ ☐ USA Ge8811 **Photo Card: Ronald with Fry Kids and balloons,** 1988. $3.00-5.00

Ge8801

Ge8802 Ge8803 Ge8804 Ge8805

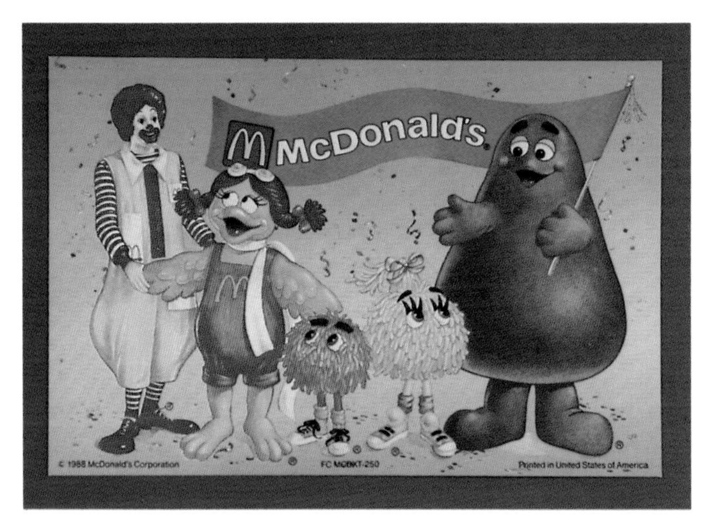

Ge8810

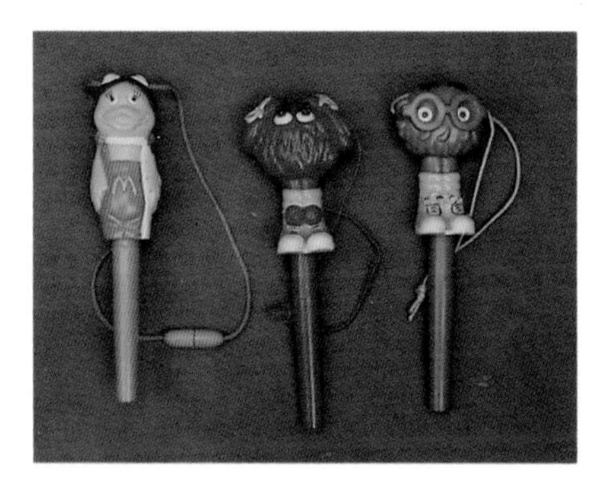

Ge8806 Ge8807 Ge8808

Ge8811

Ge8812

❑ ❑ USA GE8812 **Record: "Good Times, Great Taste"**, 45 RPM vinyl record with all the McDonald's characters on it. $2.00-3.00

❑ ❑ USA Ge8813 **Fun Times Magazine: Vol. 10 No. 1 Feb/ Mar, 1988.** $2.00-3.00

❑ ❑ USA Ge8814 **Fun Times Magazine: Vol. 10 No. 2 April/ May, 1988.** $2.00-3.00

❑ ❑ USA Ge8815 **Fun Times Magazine: Vol. 10 No. 3 Jun/ July, 1988.** $2.00-3.00

❑ ❑ USA Ge8816 **Fun Times Magazine: Vol. 10 No. 4 Aug/ Sept, 1988.** $2.00-3.00

❑ ❑ USA Ge8817 **Fun Times Magazine: Vol. 10 No. 5 Oct/ Nov, 1988.** $2.00-3.00

❑ ❑ USA Ge8818 **Fun Times Magazine: Vol. 10 No. 6 Dec/ Jan, 1988.** $2.00-3.00

❑ ❑ USA Ge8819 **Ornament: Oliver Stuffed Cat, 1988, Oliver & Company.** $4.00-6.00

❑ ❑ USA Ge8820 **Ornament: Dodger Stuffed Dog, 1988, Oliver & Company.** $4.00-6.00

Comments: Regional Distribution: USA - 1988. Happy Meal box was used when McDonald's was not selling specific Happy Meal toys/ Clean-Up weeks. Box was also used in commercials not selling specific Happy Meal toys. These are a sampling of generic premiums given away during 1988. The stuffed Oliver cat and stuffed Dodger dog were given out free with the purchase of Gift Certificates in the stores.

During April of 1988, CosMc the space character joined McDonaldland's cast of characters. He is a space alien with four to six arms, depending on the drawing. He resides in a space capsule. He appeared in the CosMc Crayola Happy Meal, April 1988, Crayola Happy Meal, 1988, and in Vol. 10 No. 2 of McDonaldland Fun Times: Welcome CosMc. Like the decreasing interest in space ventures, CosMc also disappeared from McDonaldland.

The 10,000th restaurant opened in Dale City, Virginia in 1988. McKids line of clothing and products was introduced in retail specialty stores. The retailing of McDonald's McKids merchandise expands the product line into the home.

The 10th National O/O convention was held in Washington, D.C., in conjunction with the 10,000th store opening in Dale City, Virginia. The advertising theme was: "Sharing the Dream".

Ge8816

1989

- **McChicken sandwich expands the menu selection.**

- **Big Mac sandwich is twenty-one years old - legal age!**

Beach Toy I "Collect All 4"/Test Market Happy Meal, 1989

Bags:

❑ ❑ USA Bt8955 **Hm Bag - Friendly Reflections,** 1989. $15.00-20.00

❑ ❑ USA Bt8956 **Hm Bag - Silly Story,** 1989. $15.00-20.00

❑ ❑ USA Bt8957 **Hm Bag - Splash Party**, 1989. $15.00-20.00

❑ ❑ USA Bt8958 **Hm Bag - Submarine Surprise,** 1989. $15.00-20.00

Bt8955

Bt8956

Bt8957

Bt8958

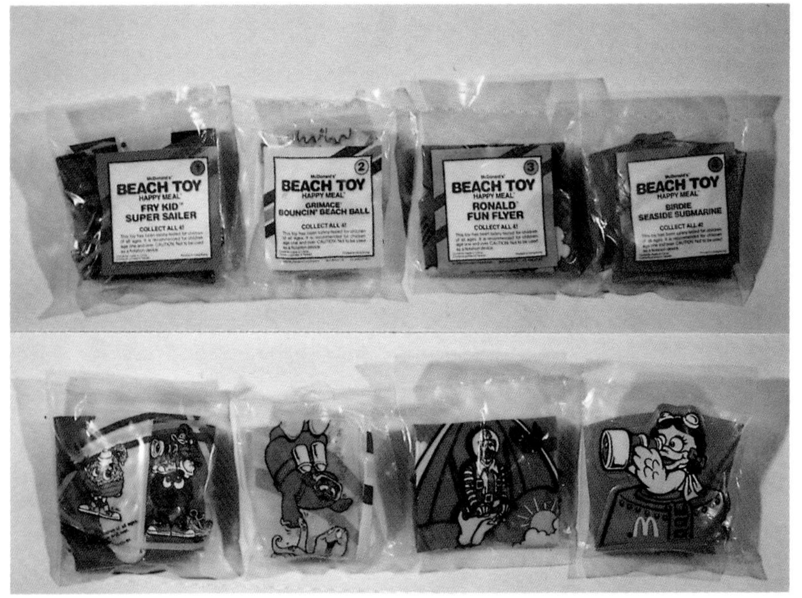

Bt8971 Bt8972 Bt8973 Bt8970

Premiums:

❑ ❑ USA Bt8971 **Set 1 Fry Kid Super Sailor,** 1989, Red-Pur Catamaran/Yel Sail. $10.00-12.00

❑ ❑ USA Bt8972 **Set 2 Grimace Bouncin Beach Ball,** 1989, Yel-Blu-Grn. $10.00-12.00

❑ ❑ USA Bt8973 **Set 3 Ronald Fun Flyer,** 1989, Turq-Org Ring. $10.00-12.00

❑ ❑ USA Bt8970 **Set 4 Birdie Seaside Sub,** 1988, Pink-Blu Inflatable. $10.00-12.00

❑ ❑ USA Bt8960 **Translite/Menu Board/Lg,** 1989. $15.00-25.00

Comments: Limited Regional Distribution: USA - June 1989 in Greenville, South Carolina and Fresno, California. "Collect All 4" was printed on the MIP polybagged package.

Bedtime/Ronald McDonald/Happy Meal, 1989

Boxes:

❑ ❑ USA Be8910 **Hm Box - Hidden Slippers,** 1988. $3.00-4.00

❑ ❑ USA Be8911 **Hm Box - Pillow Fight,** 1988. $3.00-4.00

❑ ❑ USA Be8912 **Hm Box - Scavenger Hunt,** 1988. $3.00-4.00

❑ ❑ USA Be8913 **Hm Box - Slumber Party,** 1988. $3.00-4.00

Be8910 Be8912

Be8911 Be8913

Premiums:

❑ ❑ USA Be8901 **Set 1 Toothbrush,** 1988, Yel/Ron with .85 oz. Crest Sparkle Paste. $5.00-7.00

❑ ❑ USA Be8902 **Set 2 Drinking Cup,** 1988, Ronald on Star/ 12 oz. $2.00-3.00

❑ ❑ USA Be8903 **Set 3 Foam Wash Mitt,** 1988, Ronald Scrubbing/Blu. $3.00-5.00

❑ ❑ USA Be8904 **Set 4 Nite Stand Ronald,** 1988, Glow in the Dark Star/1p. $1.50-2.50

❑ ❑ USA Be8964 **Translite/Drive-Thru/Sm,** 1988. $10.00-15.00

❑ ❑ USA Be8965 **Translite/Menu Board/Lg,** 1988. $15.00-25.00

Comments: Limited National Distribution: USA - February 3-March 2, 1989.

Be8965

Be8901 Be8902 Be8903 Be8904

Burger Six Pack to Go Promotion, 1990/1989

Boxes:

❑ ❑ USA Bu8910 **Hm Box, 1989, "McDonald's 6 2 Go" without Route 66 Sign.** $20.00-25.00

❑ ❑ USA Bu8911 **Hm Box, 1990, "Six-Pack" with Route 66 Sign.** $1.00-2.00

❑ ❑ USA Bu9064 **Translite/Lg,** 1989, "McDonald's 6 2 Go". $10.00-15.00

❑ ❑ USA Bu9065 **Translite/Lg,** 1990, Six-Pack. $10.00-15.00

Comments: Limited Regional Distribution: USA - 1989-94 in Connecticut, Florida, and Southwest USA-Route 66. Generic box held six hamburgers/cheeseburgers.

Bu8911

Bu8910

Bu9064

Chip N Dale Rescue Rangers Happy Meal, 1989

Boxes:

❑ ❑ USA Ch8960 **Hm Box - Framed,** 1989. $3.00-5.00

❑ ❑ USA Ch8961 **Hm Box - Rollin' in Dough,** 1989. $3.00-5.00

Ch8961

Ch8960

Ch8962

- ❑ ❑ USA Ch8962 **Hm Box - Yolk's on Him,** 1989. $3.00-5.00
- ❑ ❑ USA Ch8963 **Hm Box - Whale of a Time,** 1989. $3.00-5.00

U-3 Premiums:
- ❑ ❑ USA Ch8954 **U-3 Chips Rockin Racer,** 1989, Rubber/Red Rocket/Chip. $4.00-5.00
- ❑ ❑ USA Ch8955 **U-3 Gadgets Rockin Rider,** 1989, Rubber/Pnk Cup/Gadget. $4.00-5.00

Premiums:
- ❑ ❑ USA Ch8950 **Set 1 Chip's Roto-Cupter,** 1989, 3p/Red Prop/Blu-Red-Org/Yel Ruler for Blades. $3.00-4.00
- ❑ ❑ USA Ch8951 **Set 2 Dale's Roto Roadster,** 1989, 4p/Org Prop/Grn Pipe/Purp Brush Prop/Yel Copt. $3.00-4.00
- ❑ ❑ USA Ch8952 **Set 3 Gadget's Rescue Racer,** 1989, 3p/Grn Prop/Pnk Shoe Car/Blu Spatula Propeller. $3.00-4.00
- ❑ ❑ USA Ch8953 **Set 4 McJack's Propel/Phone,** 1989, 3p/Purp Prop/Turq Phone/Org Spatula Propeller. $3.00-4.00

- ❑ ❑ USA Ch8926 **Display with 4 Premiums,** 1989. $75.00-100.00

Ch8963

Ch8926

Ch8954 Ch8955 Ch8950 Ch8951 Ch8952 Ch8953

❑ ❑ USA Ch8964 **Translite/Sm,** 1989. $10.00-15.00
❑ ❑ USA Ch8965 **Translite/Lg,** 1989. $15.00-25.00

Comments: National Distribution: USA - Oct 27-Nov 23, 1989. Premium markings - "Disney China."

Craft Kits/Canceled/Happy Meal, 1989

Premiums: Craft Supplies (Projected):
❑ ❑ USA Cr8900 **Scissors - Fry Kid.** $——
❑ ❑ USA Cr8901 **Paint Brush - Hamburglar with Paint Palette.** $——
❑ ❑ USA Cr8902 **Tape Measure - Grimace.** $——
❑ ❑ USA Cr8903 **Tape Dispenser - Ronald/Birdie/Hamburglar/Grimace decal.** $——

Comments: Canceled. Had been programmed for December 5, 1989.

Ch8965

Dinosaur Talking Storybook Happy Meal, 1989

Bag:
❑ ❑ USA Di8930 **Hm Bag - Bones & Dodo/Dinosaurs,** 1989. $2.00-3.00

Premiums: Book & Tape:
❑ ❑ USA Di8900 **Book & Tape: Amazing Birthday Adventure,** 1989, tape/16 page book on Dino. $8.00-10.00
❑ ❑ USA Di8901 **Book & Tape: Creature in the Cave,** 1989, tape/16 page book on Dino. $8.00-10.00
❑ ❑ USA Di8902 **Book & Tape: Danger under the Lake,** 1989, tape/16 page book on Dino. $8.00-10.00
❑ ❑ USA Di8903 **Book & Tape: Dinosaur Baby Boom,** 1989, tape/16 page book on Dino. $8.00-10.00

❑ ❑ USA Di8964 **Translite/Sm,** 1989. $15.00-20.00
❑ ❑ USA Di8965 **Translite/Lg,** 1989. $20.00-25.00

Comments: Regional Distribution: USA - Summer 1989 In Michigan and Wisconsin.

Di8930

Di8965

Di8900-03

Fu8907

Fun with Food Happy Meal, 1989

Boxes:

❑ ❑ USA Fu8907 **Hm Box - 3 Ring Circus,** 1988. $3.00-4.00

❑ ❑ USA Fu8908 **Hm Box - In Concert,** 1988. $3.00-4.00

❑ ❑ USA Fu8909 **Hm Box - Making a Splash,** 1988. $3.00-4.00

❑ ❑ USA Fu8910 **Hm Box - Movie Making,** 1988. $3.00-4.00

Premiums:

❑ ❑ USA Fu8900 **Hamburger Guy,** 1989, 3p Top/Hmbg/Bun with Decal, Fisher-Price. $7.00-10.00

❑ ❑ USA Fu8901 **French Fry Guy,** 1989, 3p Bag with Fries and Decals, Fisher-Price. $4.00-5.00

❑ ❑ USA Fu8902 **Soft Drink Cup,** 1989, 2p Cup with Lid and Decals, Fisher-Price. $4.00-5.00

❑ ❑ USA Fu8903 **Chicken McNugget Guys,** 1989, 4p Nuggets with Container, Fisher-Price. $4.00-5.00

Fu8908

Fu8909

Fu8910

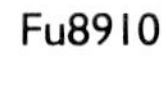

Fu8900 Fu8901 Fu8902 Fu8903

❑ ❑ USA Fu8964 **Translite/Sm,** 1989. $10.00-15.00
❑ ❑ USA Fu8965 **Translite/Lg,** 1989. $15.00-25.00

Comments: Limited Regional Distribution: USA - September 1-28, 1989.

Fu8965

Funny Fry Friends I "Collect All 4"/Test Market Happy Meal, 1991/1989

Premiums:
❑ ❑ USA Ff8900 **Gadzooks,** 1989, Blu Kid with Eyeglasses. $15.00-20.00
❑ ❑ USA Ff8901 **Matey,** 1989, Red Kid with Pirate Hat. $20.00-25.00
❑ ❑ USA Ff8902 **Tracker,** 1989, Blu Kid with Safari Hat and Snake. $15.00-20.00
❑ ❑ USA Ff8903 **Zzz's,** 1989, Turq Kid with Sleeping Cap and Bear. $15.00-20.00

Comments: Regional Distribution: USA - 1989; May/June 1991 in California, Maryland, Pennsylvania. "Collect All 4" on MIP insert card.

Fun with Food blue book.

Ff8901

Ff8903

Garfield II Happy Meal, 1989

Boxes and Bag:
❑ ❑ USA Ga8910 **Hm Box - Ahh Vacation,** 1989. $2.00-3.00
❑ ❑ USA Ga8911 **Hm Box - Cat with a Mission,** 1989. $2.00-3.00
❑ ❑ USA Ga8912 **Hm Box - Garfield Catches Lunch,** 1989. $2.00-3.00
❑ ❑ USA Ga8913 **Hm Box - Mischief this Morning,** 1989. $2.00-3.00

Ga8912 Ga8913

Ga8911 Ga8910

Ga8901 Ga8902 Ga8903 Ga8904

Ga8905 Ga8906

Ga8926

Ga8950

Ga8952

Ga8951

Ga8965

❑ ❑ USA Ga8914 **Hm Bag - Safari Garfield,** 1989. $10.00-15.00

U-3 Premiums:

❑ ❑ USA Ga8905 **U-3 Garfield Skating,** 1988, on Roller Skates. $4.00-5.00

❑ ❑ USA Ga8906 **U-3 Garfield with Pooky,** 1988, on Skateboard. $4.00-5.00

Premiums:

❑ ❑ USA Ga8901 **Set 1 Garfield on Scooter,** 1988, on Yel Scooter/Purp-Red Wheel/2p. $4.00-5.00

❑ ❑ USA Ga8902 **Set 2 Garfield in 4 Wheeler Car,** 1988, on Blu-Yel 4 Wheeler/2p. $4.00-5.00

❑ ❑ USA Ga8903 **Set 3 Garfield on Skateboard,** 1988, on Pnk Skateboard/2p. $4.00-5.00

❑ ❑ USA Ga8904 **Set 4 Garfield/Odie with Sidecar,** 1988, on Red Motorcycle/Blu Whls/2p. $4.00-5.00

❑ ❑ USA Ga8950 **Button,** 1989, Garfield with Arches. $4.00-5.00

❑ ❑ USA Ga8926 **Display/Prem,** 1989. $85.00-100.00

❑ ❑ USA Ga8941 **Danglers/4,** 1989. Each $7.00-10.00

❑ ❑ USA Ga8951 **Pin: Garfield/Odie with Arches.** $4.00-5.00

❑ ❑ USA Ga8952 **Pin: Garfield with Arches.** $4.00-5.00

❑ ❑ USA Ga8964 **Translite/Sm,** 1989. $10.00-15.00

❑ ❑ USA Ga8965 **Translite/Lg,** 1989. $15.00-25.00

Comments: National Distribution: USA - June 23-July 20, 1989. Premium markings - "United Feat. Synd. China H6." Toys came with a paper insert illustrating the toy and/or with the information printed on the bag. MIP can be either way, printed on the bag or with the insert card.

Halloween '89 Happy Meal, 1989

Premiums: Halloween Pails

❑ ❑ USA Ha8941 **Pail - McGhost,** 1989, Wht with Blk Face/ Wht Lid. $1.00-2.00

❑ ❑ USA Ha8942 **Pail - McWitch,** 1989, Grn with Grn Hat/Lid. $1.00-2.00

❑ ❑ USA Ha8964 **Translite/Sm,** 1989. $4.00-5.00

❑ ❑ USA Ha8965 **Translite/Lg,** 1989. $4.00-5.00

Comments: National Distribution: USA - October 6-31, 1989. Translite shows a third pail. USA Ha8725-27 pails were distributed again as third and/or fourth and fifth pails. USA Ha8941-42 had four grid lids/no name on back of pails.

Ha8965

Ha8942 Ha8941

Lego Motion IV Happy Meal, 1989

Boxes:

❑ ❑ USA Le8915 **Hm Box - Lake,** 1989. $2.00-4.00

❑ ❑ USA Le8916 **Hm Box - Traffic Copter/Clocks,** 1989. $2.00-4.00

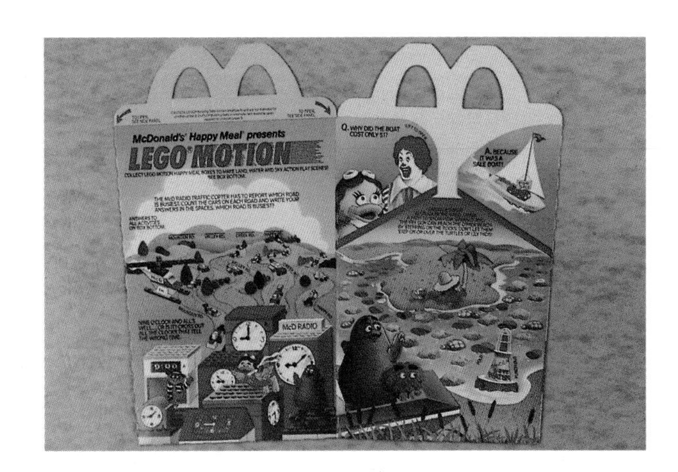

Le8916

Le8915

❑ ❑ USA Le8917 **Hm Box - Road Race,** 1989. 2.00-4.00

❑ ❑ USA Le8918 **Hm Box - Ronald in Helicopter/Blocks,** 1989. $2.00-4.00

U-3 Premiums:

❑ ❑ USA Le8908 **U-3 Giddy the Gator,** 1989, Duplo/Packaged 6p Ages 1 1/2-4. $2.00-3.00

❑ ❑ USA Le8909 **U-3 Tuttle the Turtle,** 1989, Duplo/Packaged 6p Ages 1 1/2-4. $2.00-3.00

Premiums:

❑ ❑ USA Le8900 **Set 1a Gyro Bird Helicopter,** 1989, Lego/Packaged 19p. $2.00-3.00

❑ ❑ USA Le8901 **Set 1b Turbo Force Car,** 1989, Lego/Packaged 10p. $2.00-3.00

❑ ❑ USA Le8902 **Set 2a Swamp Stinger Air Boat,** 1989, Lego/Packaged 16p. $2.00-3.00

❑ ❑ USA Le8903 **Set 2b Lightning Striker Airplane,** 1989, Lego/Packaged 14p. $2.00-3.00

Le8917

Le8918

Le8904 Le8905

Le8906 Le8907

Le8900 Le8901

Le8900 Le8901

- ❑ ❑ USA Le8904 **Set 3a Land Laser Car,** 1989, Lego/Packaged 13p. $2.00-3.00
- ❑ ❑ USA Le8905 **Set 3b Sea Eagle Seaplane,** 1989, Lego/Packaged 15p. $2.00-3.00
- ❑ ❑ USA Le8906 **Set 4a Wind Whirler Helicopter,** 1989, Lego/Packaged 17p. $2.00-3.00
- ❑ ❑ USA Le8907 **Set 4b Sea Skimmer Boat,** 1989, Lego/Packaged 17p. $2.00-3.00

- ❑ ❑ USA Le8926 **Display/Premiums,** 1989. $85.00-100.00
- ❑ ❑ USA Le8963 **Menu Board Premium Lug-On,** 1989. $7.00-10.00
- ❑ ❑ USA Le8964 **Translite/Sm,** 1989. $4.00-5.00
- ❑ ❑ USA Le8965 **Translite/Motion Wheel/Lg,** 1989. $10.00-15.00

Comments: National Distribution: USA - July 28-August 24, 1989.

Le8909 Le8908

Le8926

Le8964

Little Gardener Happy Meal, 1989

Bags:

- ❑ ❑ USA Lg8985 **Hm Bag, 1989, Birdie's Bouquet.** $1.00-2.00
- ❑ ❑ USA Lg8986 **Hm Bag, 1989, Garden Goodies.** $1.00-2.00
- ❑ ❑ USA Lg8987 **Hm Bag, 1989, Radish Contest.** $1.00-2.00
- ❑ ❑ USA Lg8988 **Hm Bag, 1989, Whose Hose.** $1.00-2.00

Lg8985 Lg8986

Lg8987 Lg8988

Lg8976 Lg8977 Lg8978

U-3 Premium:

❑ ❑ USA Lg8979 **U-3 Birdie's Shovel,** 1988, Org **without Zebra Stripes/No Seeds.** $6.00-8.00

Premiums:

❑ ❑ USA Lg8975 **Set 1 Birdie's Shovel,** 1988, Org with Burpee Marigold Seeds. $1.00 -1.50

❑ ❑ USA Lg8976 **Set 2 Fry Kids Planter,** 1988, Turq with Pur Lid/Handle. $1.00-1.50

❑ ❑ USA Lg8977 **Set 3 Grimace Rake,** 1988, Grn with Burpee Radish Seeds. $1.00-1.50

❑ ❑ USA Lg8978 **Set 4 Ron Watering Can,** 1989, Red/Yel with Handle. $1.00-1.50

❑ ❑ USA Lg8964 **Translite/Sm,** 1989. $4.00-5.00

❑ ❑ USA Lg8965 **Translite/Lg,** 1989. $7.00-10.00

Comments: National Distribution: USA - April 21-May 18, 1989. U-3 is polybagged without seeds, no U-3 zebra stripes.

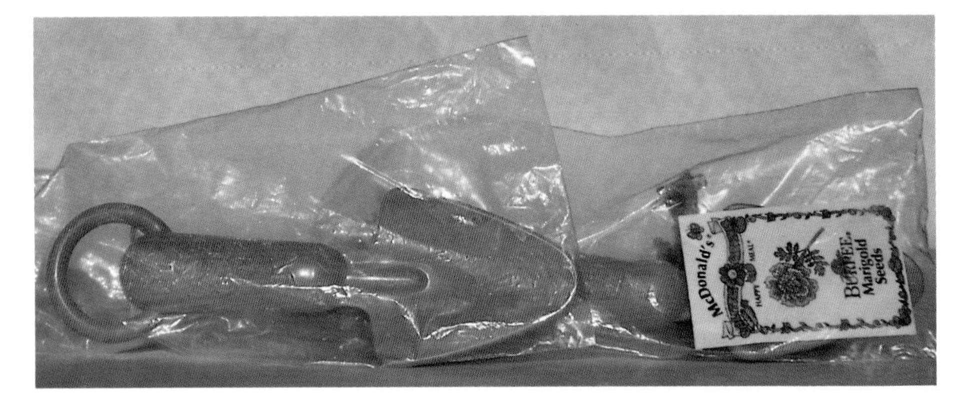

Lg8979 Lg8975

Lg8965

Little Mermaid I Happy Meal, 1989

Boxes and Bags:

❑ ❑ USA Li8910 **Hm Box - Ariel's Grotto,** 1989. $3.00-4.00

❑ ❑ USA Li8911 **Hm Box - Sea Garden,** 1989. $3.00-4.00

Li8910

Li8911

- ❑ ❑ USA Li8912 **Hm Box - Ursula's Domain,** 1989. $3.00-4.00
- ❑ ❑ USA Li8913 **Hm Box - Village Lagoon,** 1989. $3.00-4.00
- ❑ ❑ USA Li8916 **Hm Bag - Ursula's Domain,** 1989. $15.00-20.00
- ❑ ❑ USA Li8917 **Hm Bag - Village Lagoon,** 1989. $15.00-20.00

Li8913

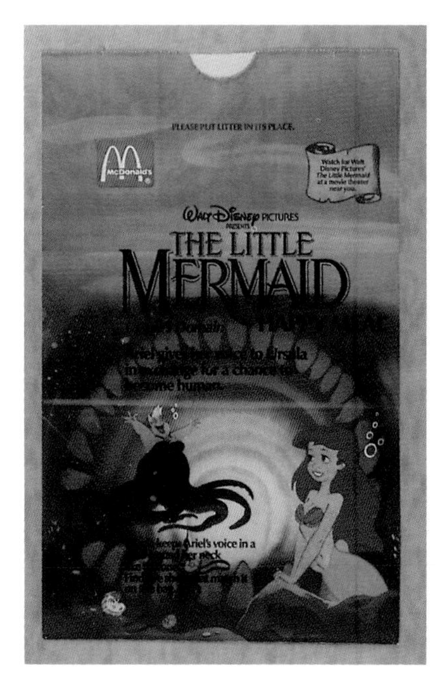

Li8916

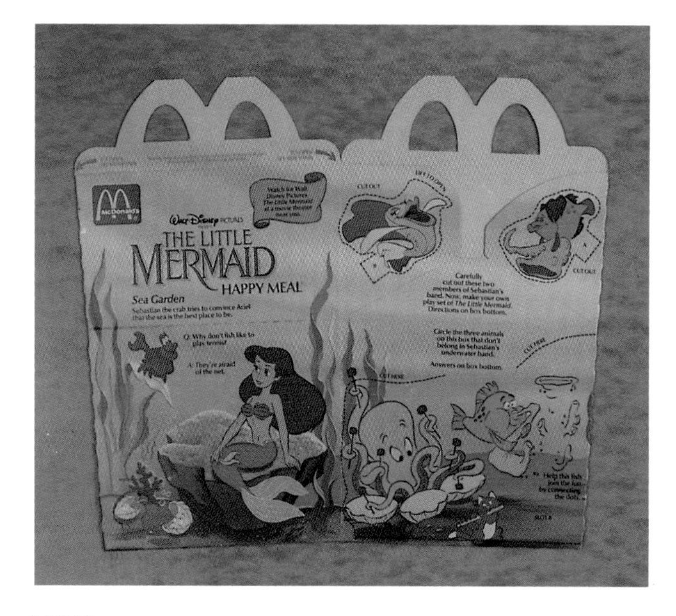

Li8912

Li8900 Li8901 Li8902 Li8903

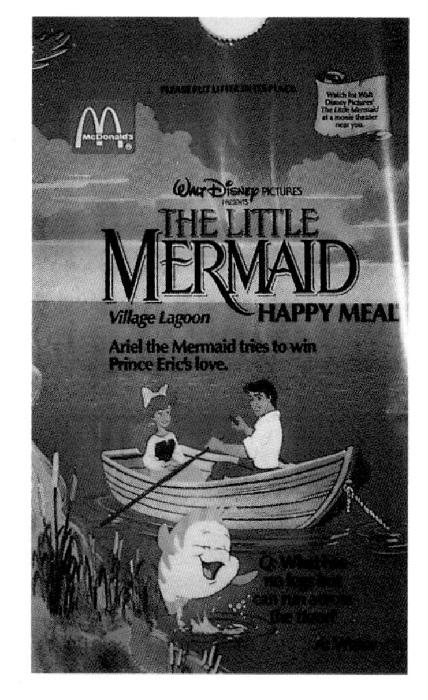

Li8917

Li8965

Premiums:

❑ ❑ USA Li8900 **Set 1 Flounder,** 1989, Fish Yel Water Squirter. $3.00-4.00

❑ ❑ USA Li8901 **Set 2 Ursula,** 1989, Octopus Blk-Pur with Suction Cup. $3.00-4.00

❑ ❑ USA Li8902 **Set 3 Prince Eric,** 1989, with Sebastian with Boat Yel. $3.00-4.00

❑ ❑ USA Li8903 **Set 4 Ariel,** 1989, Mermaid Kneeling Turq-Org. $4.00-5.00

❑ ❑ USA Li8964 **Translite/Sm,** 1989. $15.00-20.00

❑ ❑ USA Li8965 **Translite/Lg,** 1989. $20.00-25.00

Comments: National Distribution: USA - November 24-December 21, 1989. Two bags were test marketed in South Bend, Indiana. Premium markings - "Disney China."

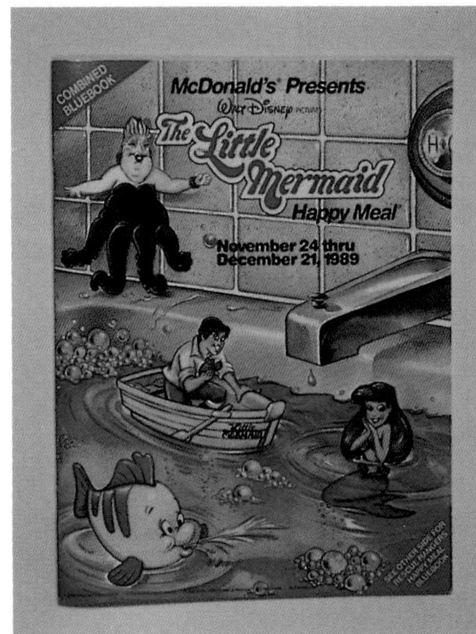

Little Mermaid blue book.

McBunny (Pails) Happy Meal, 1989

Premiums: Easter Pails

❑ ❑ USA Bp8901 **Pail: Fluffy,** 1988, Pail/Wht-Blu with Blu Lid/Yel Handle. $4.00-5.00

❑ ❑ USA Bp8902 **Pail: Pinky,** 1988, Pail/Wht-Yel with Yel Lid/Yel Handle. $4.00-5.00

❑ ❑ USA Bp8903 **Pail: Whiskers,** 1988, Pail/Wht-Grn Grn Lid/Yel Handle. $4.00-5.00

❑ ❑ USA Bp8964 **Translite/Sm,** 1988. $10.00-15.00

❑ ❑ USA Bp8965 **Translite/Lg,** 1988. $15.00-20.00

Comments: Regional Distribution: USA - Spring 1989 in Illinois, California, and Missouri. Redistributed in Alabama in the Spring of 1991. Note: Translite says, "Ears an Offer" from which the McDonald's archives derived the term: "Earsan."

Bp8965

Bp9101 Bp9102 Bp9103

Mb8915

Mickey's Birthdayland Happy Meal, 1989

Boxes:

❑ ❑ USA Mb8915 **Hm Box - Barn/Daisy Cafe,** 1988. $3.00-4.00

❑ ❑ USA Mb8916 **Hm Box - Mickey's House,** 1988. $3.00-4.00

❑ ❑ USA Mb8917 **Hm Box - Minnie's Dress Shop,** 1988. $3.00-4.00

❑ ❑ USA Mb8918 **Hm Box - Theater,** 1988. $3.00-4.00

❑ ❑ USA Mb8919 **Hm Box - Train Station,** 1988. $3.00-4.00

Mb8916

Mb8918

Mb8917

Mb8919

Mb8900 Mb8901 Mb8902 Mb8903 Mb8904

U-3 Premiums:

❑ ❑ USA Mb8905 **U-3 Donald's Jeepster,** 1988, Blu with Wht Whls/Yel Stripes. $5.00-6.00

❑ ❑ USA Mb8911 **U-3 Donald's Jeepster,** 1988, Blu with Wht Whls/Wht Stripes. $7.00-8.00

❑ ❑ USA Mb8906 **U-3 Donald's Jeepster,** 1988, Grn with Red Whls/Wht Stripes. $5.00-6.00

❑ ❑ USA Mb8907 **U-3 Goofy's Sedan,** 1988, Blu with Wht Whls. $5.00-6.00

❑ ❑ USA Mb8908 **U-3 Goofy's Sedan,** 1988, Grn with Red Whls. $5.00-6.00

❑ ❑ USA Mb8909 **U-3 Mickey's Roadster,** 1988, Red with Yel Whls. $5.00-6.00

❑ ❑ USA Mb8910 **U-3 Minnie's Convertible,** 1988, Pink with Yel Whls. $5.00-6.00

Mb8911 Mb8905 Mb8907 Mb8909

Mb8926

Mb8965

Premiums:
❑ ❑ USA Mb8900 **Set 1 Donald's Engine,** 1988, Grn. $3.00-4.00
❑ ❑ USA Mb8901 **Set 2 Minnie Convertible,** 1988, Pink. $3.00-4.00
❑ ❑ USA Mb8902 **Set 3 Goofy's Jalopy,** 1988, Blu. $3.00-4.00
❑ ❑ USA Mb8903 **Set 4 Pluto Car,** 1988, Purple Car/Pluto **without Grn Collar.** $4.00-5.00
❑ ❑ USA Mb8912 **Set 4 Pluto Car,** 1988, Purple Car/Pluto **with Grn Collar.** $3.00-4.00
❑ ❑ USA Mb8904 **Set 5 Mickey Roadster,** 1988, Red. $2.00-3.00

❑ ❑ USA Mb8926 **Display with 2 Sets/Premiums,** 1988. $200.00-250.00
❑ ❑ USA Mb8941 **Dangler,** 1988. Each $8.00-10.00
❑ ❑ USA Mb8963 **Menu Board Lug-On,** 1988. Each $15.00-20.00
❑ ❑ USA Mb8964 **Translite/Sm,** 1988. $10.00-15.00
❑ ❑ USA Mb8965 **Translite/Lg,** 1988. $15.00-20.00

Comments: National Distribution: USA - March 17-April 20, 1989. Premium markings - "Disney China."

Mix'em up Monsters Happy Meal, 1990/1989

Box:
❑ ❑ USA Mi8910 **Hm Box - Monsters on Moon,** 1988. $4.00-5.00

Premiums:
❑ ❑ USA Mi8900 **Blibble,** 1986, Green/Extended Eyes 3p. $3.00-4.00
❑ ❑ USA Mi8901 **Corkle,** 1986, Blue/Folded Arms 3p. $3.00-4.00
❑ ❑ USA Mi8902 **Gropple,** 1986, Yel/Two Heads 3p. $3.00-4.00
❑ ❑ USA Mi8903 **Thugger,** 1986, Pur/Large Tusks 3p. $3.00-4.00

❑ ❑ USA Mi8963 **Window Decal/Sm,** 1988. $7.00-10.00
❑ ❑ USA Mi8964 **Translite/Sm,** 1988. $4.00-5.00
❑ ❑ USA Mi8965 **Translite/Lg,** 1988. $7.00-10.00

Comments: Regional Distribution: USA - January/September 1989 and September 7-October 4, 1990 In Northern California and St. Louis, Missouri. Premium markings - "Current Inc China F8x."

Mi8910

Blue book picture with white connectors.

Mi8965

Muppet Kids Test Market Happy Meal, 1989

Boxes:

❑ ❑ USA Mu8910 **Hm Box - Club House,** 1989. $15.00-20.00

❑ ❑ USA Mu8911 **Hm Box - School,** 1989. $15.00-20.00

Premiums (MIP: $20.00-25.00; Loose: $15.00):

❑ ❑ USA Mu8900 **Set 1 Kermit,** 1989, Red Bike/Yel Wheel. $20.00-25.00

❑ ❑ USA Mu8901 **Set 2 Miss Piggy,** 1989, Pnk Bike/Grn Wheel/Yel. $20.00-25.00

Mu8910

Mu8911

Mu8900

Mu8901

Mu8902

Mu8903

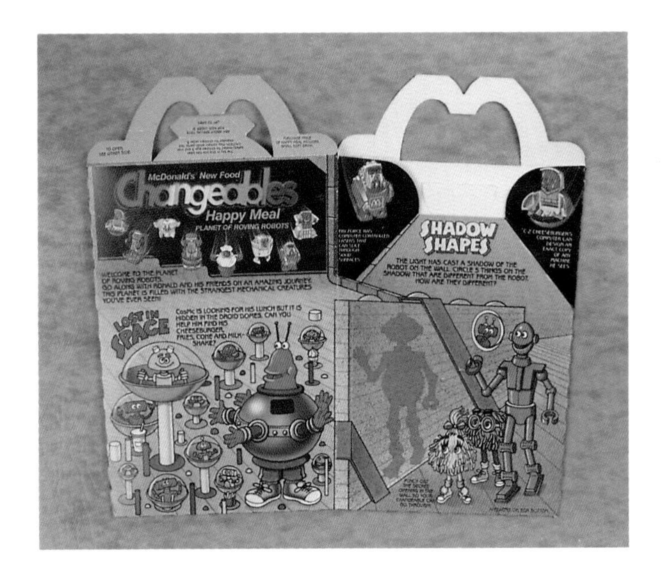

Ne8915

❏ ❏ USA Mu8902 **Set 3 Gonzo,** 1989, Yel Bike/Pnk Wheel/Red. $20.00-25.00

❏ ❏ USA Mu8903 **Set 4 Fozzie,** 1989, Grn Bike/Red Wheel. $20.00-25.00

❏ ❏ USA Mu8926 **Display/Premiums,** 1989. $225.00-275.00

❏ ❏ USA Mu8964 **Translite/Sm,** 1989. $15.00-20.00

❏ ❏ USA Mu8965 **Translite/Lg,** 1989. $25.00-35.00

Comments: Limited Regional Distribution: USA - Summer 1989 In St. Louis, Missouri test market. Discontinued after Henson dropped the kids concept. Premium markings - "Ha! 1989 China." or "Simon Mkt."

New Food Changeables Happy Meal, 1989

Boxes:

❏ ❏ USA Ne8915 **Hm Box - Lost in Space,** 1988. $2.00-3.00

❏ ❏ USA Ne8916 **Hm Box - Jeepers/Peepers,** 1988. $2.00-3.00

❏ ❏ USA Ne8917 **Hm Box - Tongue Tippers,** 1988. $2.00-3.00

❏ ❏ USA Ne8918 **Hm Box - Who's That,** 1988. $2.00-3.00

Ne8916

Ne8917

U-3 Premium:
❑ ❑ USA Ne8908 **U-3 McDonald's Pals,** 1988, Wht Cube with Char Photo. $6.00-8.00

Premiums:
❑ ❑ USA Ne8900 **Set 1a Robocakes,** 1988, Hot Cakes/Grn Hands. $2.00-4.00
❑ ❑ USA Ne8901 **Set 1b Gallacta Burger,** 1988, Quarter Pounder/Pnk Hands. $2.00-4.00
❑ ❑ USA Ne8902 **Set 2a Fry Force,** 1987, Large French Fries/Turq Hands. $2.00-4.00
❑ ❑ USA Ne8903 **Set 2b Krypto Cup,** 1988, Milk Shake Opens from Side/Blu Hands. $2.00-4.00
❑ ❑ USA Ne8904 **Set 3a Macro Mac,** 1987, Big Mac/Pnk Hands. $2.00-4.00
❑ ❑ USA Ne8905 **Set 3b Turbo Cone,** 1988, Ice Cream Cone/Pnk Hands. $2.00-4.00
❑ ❑ USA Ne8906 **Set 4a C2-Cheeseburger,** 1988, Cheeseburger/Org Hands. $2.00-4.00
❑ ❑ USA Ne8907 **Set 4b Fry-Bot,** 1988, Small Fries/Pnk Feet. $2.00-4.00

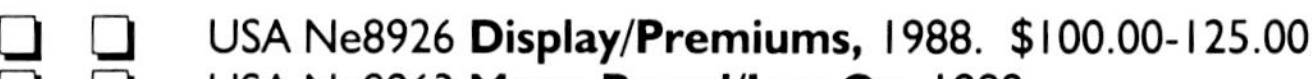

❑ ❑ USA Ne8926 **Display/Premiums,** 1988. $100.00-125.00
❑ ❑ USA Ne8963 **Menu Board/Lug-On,** 1988. $10.00-15.00
❑ ❑ USA Ne8964 **Translite/Sm,** 1988. $10.00-15.00
❑ ❑ USA Ne8965 **Translite/X-O Graphic/Lg,** 1988. $15.00-25.00

Comments: National Distribution: USA May 19-June 15, 1989. Version 1, issued in 1987 (see Changeables '87), had five premiums: Big Mac, Large French Fries, Egg McMuffin, Chicken McNuggets, Quarter Pounder. Version 2, also issued in 1987, added a Milk Shake. Version 3, issued in 1989 (New Food Changeables), reissued the Large Fries and the Big Mac from 1987 and added six new designs. The Macro Mac/Big Mac had pink painted hands.

Ne8918

Ne8900 Ne8901 Ne8902 Ne8903

Ne8904 Ne8905 Ne8906 Ne8907

Ne8926

Ne8964

Ne8908

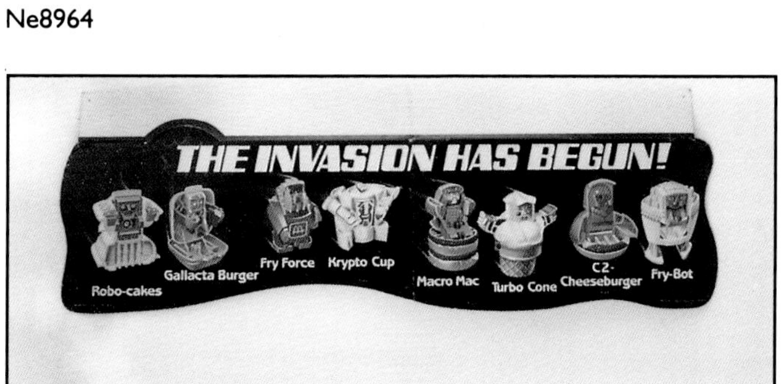

Ra8910

Raggedy Ann and Andy Happy Meal, 1989

Box:

❑ ❑ USA Ra8910 **Hm Box, 1988, Schoolhouse.** $4.00-5.00

U-3 Premium:

❑ ❑ USA Ra8905 **U-3 Camel with Wrinkled Knees,** 1989, without Teeter Totter. $8.00-10.00

Premiums (MIP: $5.00-10.00; Loose: $4.00-5.00):

❑ ❑ USA Ra8901 **Set 1 Raggedy Andy,** 1989, 3p with Slide. $7.00-10.00

❑ ❑ USA Ra8902 **Set 2 Raggedy Ann,** 1989, 5p with Swing. $7.00-10.00

❑ ❑ USA Ra8903 **Set 3 Grouchy Bear,** 1989, 3p with Carousel/ Merry-Go-Round. $5.00-6.00

❑ ❑ USA Ra8904 **Set 4 Camel with Wrinkled Knees,** 1989, 3p with Teeter Totter. $7.00-10.00

❑ ❑ USA Ra8964 **Translite/Sm,** 1989. $25.00-40.00

❑ ❑ USA Ra8965 **Translite/Lg,** 1989. $35.00-30.00

Comments: Regional Distribution: USA - September 1-28, 1989 in San Francisco, California, Portland, Oregon, Nevada, Hawaii, and southern Pennsylvania. Premium markings - "1988 Macmillian Inc China or "1989 Simon Marketing Inc China." **Loose Ragggedy Ann and Andy figurines with the pieces are selling for $4.00 - 5.00 each.**

Ra8905 Ra8901 Ra8902 Ra8903 Ra8904

Picture from Raggedy Ann and Andy blue book.

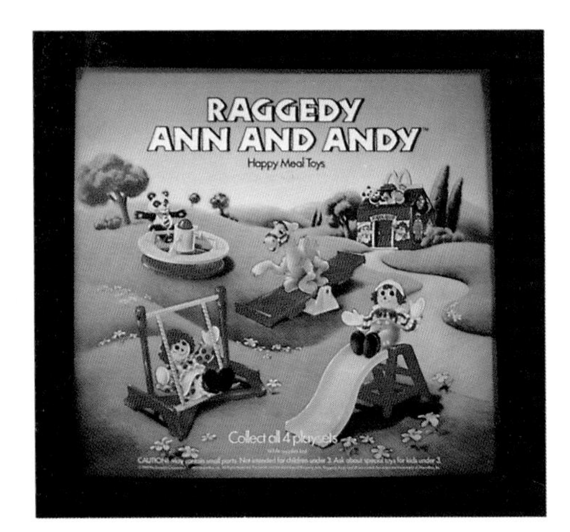

Ra8965

Rain or Shine Happy Meal, 1989

Boxes:

❑ ❑ USA Ge8900 **Hm Box - Bubbles,** 1989. $1.00-1.50

❑ ❑ USA Ge8901 **Hm Box - Umbrellas,** 1989. $1.00-1.50

Ge8901 Ge8900

❑ ❑ USA Ge8950 **Button,** 1989, Build Your Own Hm. $2.00-3.00

Ge8950

Comments: Regional Distribution: USA - 1989 during Clean-Up week. Generic premiums were used.

Read Along With Ronald Happy Meal, 1989

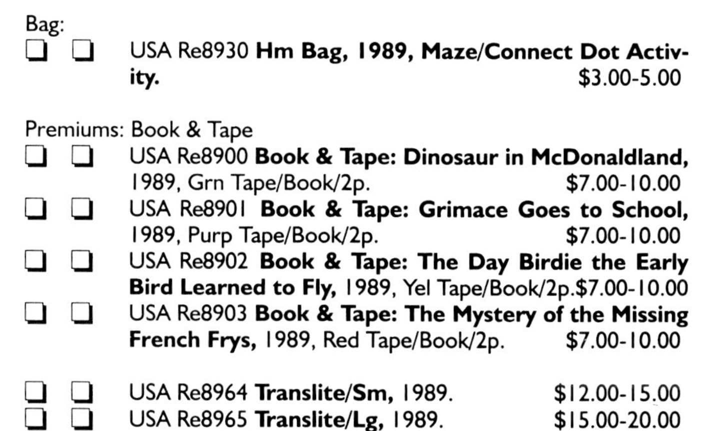

Bag:
❑ ❑ USA Re8930 **Hm Bag, 1989, Maze/Connect Dot Activity.** $3.00-5.00

Premiums: Book & Tape
❑ ❑ USA Re8900 **Book & Tape: Dinosaur in McDonaldland,** 1989, Grn Tape/Book/2p. $7.00-10.00
❑ ❑ USA Re8901 **Book & Tape: Grimace Goes to School,** 1989, Purp Tape/Book/2p. $7.00-10.00
❑ ❑ USA Re8902 **Book & Tape: The Day Birdie the Early Bird Learned to Fly,** 1989, Yel Tape/Book/2p. $7.00-10.00
❑ ❑ USA Re8903 **Book & Tape: The Mystery of the Missing French Frys,** 1989, Red Tape/Book/2p. $7.00-10.00

❑ ❑ USA Re8964 **Translite/Sm,** 1989. $12.00-15.00
❑ ❑ USA Re8965 **Translite/Lg,** 1989. $15.00-20.00

Comments: Regional Distribution: USA - Summer 1989 in New England states.

Re8900 Re8901

Re8902

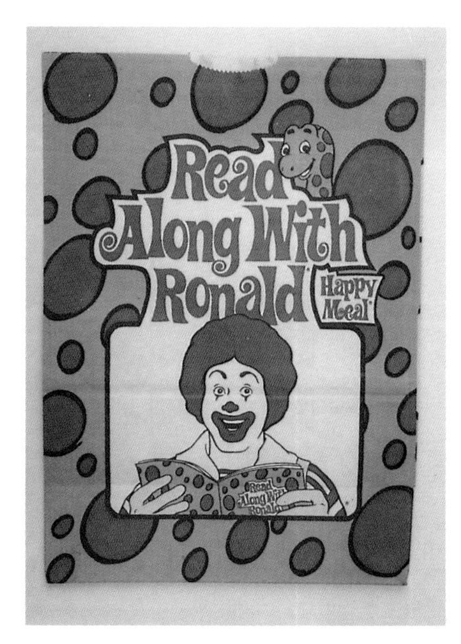

Re8930

Re8965

Re8903 Re8900

Se8930

Sea World of Texas II Happy Meal, 1989

Box:

❑ ❑ USA Se8930 **Hm Box - Ronald in a Yel Sub/ with Coupon,** 1988. $10.00-15.00

Premiums:

❑ ❑ USA Se8925 **Sea Otter,** 1988, Stuffed Animal/Brn/Wht/6". $20.00-25.00

❑ ❑ USA Sw8800 **Dolphin,** 1988, Grey/Wht with Blk Eyes 6". $20.00-25.00

❑ ❑ USA Sw8803 **Whale,** 1988, Black-Wht 6". $20.00-25.00

❑ ❑ USA Se8926 **Penguin Sun Glasses,** 1989, Blk-Wht. $40.00-50.00

❑ ❑ USA Se8927 Whale Sun Glasses, 1989, Blk-Wht. $40.00-50.00

❑ ❑ USA Se8937 Display Card Premiums, 1989. $75.00-100.00

❑ ❑ USA Se8964 Translite/Sm, 1989. $15.00-25.00

❑ ❑ USA Se8965 Translite/Lg, 1989. $20.00-35.00

Comments: Regional Distribution: USA - 1989 in San Antonio, Texas. USA Sw8800 Dolphin and USA Sw8803 Whale were distributed again with Sea World of Texas II in 1989. **Loose prices for the plush Otter, Dolphin, and Whale range $8.00 - 12.00. Loose sunglasses range $35.00 - 50.00 depending on condition.**

Se8926 Se8925 Se8927

Se8803 Se8925 Se8800 Se8927 Se8926

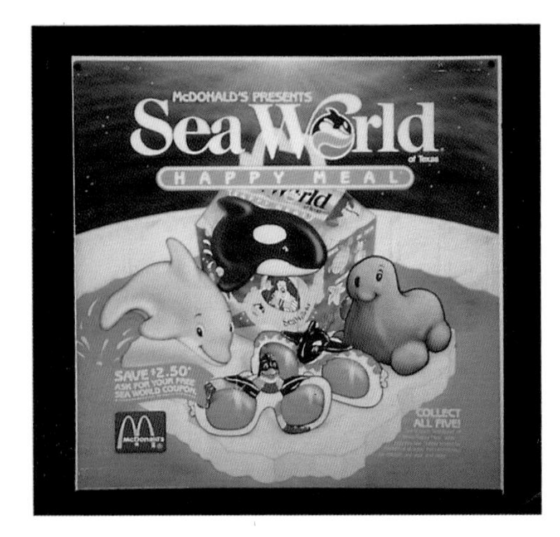

Se8965

Sun Glasses/McDonaldland Promotion, 1989

Premiums: Sunglasses:

❑ ❑ USA Su8901 **Sunglasses: Birdie the Early Bird,** 1988, Wht with Braids on Top. $5.00-7.00

❑ ❑ USA Su8902 **Sunglasses: Grimace,** 1988, Purp with Arms on Top. $5.00-7.00

Su8902 Su8901

❑ ❑ USA Su8903 **Sunglasses: Hamburglar,** 1988, Yel with Arms on Top. $5.00-7.00
❑ ❑ USA Su8904 **Sunglasses: Ronald McDonald,** 1988, Yel with Arms on Top. $5.00-7.00

❑ ❑ USA Su8964 **Translite/Sm,** 1989. $10.00-15.00

Comments: Regional Distribution: Southern USA - 1989 in Florida, Georgia, Alabama.

Su8964

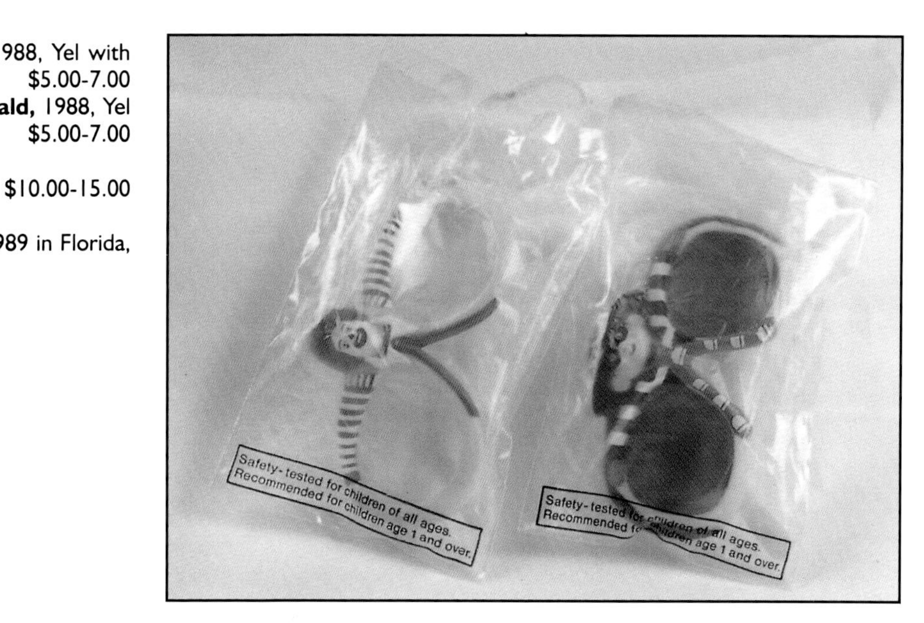

Su8904 Su8903

USA Generic Promotions, 1989

❑ ❑ USA Ge8902 **Record: Menu Song,** 45 RPM record. Record was insert in nationally distributed newspapers of February 19, 1989. Record announces McDonald's Menu Song campaign. One record announced the holder as the $1,000,000 winner. $1.00-2.00

❑ ❑ USA Ge8903 **Fun Times Magazine: Vol. 11 No. 1 Feb/ Mar, 1989.** $2.00-3.00
❑ ❑ USA Ge8904 **Fun Times Magazine: Vol. 11 No. 2 Apr/ May, 1989.** $2.00-3.00
❑ ❑ USA Ge8905 **Fun Times Magazine: Vol. 11 No. 3 Issue 3, 1989.** $2.00-3.00
❑ ❑ USA Ge8906 **Fun Times Magazine: Vol. 11 No. 4 Issue 4, 1989.** $2.00-3.00
❑ ❑ USA Ge8907 **Fun Times Magazine: Vol. 11 No. 5 Issue 5, 1989.** $2.00-3.00
❑ ❑ USA Ge8908 **Fun Times Magazine: Vol. 11 No. 6 Issue 6, 1989.** $2.00-3.00

❑ ❑ USA Ge8909 **Plate: French Fry Garden.** $4.00-6.00
❑ ❑ USA Ge8910 **Plate: Hamburger University.** $4.00-6.00

Ge8903

Ge8909

Ge8910

Ge8911

Ge8912

❑ ❑ USA Ge8911 **Plate: McNugget Band.** $4.00-6.00
❑ ❑ USA Ge8912 **Plate: Milkshake Lake.** $4.00-6.00

❑ ❑ USA Ge8913 **Ornament: Flounder - Stuffed Fish/No Date/The Little Mermaid/Yel.** $4.00-6.00
❑ ❑ USA Ge8914 **Ornament: Sebastian - Stuffed Crab/No Date/The Little Mermaid/Red.** $4.00-6.00

Comments: Regional Distribution: USA - 1989. These are a sampling of the generic premiums given away in 1989. **McChicken McDonald's and McChicken character** were short-lived ventures in 1989. Stuffed Flounder and Sebastian were given free with the purchase of Gift Certificates in the stores

Ge8913-Ge8914

For additional information on Happy Meal toys and pre-Happy Meal toys, see Joyce and Terry Losonsky's additional books:

McDonald's Pre-Happy Meal Toys from the Fifties, Sixties, and Seventies and ***McDonald's Happy Meal Toys From the Nineties,*** which trace the history of McDonald's toys and promotions from the earliest beginnings in the 1950s through mid-1998.

McDonald's Happy Meal Toys Around the World, which traces the history of Global Happy Meal toys from the beginning through June 1995.

McDonald's Collectors Club information:

Membership Secretary: Charlie Wichmann, 255 New Lenox Road, Lenox, Massachusetts, 01240 USA. E-mail: charmarmcd@aol.com
President: Sharon Iranpour, 24 San Rafael Drive, Rochester, New York 14618 USA. E-mail: siranpour@aol.com
Membership Year: January - December

Original storyboard art by Rich Seidelman.

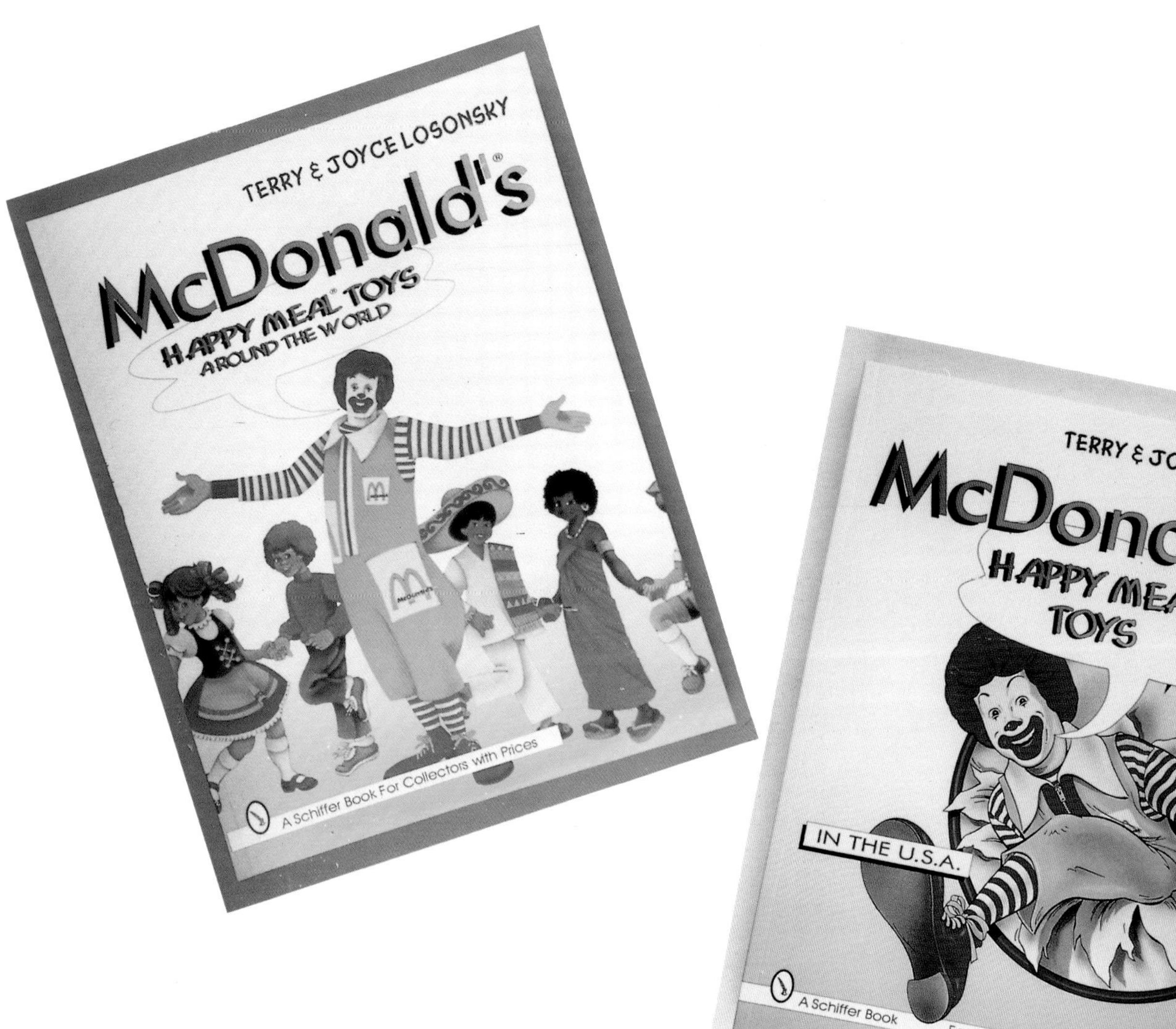
TERRY & JOYCE LOSONSKY
McDonald's
HAPPY MEAL TOYS
AROUND THE WORLD
A Schiffer Book For Collectors with Prices

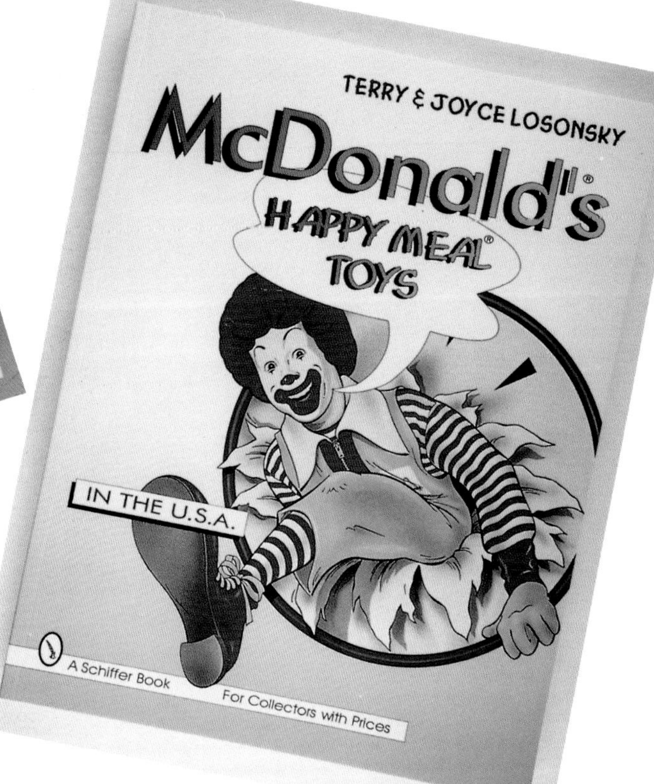
TERRY & JOYCE LOSONSKY
McDonald's
HAPPY MEAL
TOYS
IN THE U.S.A.
A Schiffer Book
For Collectors with Prices

Index

10th & 11th BILLION HAMBURGERS SOLD, 5
10th National O/O Convention held, 190
15th BILLIONTH SERVED, 5
1950s, 4, 5
1960s, 4-6
1970s, 4-6
1980s, 5, 6
1990s, 5, 6, 15, 38
20 BILLION SERVED, 5
25 BILLION SERVED, 5
3-D Happy Meal, 1981, 37, 38
3-D Paper Eye Glasses, 39
3-Window '34, 84
30 BILLION SOLD, 5
4 Knobs, 48, 69
400 HUNDRED MILLION sold, 5
45 BILLION SOLD, 6
45 RPM, 33, 52, 76, 187, 222, 245
50 BILLION SERVED, 6
500 HUNDRED MILLION, sold 5
55 BILLION SERVED, 6
56 Hi-Tail Hauler, 83
57 T-Bird, 83, 200
65 BILLION SERVED, 6
70 BILLION SERVED, 6
700 HUNDRED MILLION sold, 5
75 BILLION SERVED, 6
78 RPM, 33, 34, 76, 95
8 Windows, 48, 68, 69, 148
80 BILLION SERVED, 6
80s Firebird, 200
8th National O/O Convention, 97, 112
90 BILLION SERVED, 6
99 BILLION SERVED, 6

A

Accessory Kit, 68
Acknowledgments, 3
Adult Hand Size Puppet, 95
Adventures of Ronald McDonald Happy Meal, 1981, 40-42
Advertising, 4, 20, 35, 37, 38, 53, 55, 61, 85, 97, 138, 190, 222
Airplane, 31, 49, 72, 78, 85, 86, 104, 122, 139, 151, 201, 232
Airplane - Mac with BLACK sunglasses, 201
Airplane - Mac with BLUE sunglasses, 201
Airplane with Pilot, 122
Airplanes - *See* Airport Happy Meal, 1986, 138-139
Airport Happy Meal, 1986, 138-139
Alamo, 156
Alien Creatures, 15
Alligator, 25, 29, 174, 186, 196, 218, 219
Amazing Birthday Adventure, 227
AMC Eagle, 74 155
Amtrak, 88, 89
An American Tail Happy Meal, 1986, 140
Anniversary of Ronald McDonald House, 97
Ape, 13, 25, 218
Apollo Command Module, 159
Apple Pie Tree, 53
Apron/Uniform, 19
Archies - *See* New Archies Happy Meal, 1988, 206
Archy McDonald, 6
Argo Land Shuttle, 159
Ariel, 234, 236
As you are and eat in your car, 4
Astralia - Girl Not Holding Cone, 113
Astralia - Girl Holding Cone, 79
Astrosniks, 78, 80, 97, 98, 113
Astrosniks I '83 Happy Meal, 1983, 78-80
Astrosniks II '84 Happy Meal, 1984, 97-99
Astrosniks III '85 Happy Meal, 1985, 113
Austin, 156, 157
Axel, 144

B

Baby Piggy the Living Doll, 214
Back To Our Future, 5, 138, 162
Backpack, 111, 112
Bag, 80, 99, 109, 110, 113, 153, 185, 199, 201, 202, 209, 215, 217, 223, 227-230, 233, 235, 243
Baja Breaker, 83
Balloon, 34, 76
Bambi, 20, 190, 191
Bambi Happy Meal, 1988, 190-191
Banner - Scout Holding Yellow Flag, 113
Barney in Blue Mastodon Car, 196
Barnyard/Old McDonald's Farm Happy Meal, 1986, 138, 152-153
Baseball, 94, 181, 204, 213
Basketball, 24, 49, 73, 100, 208, 213
Basketball Hoop Shooter, 73
Beach Ball '84 - *See* Olympic Beach Ball Promotion, 1984, 97, 105
Beach Ball '85 - *See* Florida/Olympic Promotions, 1985, 113, 120
Beach Ball '86 Happy Meal, 1986, 140-141
Beach Ball, 97, 105, 113, 120, 138, 140, 185, 215, 224
Beach Ball Characters Promotion, 1985, 113
Beach Party - Grimace, 144
Beach Toy I Collect All 4/Test Market Happy Meal, 1989, 223-224
Beachcomber Happy Meal, 1986, 141
Bead Game, 26, 75, 125
Beak, 184
Bedtime/Ronald McDonald Happy Meal, 1989, 224
Beehonie Rabbit, 174
Benji Fastest Dog in the West, 62
Berenstain Bears, 138, 141, 163, 164
Berenstain Bears I '86 (Test Market) Happy Meal, 1986, 141-142
Berenstain Bears II '87 Happy Meal, 1987, 163-64
Betty, 196, 206
Betty in Orange Pterydoctil Car, 196
Big Chip, 181
Big Mac, 11, 13, 26, 27, 30-32, 41, 49, 53, 99, 118, 137, 139, 160, 167, 223, 241
Big Mac Helicopter, 139
Big Mac officially introduced, 6
Big Mac redesigned, 6
Big Mac sandwich is 21 years old, 223
Big Mac sandwich, 167, 223
Big Top Happy Meal, 1988, 191
Bigfoot/With Arches Happy Meal, 1987, 165
Bigfoot/Without Arches Happy Meal, 1987, 164-165
Billions Served on signs, 5
Bird with Eye, 103, 151
Birdie, 22, 23, 30, 31, 36-38, 41, 44, 49, 50, 73-75, 83, 88, 91, 94, 100, 105, 107, 110, 111, 113, 118, 120, 124, 125, 131, 136, 137, 139, 140, 144, 150, 160, 165, 166, 175, 176, 187, 200, 205, 208, 217, 220, 221, 224, 227, 233, 234, 243, 244
Birdie Bent Wing Brazer Airplane, 139
Birdie Glider, 136
Birdie on Raft, 165, 166
Birdie Parlor Car, 88
Birdie Seaside Sub, 224
Birdie, The Early Bird, 6, 22, 31, 36-38, 50, 111
Birdie, The Early Bird introduced, 6, 22
Birdie, The Early Bird characterization, 37-38
Birdie's Shovel, 234
Birdie's Sunshine Special Train Engine, 175
Birthday Party, 34, 77
Birthday Party Starring Ronald, 34
Bisected Arches, 4-6
Bisected Arches replaces Speedee, 5
Bisected Arches with Half Arrow Head, 5
Bisected Arches within a Ship's Wheel, 5
Black History Happy Meal, 1988, 192
Blibble, 238
Blow String Pipe, 49
Blow-Up Grimace, 111
Blue Book, 67, 208
Boat/Sail, 215
Boat with Sailor, 103, 151
Boat Whistle, 176
Boats 'N Floats, 163, 165
Boats 'N Floats Happy Meal, 1987, 165-66
Bolt, 184
Bones, 158, 206, 227
Book & Tape, 227, 243
Book, 23, 33, 34, 55, 57, 62, 67, 105, 110, 112, 114, 128, 140, 156, 157, 171, 172, 192, 208-210, 214, 227, 243
Book - Animals of the Sea, 23
Book - Animals That Fly, 23
Book - Cats in the Wild, 23
Book - The Biggest Animals, 23
Boomerang McNugget, 204
Boss Hogg Cadillac, 55
Brawn, 134
Bulldozer in Bulldozer, 172
Bumblebee, 134
Bunny Pails - *See* McBunny (Pails) Happy Meal, 1989, 236
Burger Six Pack to Go Promotion, 1990/1989, 225
Butterfly on Tail, 190, 191
Buttons, 64, 68, 70, 164, 169, 180, 190, 198, 208, 230, 243

C

C.B. - Guy with Headphones with Radio, 113
C2-Cheesburger, 241
Caboose, 64, 88
Cadillac Seville, 83
Calendar, 34, 52, 77, 96, 111, 112, 137, 246
Calendar Display, 246
Calendar of the '84 Olympic Games, 112
Call to Action, Customer Satisfaction, 5
Camel with Wrinkled Knees, 242
Camp Nippersink, 5, 55
Camp Out - Professor, 144
Can with Tin Bottom, 91
Canopy, 144
Capt Rub-A-Dub Sub, 129
Captain, 6, 26, 29, 31, 32, 41, 49, 53, 55, 73, 76,

92, 105, 110, 120, 129, 130, 137, 205
Captain Crook, 6, 29, 31, 32, 53, 55, 76, 92, 130, 137
Captain Crook redesigned, 6
Captain Crook with skull & cross bones officially introduced, 6
Car - Boss Hogg Cadillac, 55
Car - General Lee, 55
Car - Sheriff Roscoe, 55
Car - Overdrive, 134
Card, 15, 19, 23, 24, 28, 30, 43, 45, 50, 59, 63, 64, 68, 70, 72, 76, 83, 88, 91, 94, 101, 102, 107, 110, 113, 118, 123, 124, 126, 128, 132, 137, 203, 217, 221, 229, 230, 244
Cards, 22, 45, 85, 99
Cast of Characters,37, 78, 190, 222
Castle Mold, 215
Castlemakers/Sand Castle Happy Meal, 1987, 166
Cavalry Soldier, 47
Ceiling Dangler, 12, 15, 19, 28, 30, 57, 64, 70, 149
Certificate, 116
Chalk, 193
Changeables '87 Happy Meal, 1987, 166-167
Changeables '89 - *See* New Food Changeables Happy Meal, 1989, 240-241
Changeables WITH Painted Hands/Feet, 240-241
Changeables WITHOUT Painted Hands, 167
Character identification, 31, 37
Character Introduction/Redesigns, 6
Charm - Blossom, 134
Charm - Blue Belle, 134
Charm - Butterscotch, 134
Charm - Cotton Candy, 134
Charm - Minty, 134
Charm - Snuzzle, 134
Chevy Blazer, 154, 155
Chevy Blazer 4 x 4, 155
Chevy Citation "X-11" Brown, 84
Chevy S-10 Pick-Up, 155
Chevy Van, 132, 154, 155
Chicken McNugget Guys, 228
Chicken McNuggets added to menu, 78
Chimp, 178
Chip N Dale Rescue Rangers Happy Meal, 1989, 225-227
Chip N Dale, 223, 225
Chip's Roto-Cupter, 226
Chip's Rockin Racer, 226
Chocolaty Chip, 50, 111
Christmas Ornament, 96, 246
Christmas Stocking, 52, 162
Cinderella, 171, 188
Circle/Stencil, 146
Circus-3 Ring Circus Happy Meal, 1983, 80-83
Circus, 4, 10, 13, 20, 59, 78, 80, 83, 95, 153, 228
Circus Wagon Happy Meal, 1979, 10-13, 20
Cirrus Vtol, 159
CJ7 Jeep, 200
Cliffjumper, 134
Clip-on Buttons, 208
Coach, 64
Coach Car, 64
COAST to COAST, 4
Coil, 184
Collar, 238
Collectors Club Information, 247
Colorforms Happy Meal, 1986, 143-44
Coloring Board, 50
Coloring Book, 110, 112, 128, 192
Comb 25, 45, 59, 120
Come as you are and eat in your car, 4
Comic Book, 17, 20, 114
Commander Magna, 114
Commander - with Black Mask, 97, 113
Commandrons (Test Market) Happy Meal, 1985, 114
Construction Worker, 47
Construx, 138, 144
Construx Action Building System Test Market Happy Meal, 1986, 144-45
Convention, 22, 35, 71, 77, 97, 104, 112, 138, 162, 190, 222
Cookie, 25, 27, 28, 30, 31, 33, 37, 38, 50, 111, 186, 187, 194
Cookie Box 31, 38, 50, 186, 187
Cookie Boxes, 37, 50, 111, 186, 187
Cookie Cutter, 186
Cookie Mold, 25
Cookie Sampler, 27, 30
Cookies, 20, 50, 76, 110
Copter - with Helicopter Blades, 97
Corkle, 238
Corny McNugget, 204
Corvette Stingray, 83, 200
CosMc, 190, 193, 208, 222
CosMc Crayola Happy Meal, 1988, 193-94
CosMc - The little space alien joins McDonaldland, 6
Costume, 77
Cotterpin Dozer in Fork Lift, 172
Counter Card, 15, 19, 23, 28, 30, 63, 68, 70, 113, 128, 203
Cow, 47, 153
Cowboy being held up, 47
Cowgirl with Gun, 47
Cowpoke McNugget, 204
Craft Kits/Cancelled/Happy Meal, 1989, 227
Crayola/Crayon Magic I (Test Market) Happy Meal, 1986, 146-147
Crayola II Happy Meal, 1987, 168-170
Crayons, 126, 146, 169, 193
Crazy Creatures with Popoids II Happy Meal, 1985, 114-116
Creature in the Cave, 227
Crew Badge, 13, 18, 63
Crystal Ball, 90
Cup, 23, 32, 33, 51, 56, 61, 72, 76, 84, 95, 111, 112, 224, 226, 228, 236, 241
Cup - Airplane Hanger, 72
Cup - Bo, 56
Cup - Boss Hogg, 56
Cup - Country Club, 72
Cup - Daisy, 56
Cup - Figure Skating, 72
Cup - Happy Milk, 23
Cup - Jungle Gym, 72
Cup - Luke, 56
Cup - Monkey Business, 72
Cup - Sheriff, 56
Cup - Traffic Jam, 72
Cup - Uncle Jesse ,56
Cylinder, 144

D

Daisy's Jeep, 55
Daisy, 56, 204, 236
Dale's Roto Roadster, 226
Dallas Cowboys Super Box Happy Meal, 1980, 22
Danger under the Lake, 227
Datsun 200 SX, 83
Day & Night Happy Meal, 1985, 116
Delivering the Difference, 55, 77
Des Plaines, 8
Design change of Captain Crook, 6, 55
Design-O-Saurs Happy Meal, 1987, 170
Detective Kit, 90
Dick and Marice (Mac) McDonald, 8
did somebody say McDonald's?, 5
Dimples, 181
Dinah, 158
Dino, 196, 227
Dinosaur, 37, 38, 42, 43, 45, 78, 79, 113, 158, 196, 223, 227, 243
Dinosaur Baby Boom, 227
Dinosaur Days Happy Meal, 1981, 42-45
Dinosaur in McDonaldland, 243
Dinosaur Talking Storybook Happy Meal, 1989, 227
Disappearlng Hamburglar, 124
Disney Favorites Happy Meal, 1987, 171-172
Dixie Challenger, 83
Do You Believe in Magic?, 5
Dodge Rampage Pick-Up, 155
Dodger, 207, 222
Dodger the Dog with Goggles, 207
Doll, 53, 111, 112, 214, 219
Doll - Fossie, 219
Doll - Kermit, 219
Doll - Miss Piggy, 219
Doll - Ronald McDonald, 53
Doll - Grimace, 53
Doll - Ronald McDonald, 53
Dolls, 219
Dolly Dolphin, 212
Dolphin, 29, 212, 213, 244
Dolphin/Porpoise, 29
Dome Game, 26
Don't Forget to Feed the Wastebasket, 5
Donald's Engine, 238
Donald's Jeepster, 237
Doozers - *See* Fraggle Rock I Happy Meal, 1987, 172-173
Double Arch replaces Single Arch with Speedee, 5
Double Arches, 5
Dough, 127, 154, 194, 225
Dragon, 158, 196
Dress-Up McNuggets, 221
Drill - with Drill in Hands, 97
Drinking Cup, 224
Drummer McNugget, 204
Duane Pig, 174
Duck Tales I Happy Meal, 1988, 194-95
Duck Tales II Happy Meal, 1988, 195
Dukes of Hazzard Happy Meal, 1982, 55-57
Dumbo, 171
Duplo, 86, 103, 151, 232

E

E.T., 113, 117, 118
E.T. Happy Meal, 1985, 117-18
Earlybird in Car, 169
Egg McMuffin, 167, 241
Eggart Whapper, 30
Eight (8) Windows Spaceship, 48, 68, 69, 148

Eighth National O/O Convention, 97, 112
Eighties (80's) Firebird Car, 200
Elephant, 10, 25, 32, 59, 82, 177, 178
Entry Form/Star Trek, 18
Eraser, 31, 110, 183
EVERYBODY'S FAVORITE Coast to Coast, 4
Evil Grimace redesigned to Grimace (2 arms), 6
Evil Grimace, 6

F

Face Mask, 186, 218
Faller Zig Zag, 73
Farm - Ronald, 144
Farmer, 47, 65
Fast Macs, 97, 99, 113, 118
Fast Macs I '84 Promotion, 1984, 99
Fast Macs II '85 Promotion, 1985, 118-119
Feeling Good Happy Meal, 1985, 119-120
Female Gymnast, 24
Female Swimmer/Diver, 24
Fern, 158
Fievel, 140, 162
Fievel and Tiger, 140
Fievel's Boat Trip, 140
Fievel's Friends, 140
Figurine, 28, 29, 41, 80, 99, 113, 178, 212
Figurine - Big Mac, 41
Figurine - Birdie, 41
Figurine - Captain, 41
Figurine - Grimace, 41
Figurine - Hamburglar,41
Figurine - Mayor, 41
Figurine - Ronald, 41
Finger Puppets, 160, 207
Fire Chief, 200
Fire Eater, 200
Firebird Funny Car, 83
First Class, 204
Fish Mold, 215, 216
Fishin' Fun, 27
Flame, 104, 184, 202, 203
Flamingo, 178
Flat Car with 4 Fry Kids, 88
Flinger, 59
Flintstone Kids Happy Meal, 1988, 196
Floor Display, 12, 57, 98
Florida Beach Ball, 120
Florida Beach Ball/Olympic Beach Ball Promotion, 1985, 120
Flounder, 236, 246
Flower, 134, 191, 204
Floyd Male Alligator, 174
Flubber (Disney/McDonald's) introduced, 6
Flubber, 6
Fly, Fly, Birdie Launcher, 74
Fly, Fly, Birdie, 74
Flying Hamburger, 4
Flying Wheel/Frisbee, 44
Foam Wash Mitt, 224
Food, Folks, and Fun, 5
Foot Ruler, 49
Football, 22, 79, 160, 175, 213
Football/Super Spin, 160
Ford Bronco, 164, 165
Ford Pick-Up, 164, 165
Ford Ranger 23 Pick-Up, 155
Fork/Spoon Set, 45
Forty-five (45) RPM, 33, 52, 76, 187, 222, 245
Four Knobs Spaceship, 48, 69
FOUR BILLION SERVED signs, 5
Fozzie, 152, 179, 180, 240
Fozzie on Yellow Horse, 152, 180
Fraggle Rock I '87/Doozers (TM) Happy Meal, 1987, 172-173
Fraggle Rock II '88 Happy Meal, 1988, 196-198
Francis the Bulldog, 207
Fred in Green Alligator Car, 196
Free-Wheelin' Dodge Rampage, 132
Freight, 64
Freight Car, 64
French Fries, 167, 241
French Fry Faller, 81
French Fry Garden, 245
French Fry Grabber, 73
French Fry Guys, 55, 76, 228
Friend Owl, 191
Frisbee, 44, 59
Frisbee Flying Wheel, 59
Front Running Fairmont, 83
Fry-Bot, 241
Fry Force, 241
Fry Girl, 160, 175, 205, 208, 221
Fry Girl Express Train Engine, 175
Fry Girls redesigned, 6
Fry Guy Flyer Train Engine, 175
Fry Guy Flyer Airplane, 139
Fry Guy Friendly Flyer, 139, 160
Fry Guy Happy Car, 175
Fry Guy on Duck, 120
Fry Guy on Brontofry Guy Brontosaurus, 170, 186
Fry Guys, 6, 55, 76, 125, 137
Fry Kid Super Sailor, 224
Fry Kids, 50, 88, 165, 187, 211, 215, 220, 221, 234
Fry Kids Ferry, 211
Fry Kids on Raft, 165
Fry Kids Planter, 234
Fun House Mirror, 81
Fun Ruler, 94
Fun Times Magazines, 35, 36, 51, 52, 76, 77, 95, 111, 112, 137, 160, 161, 187, 188, 222, 245
Fun with Food Happy Meal, 1989, 228-229
Funny Fry Friends I "Collect all 4"/(TM) Happy Meal, 1991/1989, 223, 229

G

Gadget's Rescue Racer, 226
Gadgets Rockin Rider, 226
Gadzooks, 229
Gallacta Burger, 241
Game, 17, 26, 27, 44, 49, 60, 61, 75, 125, 164
Garfield, 190, 199, 223, 229, 230
Garfield I Test Market Happy Meal, 1988, 199
Garfield II Happy Meal, 1989, 229-230
Garfield in 4 Wheeler Car, 230
Garfield in Car, 199
Garfield/Odie with Sidecar, 230
Garfield on Big Wheels, 199
Garfield on Scooter, 199, 230
Garfield on Skateboard, 199, 230
Gears, 134
General Lee, 55
Georgette the French Poodle, 207
Ghostbusters - *See Real Ghostbusters I Happy Meal, 1987, 182-184*
Giddy the Gator, 232
Gift Wrap, 34, 35, 52
Gift Wrap Paper and Ribbon, 52
Giggles and Games Happy Meal, 1982, 57-59
Gill Face Creature, 15
Gingerbread House with Stickers, 101
Glasses, 32, 39, 91, 181, 195, 244, 245
Glider, 31, 74, 94, 136
Glo-Tron Spaceship (Test Market) Happy Meal, 1986, 148
Globe Airplane, 49
Go for Goodness at McDonald's, 4
Gobblin, 59
Gobblin Caller, 59
Gobblin Groomer, 59
Gobblins, 50, 55, 60, 61, 76
Gobblins Bowling, 61
Gobblins Horseshoes, 60
Gobblins officially introduced, 6
Gobblins renamed, 55, 76
Gobblins renamed the French Fry Guys, 55
Goblins, 11, 60
Gobo, 172, 196-198
Gobo holding Large Carrot, 197
Gobo in Carrot, 172
Going Places, 55, 59, 61, 78, 83-85
Going Places Happy Meal, 1982, 59-61
Going Places/Hot Wheels Promotion, 1983, 83-85
Golden Arches, 4, 9
Gonzo, 152, 179, 180, 240
Gonzo on Green Big Wheels, 180
Gonzo with Suspenders, 152
Good Friends Happy Meal, 1987, 173
Good Sports Happy Meal, 1984, 99-101
Good Time, Great Taste, That's Why This is My Place, 190
Good Time, Great Taste of McDonald's, 5, 190
Good Times, Great Taste, 222
Goofy's Jalopy, 238
Goofy's Sedan, 237
Great Expectations, 5
Grimace, 10, 13, 20, 24-26, 29-33, 37, 40, 41, 44, 45, 49-51, 53, 73, 75, 76, 82, 88, 92, 94, 100, 105, 107, 110-113, 120, 122, 124, 125, 129, 131, 136, 137, 139-141, 143, 144, 146, 150, 160, 165, 169, 170, 175, 176, 185, 186, 193, 200, 208, 209, 211, 215, 217, 221, 224, 227, 234, 243, 244
Grimace Ace Biplane, 139
Grimace Bouncin' Beach Ball, 224
Grimace Caboose, 88
Grimace Glider, 136
Grimace Goes to School, 243
Grimace Happy Taxi Company, 175
Grimace in a Tub, 120
Grimace in Rocket, 169
Grimace Mighty Mac Shuttle, 160
Grimace on Grimacesaur Pterodactyl, 170
Grimace Purple Streak Train Engine, 175
Grimace Rake, 234
Grimace redesigned, 6
Grimace Ski Boat, 165
Grimace Smiling Shuttle, 139, 160
Grimace still called: The Grimace, 6
Grimace Strong Gong, 82
Grimace Submarine, 211
Grimace Tubby Tugger, 129
Gropple, 238
Grouchy Bear, 242
Gus, 174, 188
Gus Father Bear, 174

Gymnastic, 73
Gyro Bird Helicopter, 232
Gyro Top, 73

H

Half Arrow Head, 5
Halloween '85 Happy Meal, 1985, 121
Halloween '86 Happy Meal, 1986, 148
Halloween '87 Happy Meal, 1987, 173
Halloween '88 Happy Meal, 1988, 199, 218
Halloween '89 Happy Meal, 1989, 231
Halloween, 113, 121, 138, 148, 163, 173, 177, 190, 199, 218, 223, 231, 246
Hamburger Guy, 228
Hamburger Man, 4, 5
Hamburger University, 78, 245
Hamburglar, 11, 25, 26, 30-32, 40, 41, 45, 49, 50, 53, 73-76, 81, 83, 92, 94, 99, 100, 105, 107, 110, 112, 118, 120, 124, 125, 129, 131, 136, 137, 144, 149, 169, 170, 173, 176, 186, 187, 200, 205, 208, 211, 217, 221, 227, 245
Hamburglar Floater, 25
Hamburglar Glider, 136
Hamburglar Hockey, 26
Hamburglar in Steam Engine, 169
Hamburglar introduced, 6
Hamburglar on Tricerahamburglar Triceratops, 170, 186
Hamburglar Pirate Ship, 211
Hamburglar redesigned to long pointed nose, 6
Hamburglar redesigned to pudgy face, 6
Hamburglar Splash Dasher, 129
Hamburglar still called: The Hamburglar, 6
Hamburglar with unusual nose officially introduced, 6
Hand Puppet, 95
Hang Glider, 74
Happy Cup, 33, 95
Happy Holidays, 97, 101
Happy Holidays Happy Meal, 1984, 101-102
Happy Hotcakes Promotion, 1980, 22-23
Happy Meal (Hm) Box, 10, 11, 13, 15, 16, 20, 24, 29, 38-40, 42, 43, 46, 57-60, 62, 65, 71, 72, 78, 80, 81, 86, 87, 89, 90, 93, 97, 99-101, 103, 106, 108, 109, 114, 116, 117, 119, 120, 122-127, 130-132, 134, 138, 139, 140-144, 146, 149, 150, 152-154, 156-158, 163, 164, 166-168, 171, 173, 174, 176, 177, 179, 181, 182, 184, 190-196, 199, 201-203, 205, 206, 208-214, 217, 218, 220, 222, 224-226, 228, 229, 231, 232, 234-236, 238-240, 242, 244
Happy Meal (Hm) Pail, 102
Happy Meal Guys, 6
Happy Meal Officially Begins, 10
Happy Pail I '83 Happy Meal, 1983, 85
Happy Pail II '84 (Olympic Theme) Happy Meal, 1984, 102
Happy Pail III '86 Happy Meal, 1986, 149
Happy Teeth Happy Meal, 1983, 86
Hardworking Burger Bulldozer, 74
Hardworking Burger Dump Truck, 74
Harmonica, 176
Hat, 47, 65, 153, 181, 182, 188, 199, 204, 229, 231
Have You Had Your Break Today?, 5
Helicopter, 86, 97, 98, 104, 122, 139, 151, 160, 232, 233
Helicopter - Big Mac, 160
Helicopter - Hello! Copter, 160
Hen, 153
High Flying Kite Happy Meal, 1986, 149-50
Hippo, 24, 25
Hobby Box Happy Meal, 1985, 121
Home of America's Favorite Hamburger, 4
Home of...America's Favorite Hamburgers...still only 15 cents, 4
Horned Cyclops, 15
Horse, 13, 47, 65, 83, 152, 158, 180
Hot Wheels, 61, 78, 83, 85, 190, 199, 200
Hot Wheels Happy Meal, 1988, 199-200
Houston, 156, 157, 166
Huey Duey Louie on Ski Boat WITH Wheels, 195
Huey Duey Louie on Ski Boat WITHOUT Wheels, 195
Huey on Skates, 195
Husband, 153

I

Icicle Stick Mold, 30
ID Bracelet, 11, 12
ID Bracelet - Big Mac, 11
ID Bracelet - Hamburglar, 11
ID Bracelet - Ronald, 12
Indian, 47, 65, 68
Indian Brave/Face to Face, 47
Insectman, 15
Iron-on, 18
Iron-On Transfer - Mr. Spock, 18
Iron-On Transfer - Lt. Ilia, 18
Iron-On Transfer - Capt Kirk, 18
Iron-On Transfer - Dr. McCoy, 18
It's a Good Time for the Great Taste of McDonald's, 5, 97, 112

J

Jab, 184
Jacque, 188
Jad, 158
Jalopy Car, 30
Jeep CJ-7 Car, 83
Jeep - Daisy's Jeep, 55
Jeep/Off Roader, 201
Jeep Renegade, 78, 132, 154, 155
Jet - on Rocket, 113
Jingle, 55, 97, 113
Jolene Female Alligator, 174
Jughead, 206
Junior - holding Ice Cream Cone, 113
Just Kermit and Me, 214

K

Kazoo, 176
Keep Your Eyes on Your Fries, 5
Kermit, 152, 179, 180, 214, 219, 239
Kermit on Red Skateboard, 152, 180
Kermit on Skates, 180
Kids Radio, 34
Kissyfur, 163, 174
Kissyfur Baby Bear, 174
Kissyfur Happy Meal, 1987, 174
Kite, 30, 138, 149, 150
Kobby, 158
Kroc, Ray, 4, 20, 77, 97, 112
Krypto Cup, 241

L

Lady and the Tramp, 171
Land Laser Car, 233
Land Lord, 83
Large Fries for Small Fries, 5, 113, 138
Laser - with Gold Gun, 113
Laser - with Gun, 79
Launchpad in Orange Plane, 195
Learn the ABC's, 110, 112
Lego, 78, 86, 97, 103, 104, 113, 122, 123, 138, 150, 151, 223, 231-233
Lego Building Sets I '83 (Test Market) Happy Meal, 1983, 86
Lego Building Sets II '84 Happy Meal, 1984, 103-104
Lego Building Sets III '86 Happy Meal, 1986, 150-151
Lego Motion IV '89 Happy Meal, 1989, 231-233
Lennie Wart Hog, 174
Lesney, 68
Let's Face It, 5
Letterland Stationery, 27
Lid, 85, 102, 111, 112, 148, 173, 215, 228, 231, 234, 236
Lids, 85, 102, 121, 141, 148, 173, 231
Light Switch Cover, 26
Lightning Striker Airplane, 232
Limited Edition Stomper 4x4 (Chev), 132
Link, 158
Lion, 11, 25, 83, 218
Little Engineer Happy Meal, 1987, 174-175
Little Gardener Happy Meal, 1989, 233-234
Little Golden Book Happy Meal, 1982, 62-63
Little Martin Jr. Coloring Book, 192
Little Mermaid, 223, 234, 246
Little Mermaid I Happy Meal, 1989, 234-236
Little Travelers with Lego Building Sets Happy Meal, 122-123
Lizard Man, 15
Lolly, 181
Look for the Drive-in with the Arches, 4
Look for the Golden Arches, 4, 9
Look for THE DRIVE-IN WITH THE (GOLDEN) ARCHES, 4
Look For The Golden Arches - The Closest Thing To Home, 4
Look Look Books Happy Meal, 1980, 23
Losonsky List #1: Jingles and Slogans, 4-5
Losonsky List #2: Sign Identification, 5-6
Losonsky List #3: Character Introduction/Redesigns, 6
Losonsky's Identification Guides, 4
Luggage Tag, 190, 200
Luggage Tag Promotion, 1988, 200
Lumpy, 181
Lunch Bag, 110, 209
Lunch Box, 121, 163, 175, 177, 190, 209
Lunch Box/Characters Promotion, 1987, 175

M

Mac Tonight (Travel Toys) Happy Meal, 1990/1989/1988, 201-202
Mac Tonight, 187, 190, 201, 202
Mac Tonight 45 RPM, 187
Mac Tonight introduced, 6
Macro Mac, 241
Magic, 5, 20, 76, 113, 123-125, 138, 146, 169, 194, 195
Magic Egg Trick, 124

Magic Marker, 169, 194
Magic Motion Map, 195
Magic Picture, 124
Magic Record #1, 76
Magic Record #2, 76
Magic Show Happy Meal, 1985, 123-124
Magic Slate Board, 125
Magic String Trick, 124
Magic Tablet, 124
Magnet, 45
Magni-Finder Glass, 91
Magnifying Glass, 195
Male Basketball Player, 24
Male Javelin Thrower, 24
Male Pole Vaulter, 24
Male Soccer Player, 24
Male Swimmer/Diver, 24
Malibu Grand Prix "Goodyear 9999" Black, 84
Manager's Guide, 19, 20, 70, 116
Map, 89, 157, 178, 195, 214
Markers/Coloring, 126
Markers/Drawing, 126
Mask, 97, 98, 113, 186, 204, 218
Matchbox (Super GT) Happy Meal, 1988, 202-203
Matchbox, 68, 190, 202
Matchbox Super GT, 190, 202
Matey, 229
Mayor McCheese, 31, 32, 53
Mayor McCheese officially introduced, 6
Mayor McCheese redesigned, 6
McBlimp, 113, 138
McBoo, 121, 148, 173
McBunny (Pails) Happy Meal, 1989, 236
McDonald's Pals, 241
McDonald brothers, 5
McDonald's and You, 5, 55
McDonald's Collectors Club Information, 247
McDonald's Fry'n the Sky, 30
McDonald's is Our Kind of Place, 4
McDonald's is Your Kind of Place, 4
McDonald's Speedee drive-ins - often imitated, never duplicated, 4
McDonald's System, Inc., 32
McDonald's--where quality starts fresh every day, 4
McDonaldland, 6, 13, 20, 27, 28, 30-32, 35-37, 53, 55, 60, 64, 71, 76, 78, 87, 90, 110-112, 131, 137, 163, 176, 177, 190, 205, 208, 222, 223, 243, 244
McDonaldland Band Happy Meal, 1987, 176-177
McDonaldland Cookies, 20, 76, 110
McDonaldland Express Happy Meal, 1982, 64
McDonaldland Fun Times Magazine, 35
McDonaldland Hockey, 60
McDonaldland Junction Happy Meal, 1983, 87-89
McDonaldland TV Lunch Box/Lunch Bunch Happy Meal, 1987, 177
McDoodle Desk & School Starter Kit, 112
McDoodler Ruler, 12
McFavorite Clown, 5
McGhost, 231
McGoblin, 121, 148, 173
McJack, 121, 226
McJack's Propel/Phone, 226
McKids Clothing, 163, 188
McNugget Band, 246
McNugget Buddies, 78, 96, 165, 190, 203, 204
McNugget Buddies Happy Meal, 1988, 203-205
McNugget Buddies introduced, 6
McNugget Buddies Life Boat, 165
McPunk'n, 121, 148, 173
McPunky, 121
McPuzzle Lock, 12
McWitch, 231
McWorld, 5
Meet Ronald McDonald, 4
Menu, 20, 37, 52, 78, 89, 91, 94, 102, 104, 108, 110, 118, 120, 126, 130, 132, 139, 148, 162, 163, 170, 172, 204, 214, 223, 224, 233, 238, 241, 245
Menu Music Chants, 52
Menu Song, 245
Mercedes 380/Silver, 84
MetroZoo Happy Meal, 1986, 177-178
Mickey Roadster, 238
Mickey's Birthdayland Happy Meal, 1989, 236-238
Mickey's Roadster, 237
Midway, 82
Mig-21, 68
Mighty Mac Robot Dozer, 160
Milkshake Lake, 246
Minitrek "Good Time Camper" White, 84
Minnie Convertible, 237-38
Mirage F1, 68
Mirror, 81, 119, 120
Miss Piggy, 179, 219, 239
Mix'em up Monsters Happy Meal, 1990/1989, 238
Mokey, 196, 197
Monkey, 24, 25, 28, 72, 81, 186, 198, 218
Monkey 3D Face Mask, 218
Moose, 206
More Than 60 Billion Served signs, 6
Motor Boat with Rubber Band, 49
Motorcycle, 201, 230
Motron, 114
Moveables/McDonaldland Happy Meal, 1988, 205
Ms. Ford Pick-Up, 164, 165
Ms. Piggy in Pink Car, 152, 180
Ms. Piggy on Skates, 180
Muppet Babies, 138, 152, 163, 179, 180, 190, 214, 219
Muppet Babies Holiday Promotion, 1988, 219
Muppet Babies I '86 (Test Market) Happy Meal, 1986, 152
Muppet Babies II '87 Happy Meal, 1987, 179-180
Muppet Kids '89 (Test Market) Happy Meal, 1989, 239-240
Museum, 182
Music Happy Meal, 1985, 124
My Little Pony I/Transformers Happy Meal, 1985, 134-136
My Little Pony, 113, 134, 136
My McDonald's, 5
Mystery, 76, 78, 89, 90, 95, 243
Mystery Happy Meal, 1983, 89-91
Mystical Scrambler Kaleidoscope, 75

N

Navigation Wrist Bracelet, 17
New Archies Happy Meal, 1988, 206
New Food Changeables Happy Meal, 1989, 240-241
Nickel, 123
Nite Stand Ronald, 224
Nobody Can Do It Like McDonald's Can, 5, 20
Nose Maze Game, 44
Note Pad/Eraser, 183

O

O/O Advertising Theme, 55, 97, 138, 190
O/O Convention, 22, 35, 77, 97, 112, 138, 162, 190, 222
O/O Theme, 22
Old McDonald's Farm, 138, 152
Old McDonald's Farm/Barnyard Happy Meal, 1986, 152-153
Old West Happy Meal, 1981, 46-47
Oliver & Company Happy Meal, 1988, 206-207
Oliver, 20, 190, 206, 207, 222
Oliver the Kitten, 207
Olympic Beach Ball Promotion, 1984, 105
Olympic Pin, 190, 208
Olympic Pin/Sports II Clip-On Buttons/ Happy Meal, 1988, 208
Olympic Sports, 97, 105, 106
Olympic Sports I Happy Meal, 1984, 106-108
Olympic Sports [Zip Action] Happy Meal (Canceled), 1984, 105-106
Olympic [Medal] Happy Meal (Canceled), 1980, 24
On the Go I '85 Happy Meal, 1985, 124-125
On the Go II '88 Lunch Box/Bags Happy Meal, 1988, 209
On the Go Transfers, 125
One of a Kind, 5, 97, 112
ONE BILLIONTH Hamburger Served, 5
ONE HUNDRED MILLION sold, 5
Ornament, 52, 77, 94, 96, 112, 137, 222, 246
Ornament - Reindeer, 137
Our World...Today, Yesterday and Tomorrow, 5, 22, 35
Over 4 Billion Served, 5
Over ONE BILLION SOLD signs, 5

P

P-911 Turbo, 200
Pail, 78, 85, 97, 102, 121, 138, 141, 148, 149, 173, 215, 216, 231, 236
Pail/Sand, 215
Pail - Fluffy, 236
Pail - Pinky, 236
Pail - Whiskers, 236
Paint Brush, 227
Paint/Number, 194
Pal [Doll], 53
Pan Pipes, 176, 177
Pancakes Wrapper, 22
Patch, 44, 116, 122, 150
Pen, 30, 74, 173, 221
Pencil, 31, 49, 110, 183, 184
Pencil Case, 110, 183, 184
Pencil Holder, 49
Pencil/Pencil Topper, 183
Pencil Sharpener, 110, 183
Pencils, 31
Penguin Sun Glasses, 244
Pennant, 26
Penny Penguin, 212
Perfido - with Red Cape, 97, 113
Peter Rabbit, 190, 209
Peter Rabbit's Happy Meal, 1988, 209-210
Phantom F4E. 68
Photo Card. 50, 76, 137, 221
Picnic Today - Hamburglar. 144

Picture Perfect Happy Meal, 1985. 125-126
Pig, 153, 174
Pink Sun Cruiser, 99, 118
Plane, 49, 195, 201
Plastic Cup, 32, 33, 51, 61, 72, 76, 95
Plate, 137, 161, 245, 246
Play Day - Birdie The Early Bird, 144
Play-Doh, 78, 91, 113, 126, 127, 137, 138, 153, 154
Play-Doh I '83 Happy Meal, 1983, 91
Play-Doh II '85 Happy Meal, 1985, 126-127
Play-Doh III '86 Happy Meal, 1986, 153-154
Playmobil I '81 (Test Market) Happy Meal, 1981, 47-48
Playmobil II '82 Happy Meal, 1982, 65-68
Pluto, Car 238
Pocket Patch, 44
Pointed Front, 48, 69
Pop Car, 74
Pop up Ring, 49, 50
Popoids, 97, 108, 109, 113-116
Popoids Crazy Creatures I Test Market Happy Meal, 1984, 108-109
Poppin', 115
Porsche 928 Turbo, 84
Poster, 70, 117, 118, 188, 207
Poster: ET, 117
Potato Dumpling, 181
Potato Head Kids I Happy Meal, 1987, 181-182
Potato Head Kids, 163, 181
Potato Puff, 181
Pre-Happy Meal Toy collecting, 4
Prince Eric, 236
Professor, 6, 27, 53, 74, 77, 82, 144, 176, 205
Punch Outs, 82, 83, 164
Puppet ,82, 95, 160
Push-Alongs, 113, 132
Puzzle Guess N' Glow, 75
Puzzle(s), 35, 75, 105, 107, 108, 164
Pyramido - Pyramid Shape with Left Hand Raised, 113

Q

QSC and Me, 5
Quacker/Whistle, 195
Quarter Pounder, 167, 241

R

Race, 83, 110, 122, 125, 151, 217, 232
Race Bait, 308 83
Race Car, 125, 151, 217
Racing - on Sled, 97
Radio Rainy Day Fun, 34
Raggedy Andy, 242
Raggedy Ann, 223, 242
Raggedy Ann and Andy, 223, 242
Raggedy Ann and Andy Happy Meal, 1989, 242
Rain or Shine Happy Meal, 1989, 242-243
Ray Kroc, 4, 20, 77, 97, 112
Read Along with Ronald Happy Meal, 1989, 243
Real Ghostbusters I Happy Meal, 1987, 182-184
Record, 33, 34, 52, 76, 95, 122, 124, 150, 187, 222, 245
Rectangle/Stencil, 146
Red Fraggle in Radish, 172
Red Holding Large Radish, 197
Red Sports Car, 99, 118
Reggie, 206
Regional Happy Meal promotions, 138, 162
Register Topper, 180, 182, 197
Rhino, 25
Riddles, 35
Right Triangle/Stencil, 147
Ring, 17, 18, 26, 45, 49, 50, 78, 80, 137, 224, 228, 246
Ring - Capt Kirk, 17
Ring - Spock, 18
Ring - Star Trek Logo, 18
Ring - U.S.S. Enterprise, 17
Ring - Play-Doh, 137
Rings, 26, 138
River Boat, 92, 129
Roadster/Race Auto, 122
Robo-Robot - Gold colored Astrosnik, 79, 113
Robocakes, 241
Rocker McNugget, 204
Rockwell, 77, 96
Ronald Acrobat, 82
Ronald and Friends in Space, 161
Ronald and Friends on a Train, 161
Ronald and Friends with Dinosaurs, 161
Ronald Clown Convention, 104
Ronald Fun Flyer, 224
Ronald Glider, 136
Ronald in Hot Air Balloon, 76
Ronald in Red Car with TEAR DROP Under Eyes, 217
Ronald in Rubber Red Car, 217
Ronald McDonald & Friends, 33
Ronald McDonald Airboat, 211
Ronald McDonald Airlines, 31
Ronald McDonald Airplane, 31
Ronald McDonald and his Flying Hamburger, 4
Ronald McDonald as national spokesman, 6
Ronald McDonald as regional spokesman, 6
Ronald McDonald Coloring Calendar, 34, 52, 112
Ronald McDonald House, 97, 112, 138
Ronald McDonald introduced in McDonaldland, 6
Ronald McDonald Secret Solver Coloring Calendar, 137, 246
Ronald McDonald Sticker Fun Coloring Calendar, 77
Ronald McDonald Visits America, 33
Ronald McDonald Visits.... Coloring Calendar, 112
Ronald McDonald with Birdie on Moon next to Spaceship, 137
Ronald McDonald with bisected arches costume, 6
Ronald McDonald with Fry Guys in Biplane, 137
Ronald McDonald with Grimace in McDonaldland Express Train, 137
Ronald McDonald with Grimace in Sailboat, 137
Ronald McDonald's costume changed to even pockets, 6
Ronald McDonald's Theme, 10, 20
Ronald McDonald, 4, 6, 10, 13, 20, 31-35, 37, 38, 40, 42, 50-53, 61, 76, 77, 95-97, 111, 112, 125, 137, 138, 211, 223, 224, 245, 246
Ronald on Fire Engine, 169
Ronald on Tractor, 169
Ronald on Tyronaldsaurus Rex Tyrannosaurus, 170
Ronald redesigned - pockets have black lines, 6
Ronald River Boat, 129
Ronald Seaplane, 139
Ronald Styro-glider, 31, 94
Ronald Train Engine, 88
Ronald Watering Can, 234
Ronald's Playplace, 163
Ronald's Railway Train Engine, 175
Ronald, 4, 6, 10-13, 20, 24-27, 29-35, 37, 38, 40-42, 44, 45, 49-53, 59, 61, 62, 73-77, 81, 82, 85, 88, 90-92, 94-97, 99-101, 104, 105, 107, 110-113, 118, 120, 122, 124, 125, 129, 131, 133, 136-141, 143, 144, 146, 150, 160, 161, 163, 169, 170, 173, 175-177, 186-188, 190, 200, 205, 208, 209, 211-213, 217, 221, 223, 224, 227, 232, 243-246
Rooster, 153
Round/Circular with Rear Engine, 68
Rub-A-Dub Sub, 92, 129
Rubber, 15, 28, 30, 41-43, 47, 49, 68, 132, 142, 156, 205, 217, 226
Ruler, 12, 49, 94, 110, 183, 184, 226
Ruler/No Metric Scale, 110
Ruler/Note Pad, 183
Runaway Robots, 163, 184, 185
Runaway Robots Happy Meal, 1987, 184-185

S

Safari Adventure Meal Happy Meal, 1980, 24-28
Sailboat, 105, 113, 120, 137, 140, 185
Sailors Happy Meal, 1988, 210-211
Sand/Fish Mold, 215
Sand Mold, 166, 216
Sand Pail, 85, 141, 149, 215
Santa Claus the Movie Happy Meal, 1985, 127-128
Sardine Can, 75
Sarge McNugget, 204
Saxophone, 176
School Days Happy Meal, 1984, 109-110
Scissors, 61, 227
Scooter, 142, 164, 199, 201, 230
Scott, Willard, 9
Scout: Holding Flag, 79
Sea Eagle Seaplane, 233
Sea Otter, 244
Sea Skimmer Boat, 233
Sea World of Ohio Happy Meal, 1988, 212
Sea World of Texas I '88 Happy Meal, 1988, 212-213
Sea World of Texas II '89 Happy Meal, 1989, 244
Seal 10, 29, 32
Sebastian, 236, 246
Seventy-eight (78) RPM, 33, 34, 76, 95
Shamu the Whale, 212
Sharing the Dream, 5, 190, 222
Shark/Great White, 29
Shark/Hammerhead, 29
Shark/Tiger, 29
Shark/Whale, 29
Sheep, 153
Sheriff, 46, 47, 55, 56, 65, 83, 200
Sheriff Patrol, 83, 200
Sheriff Patrol/Blk/Wht, 83
Sheriff Roscoe, 55, 56
Sheriff with Hands on Side, 47
Ship, 78, 86, 92, 103, 104, 113, 129, 130, 205, 211
Ship Shape I '83 Happy Meal, 1983, 92
Ship Shape II '85 Happy Meal, 1985, 129-130
Shuttle Ford, 164, 165
Signs, 138

Signs on highways changed to arches only, 6
Single Arch, 5
Sipper lid, 111, 112
Siren, 176
Skateboard, 152, 180, 199, 201, 230
Skater: on Ice Skates, 79
Ski Boat, 165, 195
Ski - with Skis/Goggles, 97
Skull, 184
Sky-Busters Happy Meal, 1982, 68
Sky Hawk A4F, 68
Slate, 58, 124, 125
Sleighfull of Surprises, 128
Slick, 181
Slogan, 4, 20, 112, 138, 190
Slugger, 181, 204
Smarty Pants, 181
Snikapotamus - Dinosaur, 79, 113
Snorkel McNugget, 204
Snow White, 188
Soap Dish, 120
Soft Drink Cup, 228
Solardyn, 114
Soldier's Horse, 47
Space Aliens, 15
Space Cruiser, 31
Space Explorer Coloring Calendar, 96
Space Raiders, 15
Space Ship, 78
Space Shuttle, 159
Space Theme Meal Happy Meal, 1979, 13-15
Space Vehicle, 159
Spacemobile Rocket, 98
Spaceship '81/Unidentified Happy Meal, 1981, 48
Spaceship '82 Happy Meal, 1982, 68-71
Spaceship, 37, 48, 55, 68-71, 78, 117, 137, 138, 148, 175, 193
Sparky McNugget, 204
Speedee, 4-6
Speedee character introduction, 6
Speedee holding 15 cent sign, 5
Speedee is the McDonald brothers Company Symbol, 5
Speedee says: look for me at McDonald's speedee drive -ins, 4
Speedee sign with Single Arch, 5
Speedee, the Hamburger Man, 4, 5
Speedy Chevy S-10 Pick-Up, 132
Speedy Chevy S-10 Pick-Up, 132
Spike, 182
Spin Pipe, 75
Spinner Baseball, 94
Spinner Bike Race, 110
Splash Dasher, 92, 129, 130
Split Window '63, 84, 200
Split Window '63 Gold, 84
Sponge, 25, 43, 44, 58, 60, 120, 213
Sport Ball (Test Market) Happy Meal, 1990/1988, 213
Sport: Holding Football, 79
Sports Car, 99, 118, 201
Sporty Chevy Van, 132
Spud, 182
Spy Glass, 195
Squad/Police Car, 99
Squire Fridell, 137, 138
Star Trek Meal Happy Meal, 1980/1979, 15-20
Stationery, 28
Stencil, 146, 147, 169
Sticker, 11, 12, 26, 48, 55-57, 64, 65, 69, 77, 79, 80, 89, 92, 100, 101, 104, 113, 129-132, 134, 143, 144, 148, 159, 165, 166, 171, 175, 177, 186, 209
Sticker Album, 131
Sticker Club Happy Meal, 1985, 130-132
Sticker - Motion (3-D), 131
Sticker - Paper, 131
Sticker Playset, 143
Sticker - Puffy, 100, 131
Sticker - Scented/Scratch & Sniff, 131
Sticker Sheet, 11, 12, 26, 48, 55-57, 64, 69, 89, 92, 129, 148, 159, 165, 166, 177, 209
Sticker - Shiny/Prismatic, 131
Stickers - Hawaii, 131
Stocking, 52, 162
Stocking - Fievel Dancing, 162
Stocking - Fievel On Sled, 162
Stomper Mini 4 x 4 I '85 Push-Alongs (Test Market) Happy Meal, 1985, 132-133
Stomper Mini 4 x 4 II '86 Happy Meal, 1986, 154-156
Stop Watch, 34
Store, 35, 46, 97, 104, 132, 141, 163, 222
Story of Texas Happy Meal, 1986, 156-157
Storybook Muppet Babies Happy Meal, 1988, 214
Storybook - See Dinosaur Talking Storybook Happy Meal, 1989, 227
Straw, 204
Street Beast, 200
Styro-glider, 31, 94
Stuffed Crab, 246
Stuffed Dog, 222
Stuffed Dolphin, 212
Stuffed Fish, 246
Stuffed Penguin, 212
Stuffed Walrus, 212
Stuffed Whale, 212
Stuffed Cat, 222
Stutz Blackhawk, 83
Success and Then Some, 4
Sugar Cookie, 33
Sunglasses, 31, 201, 223, 244, 245
Sunglasses/McDonaldland Promotion, 1989, 245-246
Super Door Alarm, 34
Super Sticker Squares, 186
Super Summer I '87 Test Market Happy Meal, 1987, 185
Super Summer II '88 Happy Meal, 1988, 215-216
Super Travelers Happy Meal, 1985, 113, 122
Surf Ski - with NO wheels, 201
Surf Ski - with WHITE wheels, 201
Swamp Stinger Air Boat, 232
Swing and Catch Game, 49

T

T-Shirt, 111, 137
Table Topper, 53
Tale of Benjamin Bunny, 209
Tale of Flopsy Bunnies, 209
Tale of Peter Rabbit, 209
Tale of Squirrel Nutkin, 209
Tanker/Boat, 122
Tape Dispenser, 227
Tape Measure, 227
Tennis Ball, 213
Tent, 80, 83, 98, 109, 127, 133, 136, 158
Tenth National O/O Convention held, 190
Texas: See Story of Texas Happy Meal, 1986, 156-157
The Captain, 92, 129, 130
The Captain redesigned, 6
The Challenge of Success, 4
The Challenging World of Number One, 5
The Country Mouse and the City Mouse, 62
The Customer is #1...What have you done lately for the Customer?, 4
The Day Birdie the Early Bird Learned to Fly, 243
The Drive-in with the arches, 4
The Eighties, 20-21
The Elves at the Top of the World, 128
The Fifties, 8
The Grimace, 6
The Hamburglar, 32
The Happy Meal Guys, 6
The Happy Meal Guys appear and disappear, 6
The Hot Stays Hot and the Cool Stays Cool, 5, 113, 138
The Legend of Gimme Gulch, 214
The Legend of Santa Claus, 128
The Monster at the End of this Book, 62
The most!, 4
The Mystery of the Missing French Frys, 243
The Poky Little Puppy, 62
The Professor, 6, 77
The Professor redesigned, 6
The Professor with long hair officially introduced, 6
The Ronald McDonald All Star Party, 76
The Seventies, 10
The Sixties, 9
The Sky's The Limit, 4
The Sword in the Stone, 172
Theater, 236
Theme - Ronald McDonald's, 10
Thirsty: Holding Drink, 79
Those who know--Go to McDonald's, 4
Three-D (3-D) Happy Meal, 1981, 38-39
Three-D Paper Eye Glasses, 39
Three-Window '34 Car, 84
Thugger, 238
Thumper, 190, 191
Thunder Streek, 200
Tic Tac Mac, 49
Tic-Tac-Teeth, 58
Tic-Tac-Top, 44
Tickle Feather, 58
Tiger, 25, 29, 43, 83, 140, 177, 178, 186, 218
Tiger 3D Face Mask, 218
Tin Bottom, 91, 127
Tinosaurs, 138, 157
Tinosaurs Happy Meal, 1986, 157-158
Tiny, 158, 246
Tiny Tim, 246
Tip 'N' Tilt, 26
Together, we've got what it takes, 5
Tom and Jerry's Party, 62
Tony and Fievel, 140
Toot Beaver ,174
Toothbrush, 86, 113, 120, 133, 224
Toothbrush Happy Meal, 1985, 133
Toothpaste, 86
Top, 44, 48, 71, 73, 92, 104, 128, 129, 149, 162, 166, 167, 190, 191, 211, 228, 244, 245
Tops, 196
Tornado, 68
Toucan 186, 218, 219
Toyota Tercel SR-5, 154, 155
Toyota Tercel SR-5 4 x 4, 155

Tracker, 229
Train, 32, 46, 64, 87, 88, 101, 137, 161, 174-177, 236
Train Cars - *See* McDonaldland Junction Happy Meal, 1983, 87-89
Train Engine, 64, 87-8, 175, 176
Train Engine Whistle, 176
Train with Stickers, 101
Transfers, 125
Transformers, 113, 134, 136
Transformers/My Little Pony I Happy Meal, 1985, 134-136
Transformers - *See* New Food Changeables Happy Meal, 1989, 240-241
Translite, 13, 15, 20, 28, 30, 39, 41, 45, 47, 48, 57, 59, 61, 63, 64, 67, 68, 70, 72, 80, 83, 85, 86, 89, 91, 92, 94, 98, 101, 102, 104, 108, 110, 116, 118-121, 123-128, 130, 132, 133, 136, 139-142, 144, 145, 147-151, 153-155, 157-159, 164-168, 170, 172-175, 177, 178, 180, 182, 184, 185, 191, 192, 194-196, 198, 200-214, 216-219, 224, 225, 227, 229-231, 233, 234, 236, 238, 240-245
Trayliner, 19, 48, 67, 80. 84, 98, 104, 108, 109, 116, 128, 130, 132, 136, 147, 157, 167, 174, 184, 204
Tree Trunk Monster, 15
Triangle/Isosceles Triangle/Stencil, 147
Triangle/Right Triangle/Stencil, 147
TriCar X8, 83
Trick, 123, 124
Truck, 55, 74, 86, 104, 200
Truck - Uncle Jesse's Pick-Up Truck, 55
Trumpet, 176
Tubby Tugger, 92, 129
Tulip, 182
Turbo Cone 241
Turbo Force Car, 232
Turbo Macs I (Test Market) Happy Meal, 1988, 217
Turismo "10" Red, 84
Turtle/Sea, 29
Tuttle the Turtle, 232
TWO BILLION SERVED signs, 5
Twoallbeefpattiesspecialsaucelettuce-cheesepickles, 5

U

U-3 (Under Age 3 Toys), 20, 59, 68, 85, 86, 103, 109, 120, 124, 129, 132, 133, 139, 143, 151, 154, 156, 164, 169, 175, 177, 180, 183, 186, 190, 191, 193, 195-197, 201, 204, 209, 211, 213, 217, 218, 226, 230, 232, 234, 237, 241, 242
Umbrella Girl, 47, 65
Uncle Jesse's Pick-Up Truck, 55
Uncle O'Grimacey, 6
Uncle O'Grimacey introduced, 6
Uncle Scrooge in Red Car, 195
Undersea Happy Meal, 1980, 28-30
Unidentified Spaceship Happy Meal, 1981, 48
Unidentified Sticker Sheet, 48
United DC-10, 68
Unpredict-A-Ball, 91
Ursula, 235, 236
USA Generic Promotions, 1980, 30-35
USA Generic Promotions, 1981, 49-54
USA Generic Promotions, 1982, 73-77
USA Generic Promotions, 1983, 94-96
USA Generic Promotions, 1984, 110-112
USA Generic Promotions, 1985, 136-138
USA Generic Promotions, 1986, 160-162
USA Generic Promotions, 1987, 186-189
USA Generic Promotions, 1988, 220-222
USA Generic Promotions, 1989, 245-246

V

Vacuum Form Train Cars - *See* McDonaldland Express Happy Meal, 1982, 64
Valentine, 31, 51
Valentine Cookie Sleeves, 31
Valentine Cup, 51
Vampire Bat Creature, 15
Veined Cranium, 15
Velocitor, 114
Veronica, 206
Video Communicator, 17
Volley McNugget, 204

W

Wacky Happy Meal, 1982, 71-72
Waco, 157
Walrus, 29, 212, 213
Wastebasket, 5
Watch/Wrist/Encoder, 195
Watering Can, 185, 215, 216, 234
Way of life coast to coast, 4
We Do It All For You, 5
Webby on Tricycle, 195
Welcome to McDonald's...The Closest Thing To Home, 4
Wembly/Boober, 197, 198
Whale, 29, 212, 213, 226, 244
Whale Sun Glasses, 244
What You Want Is What You Get [at McDonald's Today], 5
When the USA Wins You Win, 5, 97
When We Work At It--It Works, 5
Whistle, 26, 44, 49, 50, 59, 163, 176, 177, 195
White Plastic Cup, 32, 51, 76, 95
White Printing, 98
Who's in the Zoo?, 27
Wife, 153
Willard Scott, 9
Wilma in Purple Dragon Car, 196
Wind Whirler Helicopter, 233
Wing, 139, 144
Winged Amphibian, 15
Winter Worlds Happy Meal, 1983, 93-94
Woodsman with Knife in Hand, 47
Workshop of Activities, 128
Wrapper, 22

X

X-O graphic, 241
Xograph, 43

Y

Yellow Jeep, 99, 118
Yellow Plastic Cup, 32, 33
You Deserve a Break Today - So Get Up and Get Away to McDonald's, 4
You Deserve a Break Today, 4, 5, 37
You Deserve A Break Today We're Close By...Right on Your Way, 5
You, You're the One, 5
Young Astronauts I Happy Meal, 1986, 158-159
YOUR KIND OF PLACE, 4

Z

Zip Action Pull toys, 105
Zipper Pull, 221
Zoo animals, 13
Zoo Face I '87 (Test Market) Happy Meal, 1987, 186
Zoo Face II '88/Halloween '88 Happy Meal, 1988, 218-219
Zzz's, 229